Inverse Dynamics Problems

Inverse Dynamics Problems

Editor

Hamed Kalhori

MDPI • Basel • Beijing • Wuhan • Barcelona • Belgrade • Manchester • Tokyo • Cluj • Tianjin

Editor
Hamed Kalhori
University of Technology
Sydney
Australia

Editorial Office
MDPI
St. Alban-Anlage 66
4052 Basel, Switzerland

This is a reprint of articles from the Special Issue published online in the open access journal *Vibration* (ISSN 2571-631X) (available at: https://www.mdpi.com/journal/vibration/special_issues/inverse_dynamics).

For citation purposes, cite each article independently as indicated on the article page online and as indicated below:

LastName, A.A.; LastName, B.B.; LastName, C.C. Article Title. *Journal Name* **Year**, *Volume Number*, Page Range.

ISBN 978-3-0365-1066-8 (Hbk)
ISBN 978-3-0365-1067-5 (PDF)

Contents

About the Editor . . . . **vii**

Hamed Kalhori
Inverse Dynamics Problems for a Sustainable Future
Reprinted from: *Vibration* **2021**, *4*, 11, doi:10.3390/vibration4010011 . . . . **1**

Hamed Kalhori, Shabnam Tashakori and Benjamin Halkon
Experimental Study on Impact Force Identification on a Multi-Storey Tower Structure Using Different Transducers
Reprinted from: *Vibration* **2021**, *4*, 9, doi:10.3390/vibration4010009 . . . . **5**

Sergei Avdonin and Julian Edward
An Inverse Problem for Quantum Trees with Delta-Prime Vertex Conditions
Reprinted from: *Vibration* **2020**, *3*, 28, doi:10.3390/vibration3040028 . . . . **21**

Waad Subber, Sayan Ghosh, Piyush Pandita, Yiming Zhang and Liping Wang
Data-Informed Decomposition for Localized Uncertainty Quantification of Dynamical Systems
Reprinted from: *Vibration* **2021**, *4*, 4, doi:10.3390/vibration4010004 . . . . **37**

José Ramírez Senent, Jaime H. García-Palacios and Iván M. Díaz
A Shake Table Frequency-Time Control Method Based on Inverse Model Identification and Servoactuator Feedback-Linearization
Reprinted from: *Vibration* **2020**, *3*, 27, doi:10.3390/vibration3040027 . . . . **53**

Hasti Hayati, David Eager, Christian Peham and Yujie Qi
Dynamic Behaviour of High Performance of Sand Surfaces Used in the Sports Industry
Reprinted from: *Vibration* **2020**, *3*, 26, doi:10.3390/vibration3040026 . . . . **77**

Hasti Hayati, David Eager, Ann-Marie Pendrill and Hans Alberg
Jerk within the Context of Science and Engineering—A Systematic Review
Reprinted from: *Vibration* **2020**, *3*, 25, doi:10.3390/vibration3040025 . . . . **93**

Jin Yan, Simon Laflamme, Premjeet Singh, Ayan Sadhu and Jacob Dodso
A Comparison of Time-Frequency Methods for Real-Time Application to High-Rate Dynamic Systems
Reprinted from: *Vibration* **2020**, *3*, 16, doi:10.3390/vibration3030016 . . . . **133**

Ngoan T. Do, Mustafa Gül and Saeideh Fallah Nafari
Continuous Evaluation of Track Modulus from a Moving Railcar Using ANN-Based Techniques
Reprinted from: *Vibration* **2020**, *3*, 12, doi:10.3390/vibration3020012 . . . . **147**

Pasakorn Sengsri, Chayut Ngamkhanong, Andre Luis Oliveira de Melo and Sakdirat Kaewunruen
Experimental and Numerical Investigations into Dynamic Modal Parameters of Fiber-Reinforced Foamed Urethane Composite Beams in Railway Switches and Crossings
Reprinted from: *Vibration* **2020**, *3*, 14, doi:10.3390/vibration3030014 . . . . **161**

Alireza Shooshtari, Mahdi Karimi, Mehrdad Shemshadi and Sareh Seraj
Effect of Impeller Diameter on Dynamic Response of a Centrifugal Pump Rotor
Reprinted from: *Vibration* **2021**, *4*, 10, doi:10.3390/vibration4010010 . . . . **177**

Alessandro Casavola, Francesco Tedesco and Pasquale Vaglica
$\mathcal{H}_2$ and $\mathcal{H}_\infty$ Optimal Control Strategies for Energy Harvesting by Regenerative Shock Absorbers in Cars
Reprinted from: *Vibration* **2020**, *3*, 9, doi:10.3390/vibration4010009 . . . . **191**

About the Editor

Hamed Kalhori is currently a research assistant in the School of Mechanical and Mechatronic Engineering (MME) at the University of Technology Sydney (UTS) in Australia, and a lecturer (assistant professor) in Department of Mechanical Engineering at Bu-Ali Sina University in Hamedan, Iran. Hamed started his career at UTS as a research associate in the Centre for Autonomous Systems at UTS in the year 2017, and later joined MME as a lecturer. Prior to joining UTS, Hamed was a research assistant in Data61—CSIRO, working on structural health monitoring of various infrastructures across New South Wales. Hamed received his PhD in mechanical engineering from the University of Sydney in the year 2017 and received several awards during this period. Hamed's contribution to academic research has led to the publication of over 24 peer-reviewed journal articles and 16 international conference papers. Furthermore, Hamed is a reviewer for several leading scientific journals, such as *Mechanical Systems and Signal Processing, Journal of Sensors, Advanced Experimental Mechanics, Vibrations*, etc. Hamed has demonstrated excellence in his performance in teaching and learning also, evidenced by his most recent recognition. Hamed was bestowed with the 2019 University of Sydney Dean's Award for Excellence in Teaching and Tutoring. Hamed's research interests include inverse dynamics problems, linear and nonlinear vibrations, structural health monitoring, and smart materials and structures.

Editorial

Inverse Dynamics Problems for a Sustainable Future

Hamed Kalhori [1,2]

[1] School of Mechanical and Mechatronic Engineering, University of Technology Sydney, Sydney, NSW 2007, Australia; Hamed.Kalhori@uts.edu.au
[2] Department of Mechanical Engineering, Faculty of Engineering, Bu-Ali Sina University, Hamedan 65178, Iran

Citation: Kalhori, H. Inverse Dynamics Problems for a Sustainable Future. *Vibration* **2021**, *4*, 130–132. https://doi.org/10.3390/vibration4010011

Received: 7 February 2021
Accepted: 10 February 2021
Published: 12 February 2021

Inverse dynamics problems and associated aspects are all around us in everyday life but are commonly overlooked and/or not fully comprehended. However, they are of utmost significance when it comes to structural integrity and safety of engineering components we constantly interact with in our daily lives. The identification of dynamic characteristics and the estimation of applied loads to evaluate the structural health of aerospace structurers such as airliners, and/or on civil structures, such as large bridges with thousands of vehicles crossing daily, are tangible examples that show the fundamental importance of inverse dynamics problems. The inverse theory of network-like structures, with numerous practical applications in science and engineering, is an essential part of a swiftly growing area of applied mathematics. Such inverse problems encompass various applications in structural health monitoring, in water, electricity, gas, and traffic networks, in nano-electronics and quantum computing, in material science, and in biology [1].

But why solve problems inversely? What does that mean exactly? The estimation of system inputs or internal reactions by direct measurements is complicated or impossible for many real systems, either because the system input is inaccessible or unknown or simply because the nature of input is unknown and therefore cannot be instrumented. An inverse problem strategy is, therefore, a promising solution for such scenarios. Inverse problems are about identifying the cause of an effect, utilizing a set of observations and the measurement of the system response. As opposed to a forward problem yielding the system response, an inverse problem manipulates the effects considering the system's natural behavior to predict the inputs to the system. Figure 1 depicts a schematic of the forward and inverse problems.

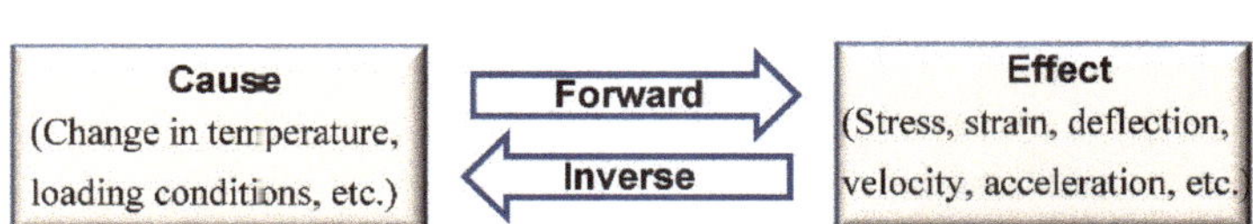

Figure 1. Forward and inverse problems.

In vibration engineering, the extraction of natural dynamic behaviors of a system, commonly referred to as an experimental modal analysis, can also be considered as an inverse problem deconvolving the input from the measured output. The modal analysis can be utilized to investigate the structural damage [2]. As a more advanced technique, operational modal analysis, OMA, seeks the system's natural behavior, manipulating the measured vibration response only [3]. Once the modal analysis is completed, it can be exploited to update the finite element model of the system to estimate the system response to more complicated loading and environmental conditions [4].

There is normally complex mathematics behind most inverse problems. Solving an inverse problem might not result in a sufficiently good outcome since the inverse problems are usually intrinsically ill-posed due to the ill-conditioned nature of the system frequency response function, making the problem sensitive to small perturbations such as measurement errors or noise. In other words, the existence, the uniqueness, or the stability of the solution might be disrupted. To avoid divergent or inaccurate results, it is, therefore, necessary to exploit a regularization method. Regularization has been the hot topic in the context of inverse problems for almost 50 years, and, more interestingly, there is still a lot of research being conducted by mathematicians and engineering scientists to develop new and more accurate regularization techniques. Inverse problems are normally challenging, as there are uncertainties that usually get amplified through the inverse process and therefore need to be properly addressed [5]. The inverse problems are performed in the time domain, frequency domain, and even the time–frequency domain [6,7].

This special issue mainly focuses on inverse problems in the context of system dynamics and vibration. Inverse dynamics, in particular, focusing on structural dynamics and/or inverse rigid body dynamics calculates the applied forces or internal forces and moments from measurements of structural vibrations and/or rigid body motions. The dynamic response can be measured using various types of contact and non-contact transducers including strain gauges, triangulation displacement sensors, laser Doppler vibrometers, accelerometers, and many other types of sensors [8].

As a widely used practical application of inverse problems in vibration engineering, impact force identification has attracted a great deal of attention [9–11]. Accidental external impact loading on structures makes them susceptible to different sorts of structural damage that reduces their load-carrying capabilities and may eventually give rise to catastrophic failure during service life. Identifying damage at the earliest possible stage by the determination of the impact location and magnitude can create a speedy structural health monitoring system. Identification of the impact loading on a structure is essential for structural integrity assessment and failure prediction. Additionally, structural condition assessment using moving sensors [12], time-varying load identification [13], moving load identification [14], bridge-weigh-in-motion systems [15], and human body and animal body inverse dynamics problems [16] are among other important applications of inverse problems in vibration engineering. Securing a more sustainable future by focusing on infrastructures' economic challenges can be achieved by the real-time monitoring of structures' health conditions through inverse algorithms.

The objective of this Special Issue was to create a forum of discussion for research scientists and engineers working in the area of inverse structural dynamics and inverse rigid body kinematics. We invited researchers to submit both original research and review articles. In total, 11 articles were accepted and published in this special issue. Herein, I would like to thank all the authors for their valuable time and effort contributing to the Special Issue.

Funding: This research received no external funding.

Institutional Review Board Statement: Not applicable.

Informed Consent Statement: Not applicable.

Conflicts of Interest: The author declares no conflict of interest.

References

1. Avdonin, S.; Edward, J. An Inverse Problem for Quantum Trees with Delta-Prime Vertex Conditions. *Vibration* **2020**, *3*, 448–463. [CrossRef]
2. Sengsri, P.; Ngamkhanong, C.; de Melo, A.L.O.; Kaewunruen, S. Experimental and numerical investigations into dynamic modal parameters of fiber-reinforced foamed urethane composite beams in railway switches and crossings. *Vibration* **2020**, *3*, 174–188. [CrossRef]
3. Sun, M.; Makki Alamdari, M.; Kalhori, H. Automated operational modal analysis of a cable-stayed bridge. *J. Bridge Eng.* **2017**, *22*, 05017012. [CrossRef]
4. Shooshtari, A.K.; Shemshadi, M.; Seraj, S. Effect of impeller diameter on dynamic response of a centrifugal pump rotor. *Vibration* **2021**, *4*, 117–129. [CrossRef]
5. Subber, W.; Ghosh, S.; Pandita, P.; Zhang, Y.; Wang, L. Data-Informed Decomposition for Localized Uncertainty Quantification of Dynamical Systems. *Vibration* **2021**, *4*, 49–63. [CrossRef]

6. Senent, J.R.; García-Palacios, J.H.; Díaz, I.M. A Shake Table Frequency-Time Control Method Based on Inverse Model Identification and Servoactuator Feedback-Linearization. *Vibration* **2020**, *3*, 425–447. [CrossRef]
7. Yan, J.; Laflamme, S.; Singh, P.; Sadhu, A.; Dodson, J. A Comparison of Time-Frequency Methods for Real-Time Application to High-Rate Dynamic Systems. *Vibration* **2020**, *3*, 204–216. [CrossRef]
8. Kalhori, H.; Tashakori, S.; Halkon, B. Experimental Study on Impact Force Identification on a Multi-Storey Tower Structure Using Different Transducers. *Vibration* **2021**, *4*, 101–116. [CrossRef]
9. Kalhori, H.; Alamdari, M.M.; Li, B.; Halkon, B.; Hosseini, S.M.; Ye, L.; Li, Z. Concurrent Identification of Impact Location and Force Magnitude on a Composite Panel. *Int. J. Struct. Stab. Dyn.* **2020**, *20*, 2042004. [CrossRef]
10. Kalhori, H.; Alamdari, M.M.; Ye, L. Automated algorithm for impact force identification using cosine similarity searching. *Measurement* **2018**, *122*, 648–657. [CrossRef]
11. Kalhori, H.; Ye, L.; Mustapha, S. Inverse estimation of impact force on a composite panel using a single piezoelectric sensor. *J. Intell. Mater. Syst. Struct.* **2017**, *28*, 799–810. [CrossRef]
12. Do, N.T.; Gül, M.; Nafari, S.F. Continuous evaluation of track modulus from a moving railcar using ANN-based techniques. *Vibration* **2020**, *3*, 149–161. [CrossRef]
13. Gursoy, E.; Niebur, D. Harmonic load identification using complex independent component analysis. *IEEE Trans. Power Deliv.* **2008**, *24*, 285–292. [CrossRef]
14. Zhu, X.; Law, S. Practical aspects in moving load identification. *J. Sound Vib.* **2002**, *258*, 123–146. [CrossRef]
15. Kalhori, H.; Alamdari, M.M.; Zhu, X.; Samali, B.; Mustapha, S. Non-intrusive schemes for speed and axle identification in bridge-weigh-in-motion systems. *Meas. Sci. Technol.* **2017**, *28*, 025102. [CrossRef]
16. Hayati, H.; Eager, D.; Peham, C.; Qi, Y. Dynamic Behaviour of High Performance of Sand Surfaces Used in the Sports Industry. *Vibration* **2020**, *3*, 410–424. [CrossRef]

Article

Experimental Study on Impact Force Identification on a Multi-Storey Tower Structure Using Different Transducers

Hamed Kalhori [1,*], Shabnam Tashakori [2,*] and Benjamin Halkon [1]

1 School of Mechanical and Mechatronic Engineering, University of Technology Sydney, Sydney, NSW 2007, Australia; Benjamin.Halkon@uts.edu.au
2 Rahesh Innovation Center, Shiraz 7188711114, Iran
* Correspondence: hamed.kalhori@uts.edu.au (H.K.); shabnam.tashakori@alum.sharif.edu (S.T.)

Abstract: This paper presents the identification of both location and magnitude of impact forces applied on different positions of a multi-storey tower structure using different types of transducers, i.e., an accelerometer, a laser Doppler vibrometer, and a triangulation displacement sensor. Herein, a model-based inverse method is exploited to reconstruct unknown impact forces based on various recorded dynamic signals. Furthermore, the superposition approach is employed to identify the impact location. Therein, it is assumed that several impact forces are applied simultaneously on potential locations of the multi-storey tower structure, while only one impact has non-zero magnitude. The purpose is then to detect the location of that non-zero impact. The influence of using different hammer tip materials for establishing the transfer function is investigated, where it is concluded that the hammer with a harder tip leads to a more accurate transfer function. An accuracy error function is proposed to evaluate the reconstruction precision. Moreover, the effect of sensor type and location on the accuracy of the reconstruction is studied, where it is shown that the proximity between the impact and sensor locations is a dominant factor in impact force reconstruction. In addition, the efficacy of using different transducers is studied for the impact localization, where it is demonstrated that reducing the degree of under-determinacy by using a combination of system responses of the same type can improve the localization accuracy.

Keywords: impact force identification; tower structure; impact localization; force history; inverse algorithm

check for updates

Citation: Kalhori, H.; Tashakori, S.; Halkon, B. Experimental Study on Impact Force Identification on a Multi-Storey Tower Structure Using Different Transducers. *Vibration* **2021**, *4*, 101–116. https://doi.org/10.3390/vibration4010009

Received: 2 December 2020
Accepted: 25 January 2021
Published: 29 January 2021

1. Introduction

Many structures are subjected to impact forces, which can be a matter of serious concern in terms of structural integrity. Measurement of these accidental impact forces is of great importance since it can help prevent system failure through evaluating the system stress and comparing it to its tolerance threshold or fatigue limit. Direct measurements of impact forces are difficult, expensive, and tedious, especially for large structures due to the difficulty of sensor installation and dynamic characteristic altering, while beforehand, localization of the impact area can make the examinations more efficient. Using system dynamic responses, captured by sensors placed distant from the impact location, the impact forces can be estimated by inverse algorithms.

The basis of inverse algorithms is to indirectly identify the impact force using responses measured at given points of the body subjected to impact. Inverse algorithms exploited in the literature can be categorized into two main classifications, namely, model-based techniques [1,2] and neural networks [3–6]. The superiority of neural networks emerges when the underlying dynamics is infeasibly complicated or inaccessible. However, as the accuracy of these techniques relies on massive training data, which is usually impractical, the model-based methods are more widely used. In model-based methods, a transfer function is found by utilizing the input and output of the system. Some examples of these methods are as follows: deconvolution technique [7–14], state variable formulation [15–20], and sum of weighted accelerations [21,22]. In [23], the inverse structural filter method, which leans on the dynamics state-space model, and the sum of the weighted accelerations technique are compared. Therein, deficiencies of the mentioned strategies are discussed and some modifications are proposed in order to enhance their performance. Among the model-based methods introduced, the deconvolution method has received significant attention in the literature. Two main attitudes of the deconvolution method are the time-domain [1,2,24,25] and the frequency-domain approach [26]. In [23], a comparison is made between the results of two time-domain strategies and those of a frequency-domain approach in order to determine the pros and cons of each method. Generally speaking, frequency-domain methods need lower computational efforts while they are usually infeasible for transient phenomena such as impact events. Solving a deconvolution problem might not result in a sufficiently good outcome since the force reconstruction problem is intrinsically ill-posed due to the ill-conditioned nature of the transfer function, i.e., the condition number of the transfer function matrix is very large, making the problem sensitive to small perturbations such as measurement errors or noise. To avoid divergent or inaccurate results, it is usually necessary to exploit a regularization method.

Several regularization techniques have been proposed in the literature. The most popular ones are Tikhonov regularization [27–31] and Singular Value Decomposition (SVD) based methods, including truncated SVD (TSVD) [27,32,33]. These two methods are compared in [34]. The theoretical backgrounds of five regularization methods, namely, generalized cross-validation, singular value decomposition, iterative method, data filtering approach, and Tikhonov regularization are introduced and main restrictions of each method are discussed in [35]. Some other exploited methods in the literature are QR factorization [36], explicit block inversion algorithms [37], Bayesian regularization [38], and the least-square QR (LSQR) iterative regularization method [39]. A combination of l_1 regularization and sparse reconstruction is proposed in [40]. In [11], a primal-dual interior point method is exploited and compared to the Tikhonov method. More recently, nonconvex sparse regularization based on generalized minimax-concave (GMC) and non-negative Bayesian learning are used in [25,41], respectively. In [42], Bayesian sparse regularization is exploited for identification and localization of multiple forces in time domain, and compared with Tikhonov regularization associated with the Generalized Cross Validation (GCV) criterion. Existing regularization methods which are proposed for force reconstruction are vector-based, while for large-scale inverse problems, matrix-based regularization has several privileges. Matrix-based regularization was recently introduced in [43] where the parameter of regularization was chosen with the Bayesian Information Criterion (BIC). Another issue that has been raised in recent years is that of moving force identification. In [44], a comparison is made between four regularization methods, i.e., (i) truncated generalized singular value decomposition (TGSVD), (ii) piecewise polynomial truncated singular value decomposition (PP-TSVD), (iii) modified preconditioned conjugate gradient (M-PCG) method, and (iv) preconditioned least-square QR-factorization (PLSQR) method, all used for reconstruction of moving forces, where it is concluded that the TGSVD method is preferred on the issue of identification accuracy. On the other hand, the M-PCG method is recommended in regard to identification efficiency.

To perform a comprehensive identification of an impact force, both its magnitude (force history) and location should be assessed. The location of the impact force is obscure in numerous cases in practice, which violates the fundamental presumption of the above mentioned methods. Various methods are introduced in the literature to localize the impact force. In [45], an experimental method is used in which an objective function is defined based on transfer functions and minimized in order to find the impact force location and in [46], a pseudo-inverse direct method is utilized to identify both the magnitude and location of the impact force. More recently, [12] pursued a similarity searching technique, and [14] introduces a superposition approach to estimate the impact location and magnitude simultaneously.

In the current paper, the identification of (i) the impact force history, and (ii) the impact location is presented. The impact force is applied on a scaled eight-storey tower structure in the laboratory. The identification is performed using recorded system outputs, i.e., the displacement, velocity, and acceleration measurements at level 3, as well as the acceleration measurement at level 8. The impact force reconstruction consists of two procedures, namely, (i) obtaining a transfer function between a reference impact force and its resulting response captured by a specific sensor, and (ii) identifying an unknown impact force using the transfer function obtained and the responses. Herein, the deconvolution technique is exploited to solve these inverse problems and the Tikhonov regularization method is used in order to deal with the ill-conditioned nature of the transfer function. To identify the impact location, the superposition approach is exploited where it is assumed that impact forces are concurrently applied on all 8 potential locations, while only one of them has a non-zero magnitude. This expresses the condition when only one impact is exerted at one of the possible locations. The actual impact location is then detected among all potential locations through an extended matrix form of the convolution equation.

The contributions of this paper are, firstly, investigating the influence of the hammer tip material on the effectiveness of the transfer function obtained, secondly, proposing an accuracy error function to evaluate the reconstruction precision, thirdly, studying the effect of sensor type and location on the accuracy of the impact force reconstruction, fourthly, using distinct sensors for the force reconstruction of different levels (i.e., using recorded signals at level 3 for the lower half of the structure and employing measurements at level 8 for the upper half), and fifthly, studying the localization accuracy based on the system responses used individually or in combination. The effectiveness of the method used for impact force reconstruction is demonstrated for all positions, with steel, soft rubber, medium rubber, and hard rubber tip hammers. The paper is organized as follows. The problem formulation is presented in Section 2. The experimental set-up is introduced in Section 3. Section 4 presents the results and discussion. Finally, the conclusions are presented in Section 5.

2. Problem Formulation

2.1. Single Impact Force Reconstruction

The impact force reconstruction consists of two procedures, namely, (i) obtaining a transfer function between a reference impact force and its resulting response captured by a specific sensor, and (ii) identifying an unknown impact force using this transfer function and the collected vibration responses. Suppose n sensors are deployed on a structure subjected to impacts to measure impact responses (e.g., displacement, velocity or acceleration) and the following assumptions hold:

- one impact is being applied at a time,
- structural responses are linear,
- the impact location is known.

Then, the relation between the impact force f applied at point x and the response r measured at point y at time t is given by a convolution integral as follows:

$$r(y,t) = \int_0^t T_s(x,y,t-\zeta) f(x,\zeta) d\zeta, \tag{1}$$

where $T_s(x,y,t-\zeta)$, $s = 1, ...n$, is the transfer function between the impact force at point x and the sth sensor at point y at time $t = \zeta$. The discretized form of the forward model (1), which is more applicable in practice, can be written as follows:

$$\mathbf{r} = \mathbf{T}_s \mathbf{f}, \tag{2}$$

with $\mathbf{r} \in R^m, \mathbf{T}_s \in R^{m \times m}, \mathbf{f} \in R^m$, where $\mathbf{r}$ is the recorded response vector, $\mathbf{f}$ is the vector of impact force which is to be reconstructed, and $\mathbf{T}_s$ is the impulse response matrix, which is a lower triangular toeplitz matrix, given by

$$\mathbf{r} = \begin{bmatrix} r(\Delta t) \\ r(2\Delta t) \\ \vdots \\ r((m-1)\Delta t) \\ r(m\Delta t) \end{bmatrix}, \mathbf{f} = \begin{bmatrix} f(\Delta t) \\ f(2\Delta t) \\ \vdots \\ f((m-1)\Delta t) \\ f(m\Delta t) \end{bmatrix},$$

$$\mathbf{T}_s = \begin{bmatrix} T_s(\Delta t) & 0 & \dots & 0 \\ T_s(2\Delta t) & T_s(\Delta t) & \dots & 0 \\ \vdots & \vdots & \ddots & \vdots \\ T_s((m-1)\Delta t) & T_s((m-2)\Delta t) & \dots & 0 \\ T_s(m\Delta t) & T_s((m-1)\Delta t) & \dots & T_s(\Delta t) \end{bmatrix}. \tag{3}$$

In (3), m is the number of samples and Δt is the time interval, which should be small enough since the above discretization assumes that the impact force f is constant within each time interval. In other words, with a higher sampling frequency, the results given by (2) are theoretically more accurate.

The solution of (2) can be theoretically obtained by using the following least squares problem:

$$min\|\mathbf{r} - \mathbf{T}_s\mathbf{f}\|_2^2, \tag{4}$$

where $\mathbf{r}$ is contaminated by experimental errors in practice. Moreover, $\mathbf{T}_s$ is a matrix with a very large condition number and hence is ill-conditioned. Therefore, the problem must be regularized. The Tikhonov regularization method alternatively searches for an approximation of $\mathbf{f}$ through the following penalized least-squares problem:

$$min\{\|\mathbf{r} - \mathbf{T}_s\mathbf{f}\|_2^2 + \delta\|I\mathbf{f}\|_2^2\}, \tag{5}$$

where $\delta \geq 0$ is the regularization parameter, determined by L-curve method, and I is the identity matrix.

2.2. Transfer Function

In order to solve (2), the transfer function $\mathbf{T}_s$ should be obtained in advance. This is achieved by using a reference impact force, its corresponding measured response, and the following relation:

$$\mathbf{r} = \mathbf{F}\mathbf{t}_s, \tag{6}$$

with $\mathbf{t}_s \in R^m, \mathbf{F} \in R^{m\times m}$, where $\mathbf{F}$ is a lower triangular toeplitz matrix, and $\mathbf{t}_s$ is the vector of transfer function, as follows:

$$\mathbf{F} = \begin{bmatrix} f(\Delta t) & 0 & \dots & 0 & 0 \\ f(2\Delta t) & f(\Delta t) & \dots & 0 & 0 \\ \vdots & \vdots & \ddots & \vdots & \vdots \\ f((m-1)\Delta t) & f((m-2)\Delta t) & \dots & f(\Delta t) & 0 \\ f(m\Delta t) & f((m-1)\Delta t) & \dots & f(2\Delta t) & f(\Delta t) \end{bmatrix}, \mathbf{t}_s = \begin{bmatrix} T_s(\Delta t) \\ T_s(2\Delta t) \\ \vdots \\ T_s((m-1)\Delta t) \\ T_s(m\Delta t) \end{bmatrix}. \tag{7}$$

The solution of (6) can be obtained by using the following least squares problem:

$$min\|\mathbf{r} - \mathbf{F}\mathbf{t}_s\|_2^2. \tag{8}$$

However, in practice, the collected impact force and the measured dynamic response are associated with noise, equivalent to high-frequency components of signals. This causes matrix $\mathbf{F}$ to, potentially, have a large condition number, making it ill-conditioned. The large condition number of $\mathbf{F}$ together with presence of noise in $\mathbf{r}$ results in deviated transfer functions. Therefore, applying regularization is deemed necessary. Employing Tikhonov regularization method, an approximation of $\mathbf{t}_s$ can be found instead, by the following penalized least-squares problem:

$$min\{\|\mathbf{r} - \mathbf{F}\mathbf{t}_s\|_2^2 + \beta\|I\mathbf{t}_s\|_2^2\}, \tag{9}$$

with $\beta \geq 0$ the regularization parameter, determined by the L-curve method.

Summarizing, impact force reconstruction consists of two steps:

1. obtaining the vector of transfer function $\mathbf{t}_s$ by solving (6) (and converting it to the triangular toeplitz transfer function matrix $\mathbf{T}_s$),
2. solving (2) for the unknown impact force $\mathbf{f}$.

As discussed, both above problems are ill-posed. Therefore, in this paper, the Tikhonov regularization method is exploited in order to avoid the sensitivity to perturbations, which can potentially make the solution unstable.

2.3. Impact Force Location

Two approaches have been employed in the literature for impact force localization: the one-to-one approach and the superposition approach [14]. In the one-to-one approach, the impact reconstruction is performed for each pair of impact and response location, while in the superposition approach, the impact forces at all possible locations are reconstructed concurrently. Generally speaking, the superposition approach considers a superposition of responses corresponding to each impact force exerted at different locations. In the following, the superposition approach is presented more in detail.

Assuming several impact forces at various points $(i = 1, \ldots, p)$ concurrently applied to a structure, the vibration response collected by a single sensor installed at position s is, therefore, a superposition of the responses generated by each individual impact force.

$$\mathbf{r} = \sum_{i=1}^{p} \mathbf{T}_s^i \mathbf{f}_i, \tag{10}$$

where $\mathbf{f}_i$ is the impact force applied on the location i, $i = 1, ..., p$, and $\mathbf{T}_s^i$ is the transfer function between the location i and the sth sensor location. Equation (10) can be written in matrix-vector form as follows:

$$\mathbf{r} = \begin{bmatrix} \mathbf{T}_s^1 & \mathbf{T}_s^2 & \ldots & \mathbf{T}_s^p \end{bmatrix} \begin{bmatrix} \mathbf{f}_1 \\ \mathbf{f}_2 \\ \vdots \\ \mathbf{f}_p \end{bmatrix}. \tag{11}$$

The procedure for creating the transfer functions were already discussed in Section 2.2. For brevity, (11) is presented by $\mathbf{r} = \mathbf{T}_s\mathbf{f}$. As previously pointed out, matrix $\mathbf{T}_s$ is ill-conditioned and vector $\mathbf{r}$ is contaminated with noise, necessitating applying regularization to solve for $\mathbf{f}$. Similar to (9), Tikhonov regularization is implemented. It is worth mentioning that (11) is severly under-determined, as there is one equation with p unknown forces. Now, let us make an important assumption that is the magnitude of all impact forces but one is actually equal to zero. This condition entails that an impact occurs at only one location. The purpose is, therefore, to detect the actual impact location among all potential locations, together with its force history. Using this approach, a reconstructed impact force is obtained for each potential impact location. In other words, p impact forces, i.e., $\mathbf{f}_1, \mathbf{f}_2, \ldots \mathbf{f}_p$, are reconstructed, keeping in mind that there is only one actual non-zero impact force. The reconstructed impact forces at spurious locations are expected to have zero magnitude as no impact has actually occurred at these locations. However, there might be some non-zero reconstructed impact forces at spurious locations. The reconstructed force at each location is qualitatively assessed, addressing key characteristics of a normal impact force such as the shape and the maximum amplitude of the first peak if applicable. A normal impact force has typically a smooth half-sine shape. More comprehensive description of this method as well as various case studies can be found in previous works of the authors [10,12,14]. Figure 1 shows a schematic of the problem.

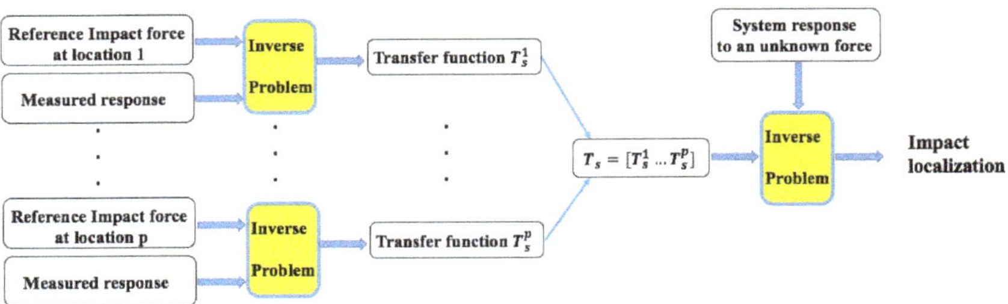

Figure 1. Schematic of impact force localization using the superposition approach.

3. Experimental Set-Up

To investigate the impact force reconstruction experimentally, the structure shown in Figure 2 is used with the parameters given in Table 1. The structure is a hollow rectangular steel beam, fixed at the bottom and free at the top, on which eight lumped masses are clamped at equally distributed distances. In the following, these masses are called level 1 to 8 with level 1 at the bottom and level 8 at the top.

Four sensors, namely, two DC-response MEMS accelerometers (Measurement Specialties 4000A-005) at level 3 and level 8, a laser Doppler vibrometer (Polytec PDV-100) at level 3, and a laser triangulation sensor (Micro-Epsilon optoNCDT 1302, ILD 1302-50) at level 3 were employed to gauge system responses, as shown in Figure 2.

The impact forces were applied by a modally tuned instrumented impact hammer that provided the measurement of the actual dynamic load. This hammer was used with different tips (i.e., steel, hard rubber, medium rubber, and soft rubber tips) in order to simulate various hit modes.

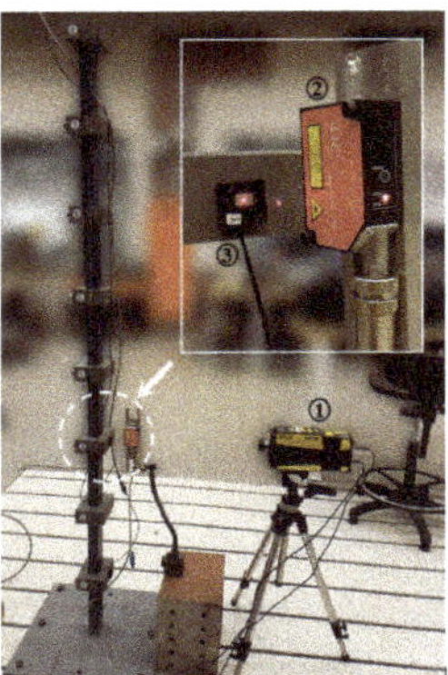

Figure 2. Experimentalset-up showing the multi-storey tower and primary response transducers including (1) laser Doppler vibrometer, (2) laser triangulation sensor, and (3) accelerometer.

Table 1. Experimental set-up parameters.

Parameter	Value
Beam length	2 m
Beam cross-section	65.3×35 mm^2
Beam thickness	2.5 mm
Lumped masses dimension	$128\times 98 \times 50$ mm^3
Lumped masses weight	4 kg
Lumped masses distances	250 mm

4. Results and Discussion

4.1. Effect of Regularization

As stated previously, identification of transfer functions as well as reconstruction of impact forces are both ill-posed problems. Solving (6) using the least-square technique without applying regularization, given in (8), led to transfer functions with magnitudes in the order of 10^3, while solving the problem considering regularization, defined in (9), gave the transfer functions with magnitudes in the order of 10^{-5}. Figure 3 depicts the reconstructed impact force implementing the transfer function obtained with and without regularization. In this figure, the impact force was applied at level 5, and the vibration response was measured in level 8. As seen in Figure 3, the reconstruction process utilizing the transfer function obtained without regularization yields a reconstructed impact force with virtually zero amplitude. This is because the transfer function has large singular values. The large singular values are inverted through the inverse algorithm, causing the reconstructed force to approach zero.

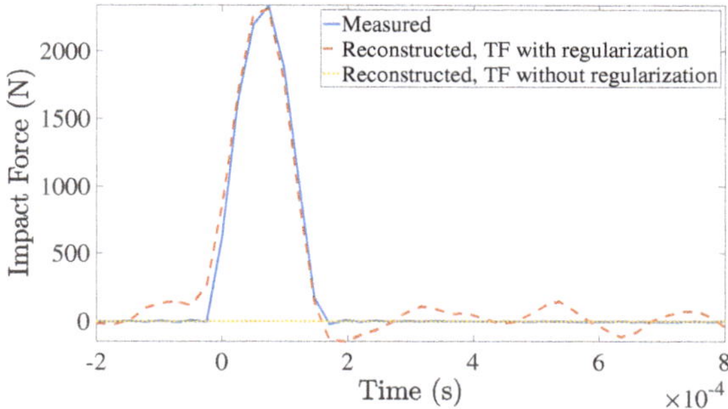

Figure 3. The effect of regularization in establishing the transfer function and reconstructing the impact force.

4.2. Establishing the Transfer Function

Different hammer tips can produce different half-sine shapes of impact force, with different rising patterns and time durations. As shown in Figure 4, harder tips produce sharper graphs of impact force, i.e., closer to the graph of Dirac delta function. The sharpest graph shows the results of the hammer with the hardest tip (steel tip), which, theoretically, should result in the most accurate transfer function. This result is also shown illustratively in Figure 5 in which the impact force applied by the soft rubber tip at level 4 is reconstructed by using the measured acceleration at level 8 and different transfer functions established, namely, by the steel tip, the hard rubber tip, the medium rubber tip, and the soft rubber tip itself. As can be seen, the transfer function obtained by the steel tip gives the most accurate force reconstruction result. Quantitatively, the correlation coefficient between the measured impact force and the corresponding reconstructed forces, shown in Figure 5, are 0.9917, 0.9829, 0.9917, and 0.9752, respectively, for the steel tip, the hard rubber tip, the medium rubber tip, and the soft rubber tip. Additionally, the percentage of peak errors are, respectively, −1.12%, 7.50%, 11.29%, and −1.58%. Interestingly, the best force reconstruction is not necessarily achieved when the transfer function is obtained with the same tip as the tip generating force. Even if that were the case, it would not be applicable as the material of the object impacting the structure is usually not known or predictable in practice.

To show the effectiveness of the force reconstruction, two quantities, correlation coefficient and peak error, should be considered simultaneously. In other words, the force reconstruction with a higher correlation coefficient (i.e., closer to 1), and concurrently, lower peak error (i.e., closer to 0%) is more desirable. Therefore, we introduce the following reconstruction accuracy error in order to use only one variable:

$$e = \sqrt{(correlation\ coefficient - 1)^2 + (peak\ error)^2}. \tag{12}$$

In the worst case scenario, the maximum value of the accuracy error is $\sqrt{2}$. On the other hand, when e is closer to zero, the reconstruction is more precise. For instance, in Figure 5, the reconstruction accuracy errors for steel tip, hard rubber tip, and medium rubber tip are 0.0139, 0.0769, and 0.1132, respectively, which shows the reconstruction precision in the case of steel tip as its corresponding accuracy error is closer to zero. Similar results were observed for impact at other levels. It is worth mentioning that neither the correlation coefficient nor the peak error can singly lead to this conclusion. As concluded from the above discussion, from now on, the transfer functions are established by using the steel tip hammer.

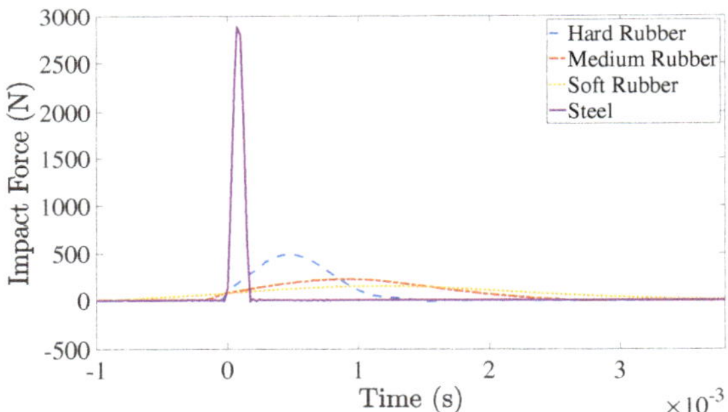

Figure 4. Impact force graphs produced using different hammer tips.

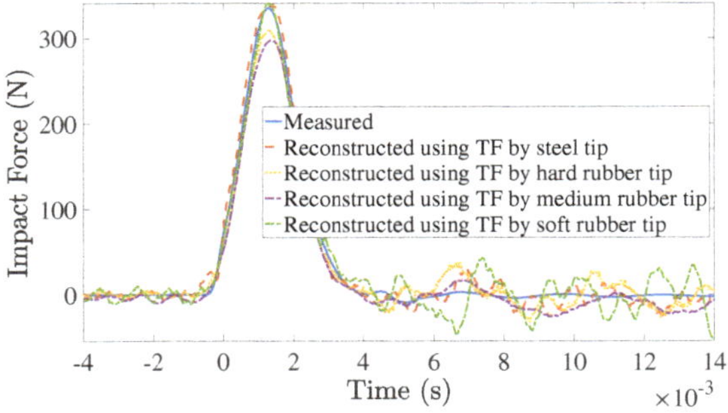

Figure 5. Impact force reconstruction using transfer functions (TF) established by different hammer tips.

4.3. Influence of Sensor Type and Location

As previously pointed out in Section 3, four transducers are mounted on the experimental set-up, measuring the displacement, velocity, and acceleration at level 3, as well as the acceleration at level 8. These measurements can be utilized in combination or individually for the force reconstruction.

Figure 6 shows the reconstruction of the impact forces applied by steel tip hammer implementing different system responses (i.e., the velocity at level 3, acceleration at level 3, and acceleration at level 8). Since the results of using displacement at level 3 were not satisfactory at all, these are not shown for better clarity. To investigate this comparison quantitatively, Table 2 shows the values of the accuracy error for each condition, as well.

As shown in Figure 6 and Table 2, the distance between the impact location and the sensor location is a dominant factor in impact force reconstruction. More specifically, employing velocity measurement at level 3 leads to better reconstruction results for impacts at lower levels (1, 2, 3, and 4). On the other hand, to reconstruct impact forces applied at higher levels (5, 6, 7, and 8), using the acceleration measurement at level 8 is more effective. Note that the minimum accuracy error in each row is colored in Table 2.

It is observed that making the problem over-determined (e.g., employing a combination of velocity at level 3 and acceleration at level 8) does not necessarily improve the reconstruction. Therefore, in this paper, the problem is kept even-determined in order not to use extra ineffective computation costs. In this regard, the impact forces will be reconstructed by using the velocity measurement at level 3 when the impact location is in the lower half of the structure and employing the acceleration measurement at level 8, otherwise.

Table 2. Accuracy errors of impact force reconstruction using different system responses.

Impact Location	Measured Response		
	Vel. at Level 3	Acc. at Level 3	Acc. at Level 8
Level 1	0.0219	0.2448	0.0956
Level 2	0.0141	0.0182	0.0682
Level 3	0.0053	0.0551	0.0323
Level 4	0.0140	0.0342	0.0290
Level 5	0.1644	0.0908	0.0157
Level 6	0.5969	0.0223	0.0157
Level 7	0.3779	0.0518	0.0125
Level 8	0.7738	0.3177	0.0108

Figure 7 illustrates the reconstruction accuracy errors for impact forces applied on different levels of the structure. As shown in Figure 7a, when using the velocity measurement at level 3, the minimum of the accuracy error occurs when the impact force is applied at level 3. Similarly, when it comes to using the acceleration measurement at level 8, the accuracy error is the minimum if the impact location is also at level 8, as illustrated in Figure 7c. Concerning the acceleration measurement at level 3, it gives its most accurate result for mid-levels (not end-levels), as can be seen in Figure 7b. Figure 7d shows the accuracy error when using the velocity at level 3 for the lower half of the structure (i.e., level 1 to 4), and employing the acceleration at level 8 for the upper half (i.e., level 5 to 8). As can be seen both in Figure 7d and Table 2, the reconstruction is poorer when the impact force is applied at level 1. The reason is that this level is very close to the fixed support, which prevents the proper stimulation of vibration modes and hence the signal can not satisfactorily be captured by sensors placed distant from this location. On the other hand, the most accurate results are obtained at levels 3 and 8, exactly where the transducers are placed, which demonstrates the effect of proximity of impact location and sensor location.

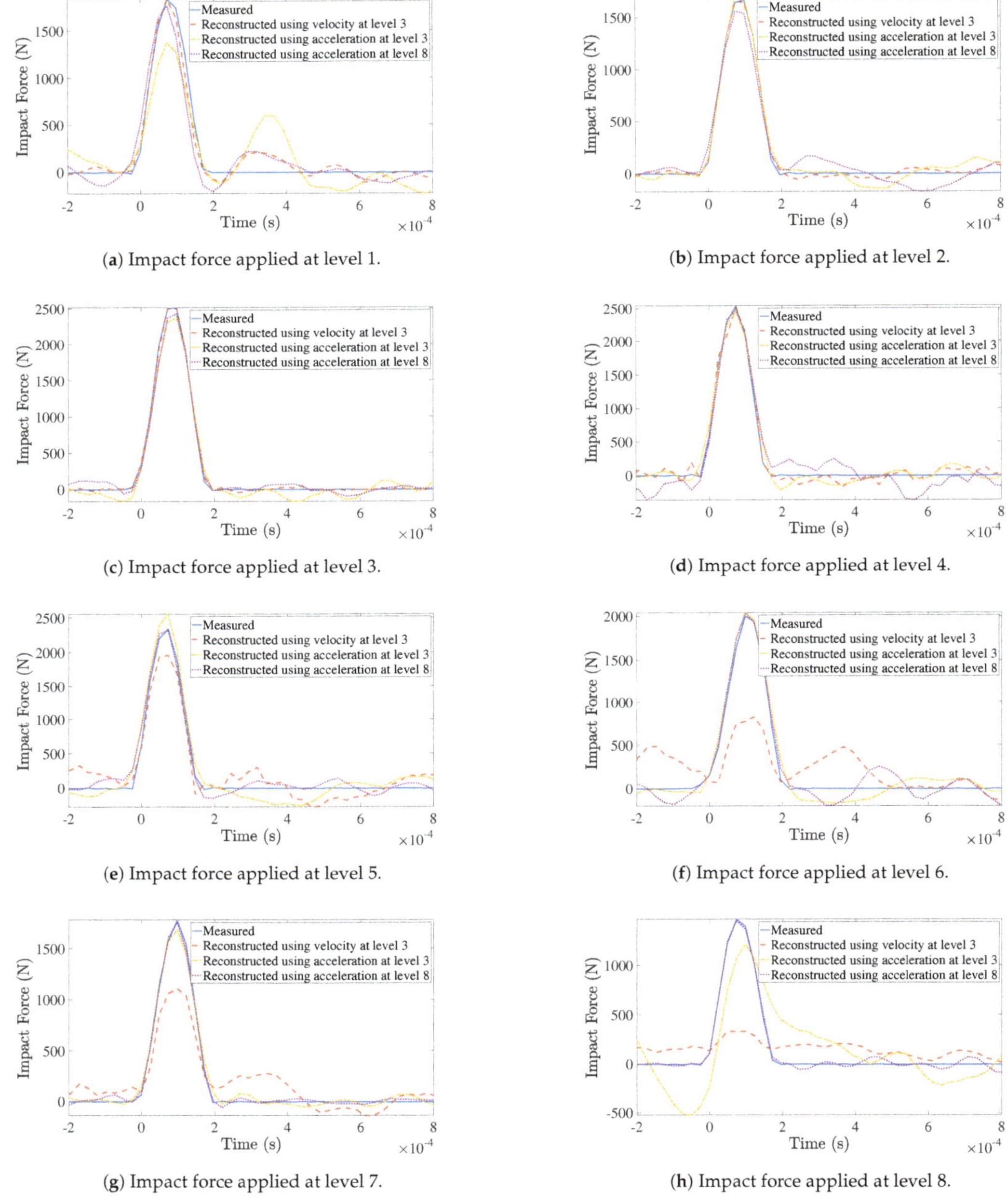

(**a**) Impact force applied at level 1.

(**b**) Impact force applied at level 2.

(**c**) Impact force applied at level 3.

(**d**) Impact force applied at level 4.

(**e**) Impact force applied at level 5.

(**f**) Impact force applied at level 6.

(**g**) Impact force applied at level 7.

(**h**) Impact force applied at level 8.

Figure 6. Using different transducers for impact force reconstruction at different locations.

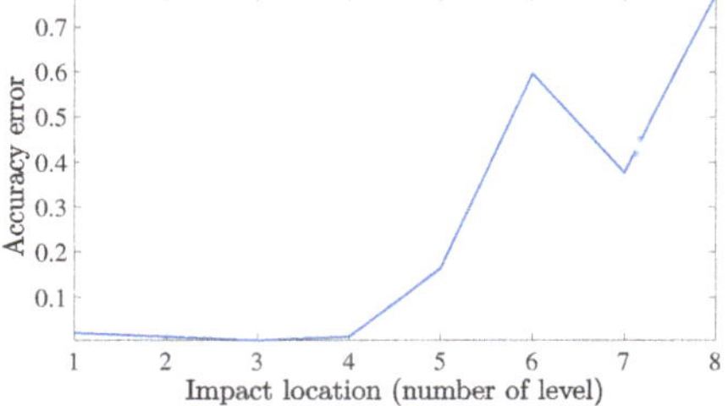

(a) Using velocity measurement at level 3.

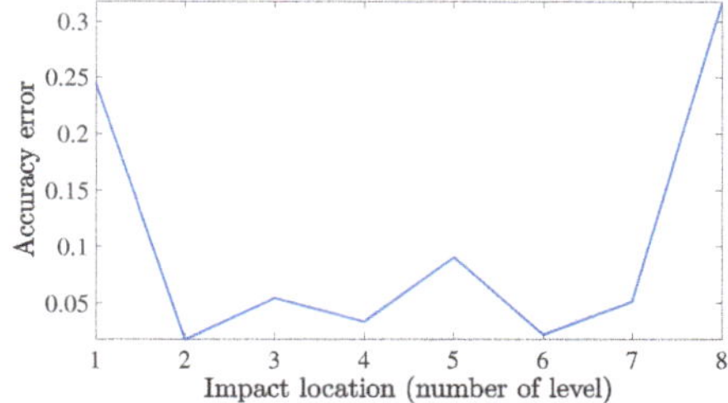

(b) Using acceleration measurement at level 3.

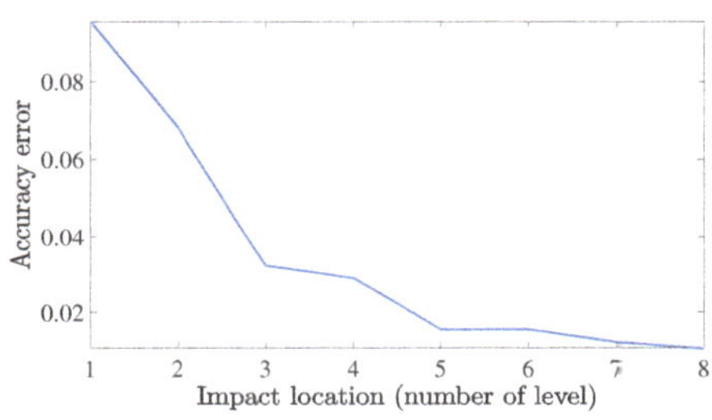

(c) Using acceleration measurement at level 8.

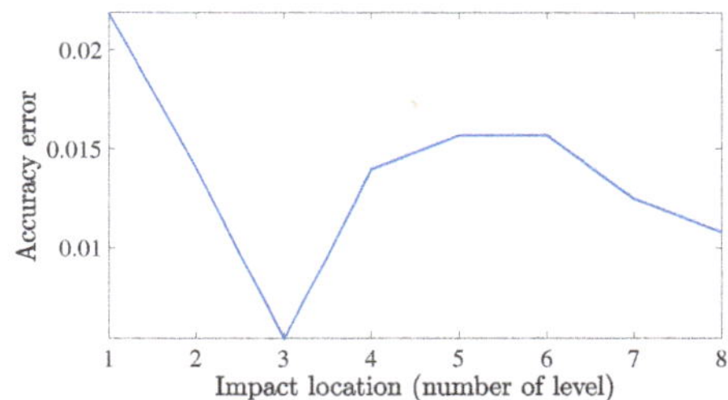

(d) Minimum accuracy error accessible using either of the sensor measurements.

Figure 7. Relation between the sensor location and impact force reconstruction accuracy error.

In order to complement the above discussion, the impact forces applied by different rubber tip hammers (i.e., soft, medium, and hard rubber tips) are reconstructed in the following. Note that based on the conclusion made in Section 4.2, the transfer functions are obtained by using the steel tip hammer. Additionally, based on the conclusion made earlier in the current subsection, the velocity measurement at level 3 and the acceleration measurement at level 8 are employed for lower half and upper half of the structure, respectively. Table 3 shows the accuracy errors of the reconstruction for each rubber hammer tip at different levels. These errors could be reduced by manually changing the regularization parameter, however, it was not the purpose of the current paper. As shown, the accuracy error is acceptable in most of the cases, which demonstrates the efficacy of the transfer function obtained and the responses used.

Table 3. Accuracy errors of the reconstruction of impact forces applied by rubber tip hammers at different levels.

Hammer Tip	Impact Location							
	1	2	3	4	5	6	7	8
Soft rubber	0.0974	0.0516	0.0339	0.0576	0.0482	0.0642	0.0563	0.0604
Medium rubber	0.0863	0.0502	0.0590	0.0563	0.0967	0.0902	0.1031	0.0932
Hard rubber	0.0472	0.0239	0.0558	0.0404	0.0558	0.3282	0.0464	0.0421

4.4. Impact Force Location

In the following results, it is assumed that eight impact forces are applied concurrently at levels 1 to 8, while the magnitude of only one impact force is non-zero. Herein, the superposition approach introduced in Section 2.3 is employed for location identification. Different scenarios were tested, namely, (i) using each of the available measurements singly, and (ii) different combinations of two system responses. Among all, the combination of the acceleration at level 3 and the acceleration at level 8 leads to the most satisfactory impact localization. This is shown quantitatively in Table 4, where the accuracy errors corresponding to the reconstruction of the actual impact forces are presented for all above-mentioned scenarios. The minimum possible accuracy error for each impact location is colored in the table. As demonstrated, for most levels, the minimum occurs when the combination of acceleration at level 3 and acceleration at level 8 is employed. It is concluded that reducing the degree of under-determinacy can improve the localization accuracy. Moreover, it seems that when the two measurements, selected in combination, are of the same type, the impact force can be localized more accurately. Therefore, the actual impact location can be detected through the following relation:

$$\begin{bmatrix} \mathbf{r}_3 \\ \mathbf{r}_8 \end{bmatrix} = \begin{bmatrix} \mathbf{T}_3^1 & \mathbf{T}_3^2 & \dots & \mathbf{T}_3^8 \\ \mathbf{T}_8^1 & \mathbf{T}_8^2 & \dots & \mathbf{T}_8^8 \end{bmatrix} \begin{bmatrix} \mathbf{f}_1 \\ \mathbf{f}_2 \\ \vdots \\ \mathbf{f}_8 \end{bmatrix}, \tag{13}$$

where $\mathbf{r}_3$ and $\mathbf{r}_8$ are the acceleration response at level 3 and 8, respectively, and $\mathbf{T}_j^i$ is the transfer function between the impact location i, $i = 1, ..., 8$, and measurement location j, $j = 3, 8$. As presented in Section 2.3, (13) is solved for $\mathbf{f}_i$, $i = 1, ..., 8$, where the magnitude of one of these reconstructed forces is significantly greater that others which specifies the actual impact location.

Figure 8 shows the reconstruction of impact forces at all possible locations when the actual impact force is applied at levels 1 to 8, individually, and a combination of the acceleration at level 3 and the acceleration at level 8 is considered as the system response. It demonstrates the efficacy of the approach as the reconstructed impact force associated to the true impact location has a smooth half-sine shape with a higher peak amplitude than other possible locations, as expected.

Table 4. Accuracy errors of the reconstruction of actual impact force using different traducers at each individual level.

Measured Response	Impact Location							
	1	2	3	4	5	6	7	8
Vel. at level 3	0.9581	0.9370	0.2522	0.9395	0.9608	1.0389	0.9821	1.0117
Acc. at level 3	0.9303	0.8922	0.4225	0.8439	0.9145	0.9701	0.9706	0.9883
Acc. at level 8	1.0055	0.9861	0.9868	1.0233	0.9899	0.9831	0.9164	0.1559
Vel. at l3 and Acc. at l8	1.0093	0.9842	0.9934	1.0149	0.9992	1.0375	0.9067	0.1163
Acc. at l3 and Acc. at l8	0.8947	0.8839	0.3038	0.7664	0.8968	0.9948	0.8844	0.0991

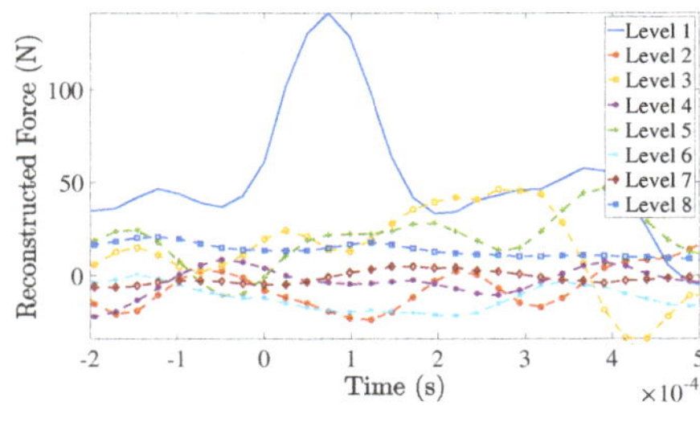

(a) True impact location is level 1.

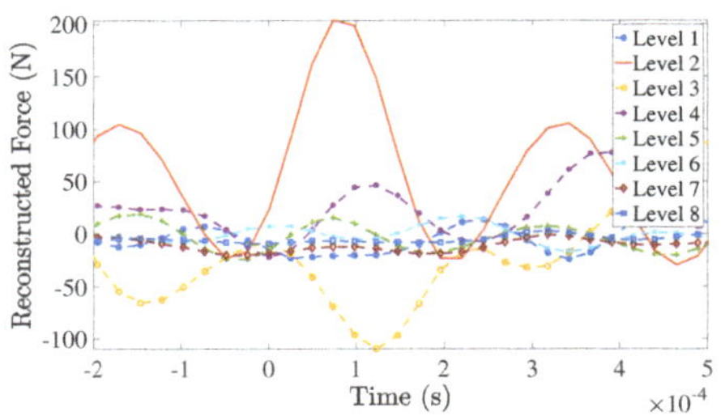

(b) True impact location is level 2.

Figure 8. *Cont.*

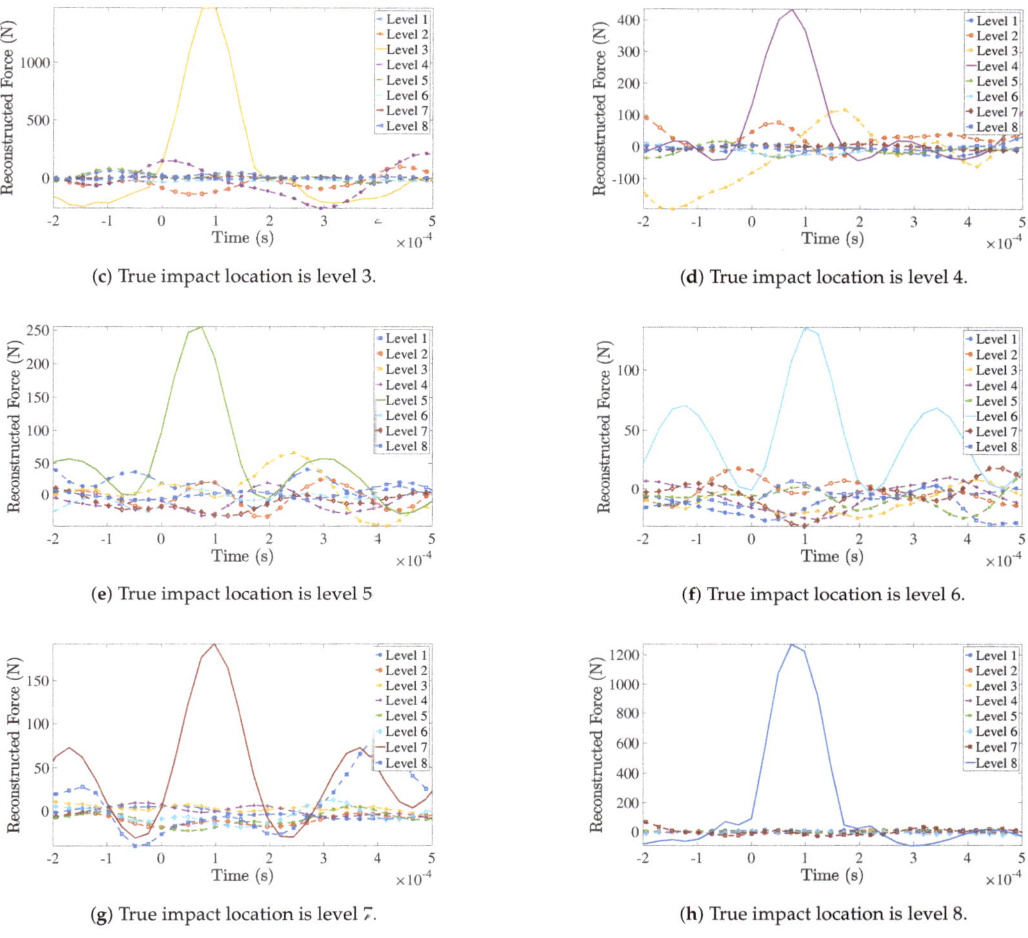

(**c**) True impact location is level 3.

(**d**) True impact location is level 4.

(**e**) True impact location is level 5

(**f**) True impact location is level 6.

(**g**) True impact location is level 7.

(**h**) True impact location is level 8.

Figure 8. Identification of the impact location.

5. Conclusions

Inverse identification of an impact force acting on a multi-storey tower structure was studied experimentally using dynamic signals measured by different transducers. Herein, both the magnitude and location of the impact force were investigated. It was shown that using the hammer with the hardest tip can lead to a more accurate transfer function, where an accuracy error function was proposed to evaluate the reconstruction precision as a function of the correlation coefficient and the peak error. Moreover, it was observed that the proximity between the impact and sensor locations is a dominant factor in impact force reconstruction. Therefore, the velocity measurement at level 3 was used for the lower half of the structure and the acceleration measurement at level 8 was employed for the upper half and the effectiveness of this idea for impact force reconstruction at all positions was demonstrated both for steel tip hammer and rubber tip hammers. For impact localization, the superposition method was exploited, where the effect of different transducers was studied. It was concluded that reducing the degree of under-determinacy by using a combination of system responses of the same type can improve the localization accuracy.

Therefore, a combination of the acceleration at level 3 and the acceleration at level 8 was employed for the localization.

As a potential real-world application of this study, identification of impact forces on bridge structures can be of great interest to the bridge owners and engineers. The bridge can be modelled as a multi-degree of freedom system with the expansion joints of the bridge deck taken as the potential impact locations. Measurement of the vibration response generated by the impact of heavy trucks can be carried out using accelerometers or contactless sensors such LDVs installed distant from the impact location. Another possible application of the current study is in oil-well drilling industry. During the whirling motion, the rotating drill string strikes the borehole wall, generating shocks from lateral vibrations. The location and magnitude of these impact forces are unknown as it is indeed impossible to place sensors on the string. However, using top-side measurements and inverse algorithms, the impact force can be identified, which helps in stability analysis and controller design for such structures.

Author Contributions: Conceptualization, H.K.; Formal analysis, S.T.; methodology, H.K. and S.T.; software, H.K. and B.H.; validation, H.K. and S.T.; supervision, H.K.; writing—original draft preparation, H.K. and S.T.; writing—review and editing, H.K. and S.T.; project administration, H.K. and B.H.; funding acquisition, B.H. All authors have read and agreed to the published version of the manuscript.

Funding: This research received no external funding.

Institutional Review Board Statement: Not applicable.

Informed Consent Statement: Not applicable.

Conflicts of Interest: The authors declare no conflict of interest.

References

1. Qiao, B.; Mao, Z.; Liu, J.; Zhao, Z.; Chen, X. Group sparse regularization for impact force identification in time domain. *J. Sound Vib.* **2019**, *445*, 44–63. [CrossRef]
2. Qiu, B.; Zhang, M.; Li, X.; Qu, X.; Tong, F. Unknown impact force localisation and reconstruction in experimental plate structure using time-series analysis and pattern recognition. *Int. J. Mech. Sci.* **2020**, *166*, 105231. [CrossRef]
3. Sung, D.U.; Oh, J.H.; Kim, C.G.; Hong, C.S. Impact monitoring of smart composite laminates using neural network and wavelet analysis. *J. Intell. Mater. Syst. Struct.* **2000**, *11*, 180–190. [CrossRef]
4. LeClerc, J.; Worden, K.; Staszewski, W.J.; Haywood, J. Impact detection in an aircraft composite panel—A neural-network approach. *J. Sound Vib.* **2007**, *299*, 672–682. [CrossRef]
5. Sarego, G.; Zaccariotto, M.; Galvanetto, U. Artificial neural networks for impact force reconstruction on composite plates. In Proceedings of the 2017 IEEE International Workshop on Metrology for AeroSpace (MetroAeroSpace), Padua, Italy, 21–23 June 2017; pp. 211–216.
6. Sarego, G.; Zaccariotto, M.; Galvanetto, U. Artificial neural networks for impact force reconstruction on composite plates and relevant uncertainty propagation. *IEEE Aerosp. Electron. Syst. Mag.* **2018**, *33*, 38–47. [CrossRef]
7. Jacquelin, E.; Bennani, A.; Hamelin, P. Force reconstruction: Analysis and regularization of a deconvolution problem. *J. Sound Vib.* **2003**, *265*, 81–107. [CrossRef]
8. Kalhori, H.; Ye, L.; Mustapha, S.; Li, J. Impact force reconstruction on a concrete deck using a deconvolution approach. In Proceedings of the 8th Australasian Congress on Applied Mechanics: ACAM 8, Engineers Australia, Melbourne, Australia, 23–26 November 2014; p. 763.
9. Kalhori, H.; Ye, L.; Mustapha, S.; Li, J.; Li, B. Reconstruction and analysis of impact forces on a steel-beam-reinforced concrete deck. *Exp. Mech.* **2016**, *56*, 1547–1558. [CrossRef]
10. Kalhori, H.; Ye, L.; Mustapha, S. Inverse estimation of impact force on a composite panel using a single piezoelectric sensor. *J. Intell. Mater. Syst. Struct.* **2017**, *28*, 799–810. [CrossRef]
11. Qiao, B.; Zhang, X.; Gao, J.; Liu, R.; Chen, X. Sparse deconvolution for the large-scale ill-posed inverse problem of impact force reconstruction. *Mech. Syst. Signal Process.* **2017**, *83*, 93–115.
12. Kalhori, H.; Alamdari, M.M.; Ye, L. Automated algorithm for impact force identification using cosine similarity searching. *Measurement* **2018**, *122*, 648–657. [CrossRef]
13. Tran, H.; Inoue, H. Development of wavelet deconvolution technique for impact force reconstruction: Application to reconstruction of impact force acting on a load-cell. *Int. J. Impact Eng.* **2018**, *122*, 137–147. [CrossRef]
14. Kalhori, H.; Alamdari, M.M.; Li, B.; Halkon, B.; Hosseini, S.M.; Ye, L.; Li, Z. Concurrent Identification of Impact Location and Force Magnitude on a Composite Panel. *Int. J. Struct. Stab. Dyn.* **2020**, *20*, 2042004. [CrossRef]

15. Hollandsworth, P.; Busby, H. Impact force identification using the general inverse technique. *Int. J. Impact Eng.* **1989**, *8*, 315–322. [CrossRef]
16. Law, S.S.; Fang, Y. Moving force identification: Optimal state estimation approach. *J. Sound Vib.* **2001**, *239*, 233–254. [CrossRef]
17. Lourens, E.; Reynders, E.; De Roeck, G.; Degrande, G.; Lombaert, G. An augmented Kalman filter for force identification in structural dynamics. *Mech. Syst. Signal Process.* **2012**, *27*, 446–460.
18. Ding, Y.; Law, S.; Wu, B.; Xu, G.; Lin, Q.; Jiang, H.; Miao, Q. Average acceleration discrete algorithm for force identification in state space. *Eng. Struct.* **2013**, *56*, 1880–1892.
19. Wang, T.; Wan, Z.; Wang, X.; Hu, Y. A novel state space method for force identification based on the Galerkin weak formulation. *Comput. Struct.* **2015**, *157*, 132–141. [CrossRef]
20. Liu, J.; Xie, J.; Li, B.; Hu, B. Regularized Cubic B-Spline Collocation Method With Modified L-Curve Criterion for Impact Force Identification. *IEEE Access* **2020**, *8*, 36337–36349. [CrossRef]
21. Kreitinger, T.; Wang, M.; Schreyer, H. Non-parametric force identification from structural response. *Soil Dyn. Earthq. Eng.* **1992**, *11*, 269–277.
22. Carne, T.G.; Mayes, R.L.; Bateman, V.I. *Force Reconstruction Using the Sum of Weighted Accelerations Technique—Max-Flat Procedure*; Technical Report; Sandia National Labs: Albuquerque, NM, USA, 1993.
23. Allen, M.; Carne, T. Comparison of inverse structural filter (ISF) and sum of weighted accelerations (SWAT) time domain force identification methods. In Proceedings of the 47th AIAA/ASME/ASCE/AHS/ASC Structures, Structural Dynamics, and Materials Conference 14th AIAA/ASME/AHS Adaptive Structures Conference 7th, Newport, RI, USA, 1–4 May 2006; p. 1885.
24. Shiozaki, H.; Geluk, T.; Daenen, F.; Iwanaga, Y.; Van Herbruggen, J. *Time-Domain Transfer Path Analysis for Transient Phenomena Applied to Tip-In/Tip-Out (Shock & Jerk)*; Technical Report, SAE Technical Paper; SAE: Warrendale, PA, USA, 2012.
25. Yan, G.; Sun, H. A non-negative Bayesian learning method for impact force reconstruction. *J. Sound Vib.* **2019**, *457*, 354–367. [CrossRef]
26. Liu, Y.; Shepard, W.S., Jr. Dynamic force identification based on enhanced least squares and total least-squares schemes in the frequency domain. *J. Sound Vib.* **2005**, *282*, 37–60. [CrossRef]
27. Thite, A.; Thompson, D. The quantification of structure-borne transmission paths by inverse methods. Part 2: Use of regularization techniques. *J. Sound Vib.* **2003**, *264*, 433–451. [CrossRef]
28. Nordström, L.J. A dynamic programming algorithm for input estimation on linear time-variant systems. *Comput. Methods Appl. Mech. Eng.* **2006**, *195*, 6407–6427. [CrossRef]
29. Ronasi, H.; Johansson, H.; Larsson, F. A numerical framework for load identification and regularization with application to rolling disc problem. *Comput. Struct.* **2011**, *89*, 38–47. [CrossRef]
30. Feng, D.; Feng, M.Q. Identification of structural stiffness and excitation forces in time domain using noncontact vision-based displacement measurement. *J. Sound Vib.* **2017**, *406*, 15–28. [CrossRef]
31. Pan, C.D.; Yu, L.; Liu, H.L. Identification of moving vehicle forces on bridge structures via moving average Tikhonov regularization. *Smart Mater. Struct.* **2017**, *26*, 085041. [CrossRef]
32. Leclere, Q.; Pezerat, C.; Laulagnet, B.; Polac, L. Indirect measurement of main bearing loads in an operating diesel engine. *J. Sound Vib.* **2005**, *286*, 341–361. [CrossRef]
33. Liu, J.; Sun, X.; Han, X.; Jiang, C.; Yu, D. A novel computational inverse technique for load identification using the shape function method of moving least square fitting. *Comput. Struct.* **2014**, *144*, 127–137. [CrossRef]
34. Miao, B.; Zhou, F.; Jiang, C.; Chen, X.; Yang S. A comparative study of regularization method in structure load identification. *Shock Vib.* **2018**, *2018*. [CrossRef]
35. Uhl, T. The inverse identification problem and its technical application. *Arch. Appl. Mech.* **2007**, *77*, 325–337. [CrossRef]
36. Nordberg, T.P.; Gustafsson, I. Using QR factorization and SVD to solve input estimation problems in structural dynamics. *Comput. Methods Appl. Mech. Eng.* **2006**, *195*, 5891–5908. [CrossRef]
37. Nordberg, T.P.; Gustafsson, I. Dynamic regularization of input estimation problems by explicit block inversion. *Comput. Methods Appl. Mech. Eng.* **2006**, *195*, 5877–5890. [CrossRef]
38. Feng, D.; Sun, H.; Feng, M.Q. Simultaneous identification of bridge structural parameters and vehicle loads. *Comput. Struct.* **2015**, *157*, 76–88. [CrossRef]
39. Liu, J.; Meng, X.; Jiang, C.; Han, X.; Zhang, D. Time-domain Galerkin method for dynamic load identification. *Int. J. Numer. Methods Eng.* **2016**, *105*, 620–640. [CrossRef]
40. Qiao, B.; Zhang, X.; Wang, C.; Zhang, H.; Chen, X. Sparse regularization for force identification using dictionaries. *J. Sound Vib.* **2016**, *368*, 71–86. [CrossRef]
41. Liu, J.; Qiao, B.; He, W.; Yang, Z.; Chen, X. Impact force identification via sparse regularization with generalized minimax-concave penalty. *J. Sound Vib.* **2020**, *484*, 115530. [CrossRef]
42. Samagassi, S.; Jacquelin, E.; Khamlichi, A.; Sylla, M. Bayesian sparse regularization for multiple force identification and location in time domain. *Inverse Probl. Sci. Eng.* **2019**, *27*, 1221–1262. [CrossRef]
43. Pan, C.; Ye, X.; Zhou, J.; Sun, Z. Matrix regularization-based method for large-scale inverse problem of force identification. *Mech. Syst. Signal Process.* **2020**, *140*, 106698 [CrossRef]
44. Chen, Z.; Chan, T.H.; Yu, L. Comparison of regularization methods for moving force identification with ill-posed problems. *J. Sound Vib.* **2020**, *478*, 115349. [CrossRef]

45. Boukria, Z.; Perrotin, P.; Bennani, A. Experimental impact force location and identification using inverse problems: Application for a circular plate. *Int. J. Mech.* **2011**, *5*, 48–55.
46. Hundhausen, R.J.; Adams, D.E.; Derriso, M. Impact loads identification in standoff metallic thermal protection system panels. *J. Intell. Mater. Syst. Struct.* **2007**, *18*, 531–541. [CrossRef]

Article

An Inverse Problem for Quantum Trees with Delta-Prime Vertex Conditions

Sergei Avdonin [1,2,†] **and Julian Edward** [3,*,†]

1 Department of Mathematics and Statistics, University of Alaska Fairbanks, Fairbanks, AK 99775, USA; saavdonin@alaska.edu
2 Moscow Center for Fundamental and Applied Mathematics, Moscow 119333, Russia
3 Department of Mathematics and Statistics, Florida International University, Miami, FL 33199, USA
* Correspondence: edwardj@fiu.edu; Tel.: +1-305-562-5630
† These authors contributed equally to this work.

Received: 30 September 2020; Accepted: 16 November 2020; Published: 17 November 2020

Abstract: In this paper, we consider a non-standard dynamical inverse problem for the wave equation on a metric tree graph. We assume that the so-called delta-prime matching conditions are satisfied at the internal vertices of the graph. Another specific feature of our investigation is that we use only one boundary actuator and one boundary sensor, all other observations being internal. Using the Neumann-to-Dirichlet map (acting from one boundary vertex to one boundary and all internal vertices) we recover the topology and geometry of the graph together with the coefficients of the equations.

Keywords: inverse problems; quantum graphs; delta-prime vertex conditions

---

1. Introduction

This paper concerns inverse problems for differential equations on quantum graphs. Under quantum graphs or differential equation networks (DENs) we understand differential operators on geometric graphs coupled by certain vertex matching conditions. Network-like structures play a fundamental role in many problems of science and engineering. The range for the applications of DENs is enormous. Here is a list of a few.

–*Structural Health Monitoring.* DENs, classically, arise in the study of stability, health, and oscillations of flexible structures that are made of strings, beams, cables, and struts. Analysis of these networks involve DENs associated with heat, wave, or beam equations whose parameters inform the state of the structure, see, e.g., [1].

–*Water, Electricity, Gas, and Traffic Networks.* An important example of DENs is the Saint-Venant system of equations, which model hydraulic networks for water supply and irrigation, see, e.g., [2]. Other important examples of DENs include the telegrapher equation for modeling electric networks, see, e.g., [3], the isothermal Euler equations for describing the gas flow through pipelines, see, e.g., [4], and the Aw-Rascle equations for describing road traffic dynamics, see e.g., [5].

–*Nanoelectronics and Quantum Computing.* Mesoscopic quasi-one-dimensional structures such as quantum, atomic, and molecular wires are the subject of extensive experimental and theoretical studies, see, e.g., [6], the collection of papers in [7–9]. The simplest model describing conduction in quantum wires is the Schrödinger operator on a planar graph. For similar models appear in nanoelectronics, high-temperature superconductors, quantum computing, and studies of quantum chaos, see, e.g., [10–12].

–*Material Science.* DENs arise in analyzing hierarchical materials like ceramic and metallic foams, percolation networks, carbon and graphene nano-tubes, and graphene ribbons, see, e.g., [13–15].

–*Biology*. Challenging problems involving ordinary and partial differential equations on graphs arise in signal propagation in dendritic trees, particle dispersal in respiratory systems, species persistence, and biochemical diffusion in delta river systems, see, e.g., [16–18].

Quantum graph theory gives rise to numerous challenging problems related to many areas of mathematics from combinatoric graph theory to PDE and spectral theories. A number of surveys and collections of papers on quantum graphs appeared in previous years; we refer to the monograph by Berkolaiko and Kuchment, [19], for a complete reference list. The inverse theory of network-like structures is an important part of a rapidly developing area of applied mathematics—analysis on graphs. It is tremendously important for all aforementioned applications. In this paper, we solve a non-standard dynamical inverse problem for the wave equation on a metric tree graph.

Let $\Omega = \{V,\, E\}$ be a finite compact and connected metric tree (i.e., graph without cycles), where V is a set of vertices and E is a set of edges. We recall that a graph is called a *metric graph* if every edge $e_j \in E$, $j = 1, \ldots, N$, is identified with an interval (a_{2j-1}, a_{2j}) of the real line with a positive length l_j. We denote the boundary vertices (i.e., vertices of degree one) by $\Gamma = \{\gamma_0, ..., \gamma_m\}$, and interior vertices (whose degree is at least 2) by $\{v_{m+1}, ..., v_N\}$. The vertices can be regarded as equivalence classes of the edge end points a_j. For each vertex v_k, denote its degree by Υ_k. We write $j \in J(v)$ if $e_j \in E(v)$, where $E(v)$ is the set of edges incident to v.

The graph Ω determines naturally the Hilbert space of square integrable functions $\mathcal{H} = L^2(\Omega)$. We define its subspace $\mathcal{H}^1$ as the space of functions y on Ω such that $y|_e \in H^1(e)$ for every $e \in E$ and $y|_{\Gamma\setminus\{\gamma_0\}} = 0$, and let $\mathcal{H}^{-1}$ be the dual space to $\mathcal{H}^1$. When convenient, we denote the restriction of a function w on Ω to e_j by w_j. For any vertex v_k and function $w(x)$ on the graph, we denote by $\partial w_j(v_k)$ the derivative of w_j at v_k in the direction pointing away from the vertex.

Our system is described by the following initial boundary value problem (IBVP) with so-called delta-prime compatibility conditions at each internal vertex v_k:

$$
\begin{aligned}
u_{tt} - u_{xx} + qu &= 0,\; (x,t) \in (\Omega \setminus V) \times [0,T], &(1)\\
u|_{t=0} = u_t|_{t=0} &= 0,\; x \in \Omega, &(2)\\
\partial u_i(v_k,t) &= \partial u_j(v_k,t),\; i,j \in J(v_k),\; v_k \in V \setminus \Gamma,\; t \in [0,T], &(3)\\
\sum_{j \in J(v_k)} u_j(v_k,t) &= 0,\; v_k \in V \setminus \Gamma,\; t \in [0,T], &(4)\\
\frac{\partial u}{\partial x}(\gamma_0,t) &= f(t),\; t \in [0,T], &(5)\\
u(\gamma_k,t) &= 0,\; k = 1,\ldots,m,\; t \in [0,T]. &(6)
\end{aligned}
$$

Here, T is arbitrary positive number, $q_j \in C([a_{2j-1}, a_{2j}])$ for all j, and $f \in L^2(0,T)$. The physical interpretation of conditions (3) and (4), and some other matching conditions was discussed in [20].

The well-posedness of this system is discussed in Section 2; it will be proved that $u \in C([0,T]; \mathcal{H}^1) \cap C^1([0,T]; \mathcal{H})$. In what follows, we refer to γ_0 as the root of Ω and f as the control.

We now pose our inverse problem. Assume an observer knows the boundary condition (5), and that (6) holds at the other boundary vertices, and that the graph is a tree. The unknowns are the number of boundary vertices and interior vertices, the adjacency relations for this tree, i.e., for each pair of vertices, whether or not there is an edge joining them, the lengths $\{\ell_j\}$, and the function q. We wish to determine these quantities with a set of measurements that we describe now. We can suppose v_N is the interior vertex adjacent to γ_0 with e_1 the edge joining the two, see Figure 1. Our first measurement is then the following measurement at γ_0:

$$
(R_{0,1}f)(t) := u_1^f(\gamma_0, t). \tag{7}
$$

We show that from operator $R_{0,1}$ one can recover ℓ_1 and the degree Υ_N of v_N. Then by a well known argument, see [21], one can then determine q_1.

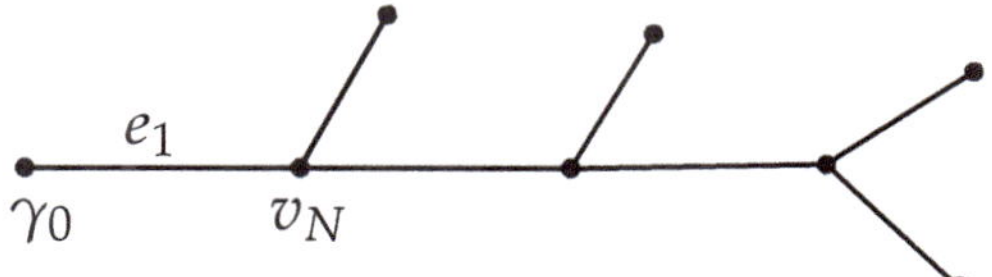

Figure 1. A metric tree.

Having established these quantities, in our second step, we propose to place sensors on the edges incident to v_N, and using these measurements together with $R_{0,1}$ to determine the data associated to these edges. Note that the one control remains at γ_0. The goal is to repeat these steps until all data associated to the graph have been determined. To define the interior measurements we require more notation. For each interior vertex v_k we list the incident edges by $\{e_{k,j} \, : \, j = 1, ..., \Upsilon_k\}$. Here $e_{k,1}$ is chosen to be the edge lying on the unique path from γ_0 to v_k, and the remaining edges are labeled randomly, see Figure 2. Then the sensors measure

$$(R_{k,j}f)(t) := u_j^f(v_k, t), \; k = m+1, ..., N, \; j = 2, ..., \Upsilon_k - 1. \tag{8}$$

We show that we do not need sensors at $e_{k,1}, e_{k,\Upsilon_k}$. Thus the total number of sensors is $1 + \sum_{j=m+1}^{N}(\Upsilon_j - 2)$. It is easy to check that this number is equal to $|\Gamma| - 1$. We denote by R^T the $(|\Gamma| - 1)$-tuple $(R_{0,1}, R_{N,2}, R_{N,3}, ...)$ acting on $L^2(0, T)$.

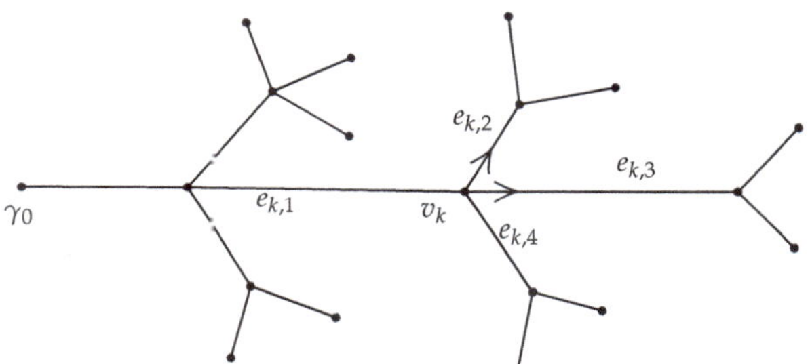

Figure 2. Sensors at vertex v_k marked by arrows.

2. Results

Let ℓ be equal to the maximum distance between γ_0 and any other boundary vertex. Our main result is the following

Theorem 1. *Assume $q_j \in C([a_{2j-1}, a_{2j}])$ for all j. Suppose $T > 2\ell$. Then from R^T one can determine the number of interior and boundary vertices, the adjacency relations of the tree, q, and the lengths of the edges.*

3. Discussion

We now compare this result to others in the literature. We are unaware of any works treating the inverse problem on general tree graphs with delta-prime conditions on the internal vertices. The most common conditions for internal vertices are continuity together with Kirchhoff–Neumann condition: $\sum_{j \in J(v_k)} \partial u_j(v_k, t) = 0$ and all references in this paragraph assume these conditions. In [21], the authors assume that controls and measurements take place at all boundary vertices but one. The authors use an iterative method called "leaf peeling", where the response operator on Ω is used first to determine the data on the edges adjacent to the boundary, and then to determine the response operator associated to a proper subgraph. In [21], the leaf peeling argument includes spectral methods that require knowing R^T for all T. In [22], the methods of [21] are extended to the case where masses are placed at internal vertices, see also [23]; however these methods still require knowledge of R^T for all T. Also in [22], it is proven that that for a single string of length ℓ with N attached masses and $T > 2\ell$, $R^T_{0,1}$ is sufficient to solve the inverse problem. In particular, [23] uses a spectral variant of the boundary control method, together with the relationship between the response operator and the connecting operator. In [24,25], a dynamical leaf peeling argument is developed for a tree with no masses and with response operators at all but one boundary points, allowing for the solution of the inverse problem for finite T sufficiently large. An important ingredient in their leaf peeling is determining the response operators associated with subtrees, called "reduced response operators", from the response operator associated to the original tree. In all of these papers, it is assumed that there are no interior measurements. In [26], the iterative methods from [24,25,27] are adapted to a tree with masses placed at internal vertices, with a single control at the roof and measurements there and at internal vertices. For other works on quantum graphs, see [1,16,19,28–31].

A special feature of the present paper is that we use only one control together internal observations. This may be useful in some physical settings where some or most boundary points are inaccessible. Another potential advantage of the method presented here is that we recover all parameters of the graphs, including its topology, from the $(|\Gamma| - 1)$-tuple response operator acting on $L^2(0,T)$. In previous papers, the authors recovered the graph topology from a larger number of measurements: the $(|\Gamma|-1) \times (|\Gamma|-1)$ matrix (boundary) response operator or, equivalently, from $(|\Gamma|-1) \times (|\Gamma|-1)$ Titchmarsh–Weyl matrix function. In [32], the inverse problems on a star graph for the wave equation with general self-adjoint matching conditions was solved by the $(|\Gamma|-1) \times (|\Gamma|-1)$ matrix boundary response operator.

4. Materials and Methods

4.1. Preliminaries

In what follows, we use the notations

$$\mathcal{F}^n = \{f \in \mathcal{H}^n(\mathbb{R}) : f(t) = 0 \text{ if } t \leq 0\}, \tag{9}$$

where $\mathcal{H}^n(\mathbb{R})$ are the standard Sobolev spaces. We define the Heaviside function by $H(t) = 1$ for $t > 0$, and $H(t) = 0$ for $t < 0$. Then, we define $H_n \in \mathcal{F}^n$ as the unique solution to

$$\frac{d^n}{dt^n} H_n = H;$$

at times we use $H_{-1}(t)$, resp. $H_{-2}(t)$ for $\delta(t)$, resp. $\delta'(t)$. Here $\delta(t)$ denotes the Dirac delta function supported at $t = 0$. In this section and those that follow, we drop the superscript T from R^T when convenient.

Consider a star shaped graph with edges $e_1, ..., e_N$. For each j, we identify e_j with the interval $(0, \ell_j)$ and the central vertex with $x = 0$, see Figure 3.

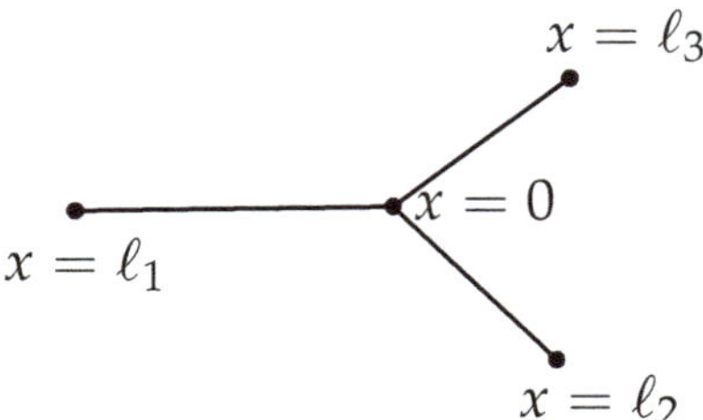

Figure 3. Star with coordinate system: e_j identified with $[0, \ell_j]$.

Recall the notation $q_j = q|_{e_j}$, and $u_j(*,t) = u(*,t)|_{e_j}$. Thus, we consider the system

$$
\begin{aligned}
\frac{\partial^2 u}{\partial t^2} - \frac{\partial^2 u}{\partial x^2} + qu &= 0,\ x \in e_j,\ j = 1, ..., N,\ t \in \times[0,T], &(10)\\
u|_{t=0} = u_t|_{t=0} &= 0, &(11)\\
\partial u_i(0,t) &= \partial u_j(0,t),\ i \neq j,\ t \in [0,T], &(12)\\
\sum_{j=1}^{N} u_j(0,t) &= 0,\ t \in [0,T], &(13)\\
\partial u_1(\ell_1,t) &= f(t),\ t \in [0,T], &(14)\\
u_j(\ell_j,t) &= 0,\ j = 2, ..., N,\ t \in [0,T]. &(15)
\end{aligned}
$$

Let u^f solve (10)–(15), and set

$$g_j(t) = u_j^f(0,t),\ j = 1, ..., N. \tag{16}$$

For (11), it is standard that the waves have unit speed of propagation on the interval, so $g_j(t) = 0$ for $t < \ell_1$ and all j. It will be useful first to consider the vibrating string on an interval.

4.2. Representation of Solution on an Interval and Reduced Response Operator

We adapt a representation of $u^f(x,t)$ developed in [27], where only Dirichlet control and boundary conditions were considered. Fix $j \in \{1, ..., N\}$. We extend q_j to $(0,\infty)$ as follows: first evenly with respect to $x = \ell_j$, and then periodically. Thus $q_j(2k\ell_j \pm x) = q_j(x)$ for all positive integers k.

Define w_j to be the solution to the Goursat-type problem

$$
\begin{cases}
\frac{\partial w^2}{\partial s^2}(x,s) - \frac{\partial w^2}{\partial x^2}(x,s) + q_j(x)w(x,s) = 0, & 0 < x < s < \infty,\\
w_x(0,s) = 0,\ w(x,x) = -\frac{1}{2}\int_0^x q_j(\eta)d\eta,\ x > 0. &
\end{cases}
$$

A proof of solvability of this problem can be found in [33].

Consider the IBVP on the interval $(0,\ell_j)$:

$$
\begin{aligned}
\tilde{u}_{tt} - \tilde{u}_{xx} + q_j(x)\tilde{u} &= 0,\ 0 < x < \ell_j,\ t \in (0,T), &(17)\\
\tilde{u}(x,0) = \tilde{u}_t(x,0) &= 0,\ 0 < x < \ell_j, &(18)\\
\partial\tilde{u}(0,t) &= p(t), &(19)\\
\tilde{u}(\ell_j,t) &= 0,\ t > 0. &(20)
\end{aligned}
$$

Let $P(t) = -\int_0^t p(s)ds$. Then, the solution to (17)–(20) on e_j can be written as

$$
\begin{aligned}
\tilde{u}_j^p(x,t) = &\sum_{n\geq 0:\ 0\leq 2n\ell_j+x\leq t} (-1)^n \left(P(t-2n\ell_j-x) + \int_{2n\ell_j+x}^{t} w_j(2n\ell_j+x,s)P(t-s)ds \right)\\
&+ \sum_{n\geq 1:\ 0\leq 2n\ell_j-x\leq t} (-1)^n \left(P(t-2n\ell_j+x) + \int_{2n\ell_j-x}^{t} w_j(2n\ell_j-x,s)P(t-s)ds \right).
\end{aligned} \tag{21}
$$

In what follows, we only consider $t \leq T$ for some finite T, so all sums will be finite.

Let us now change the condition (20) to $\tilde{u}_x(\ell_j,t)=0$. In this case, the solution becomes

$$\begin{aligned}\tilde{u}_j^p(x,t) &= \sum_{n\geq 0:\ 0\leq 2n\ell_j+x\leq t}\left(P(t-2n\ell_j-x)+\int_{2n\ell_j+x}^{t} w_j(2n\ell_j+x,s)P(t-s)ds\right)\\ &+\sum_{n\geq 1:\ 0\leq 2n\ell_j-x\leq t}\left(P(t-2n\ell_j+x)+\int_{2n\ell_j-x}^{t} w_j(2n\ell_j-x,s)P(t-s)ds\right).\end{aligned}$$

To represent the solution of the wave equation on the edge e_1 in a star graph, we must account for the control at $x=\ell_1$. Thus it will also be useful to represent the solution of a wave equation on an interval when the control is on the right end. Consider the IBVP:

$$\begin{aligned} v_{tt}-v_{xx}+q_1(x)v &= 0,\ 0<x<\ell_1,\ t>0,\\ v(x,0)=v_t(x,0) &= 0,\ 0<x<\ell_1,\\ \partial v(0,t) &= 0,\\ \partial v(\ell_1,t) &= f(t),\ t>0. \end{aligned} \tag{22}$$

Set $\tilde{q}_1(x)=q_1(\ell_1-x)$, and extend $\tilde{q}_1$ to $[0,\infty)$ by $\tilde{q}_1(2k\ell_1\pm x)=\tilde{q}_1(x)$. Define k_1 to be the solution to the Goursat-type problem

$$\begin{cases} \frac{\partial k^2}{\partial s^2}(x,s)-\frac{\partial k^2}{\partial x^2}(x,s)+\tilde{q}_1(x)k(x,s)=0, & 0<x<s,\\ k_x(0,s)=0,\ k(x,x)=-\frac{1}{2}\int_0^x\tilde{q}_1(\eta)d\eta,\ x<\ell_j. \end{cases}$$

Let $F(t)=-\int_0^t f(s)ds$. One can then verify that

$$\begin{aligned} v^f(x,t) &= F(t-\ell_1+x)+\int_{\ell_1-x}^{t}k_1(\ell_1-x,s)F(t-s)\,ds\\ &+F(t-\ell_1-x)+\int_{\ell_1+x}^{t}k_1(\ell_1+x,s)F(t-s)\,ds\\ &+F(t-3\ell_1+x)+\int_{3\ell_1-x}^{t}k_1(3\ell_1-x,s)F(t-s)\,ds\\ &+F(t-3\ell_1-x)+\int_{3\ell_1+x}^{t}k_1(3\ell_1+x,s)F(t-s)\,ds\\ &\ldots \end{aligned} \tag{23}$$

Thus

$$v^f(0,t)=2\sum_{n=1}^{\infty}\left(F(t-(2n-1)\ell_1)+\int_{(2n-1)\ell_1}^{t}k_1((2n-1)\ell_1,s)F(t-s)\right). \tag{24}$$

We now show that the system (10)–(15) is well-posed. Recall that $\mathcal{F}^1$ was defined in (9), and $g_j(t)=u_j^f(0,t)$.

Theorem 2.

(a) If $f\in L^2(0,T)$, then there exists a unique solution $u(x,t)$ solving the system (10)–(15), and mapping $t\mapsto u^f(x,t)$ is in $C(0,T;\mathcal{H}^1)\cap C^1(0,T;\mathcal{H})$.

(b) For each $j=1,...,N$, the mapping $f\mapsto g_j$ is a continuous mapping $L^2(0,T)\mapsto\mathcal{F}^1$.

Proof. On $[0,\ell_j]$ with $j\geq 2$, the wave will be generated by the "control" $\partial(u_j^f)(0,t)$, whereas on $[0,\ell_1]$ the wave is generated by the two controls $\partial(u_1^f)(0,t)$, $\partial(u_1^f)(\ell_1,t)=f(t)$. We assume first that $f\in C_0^2(0,T)$.

Let

$$p(t) := (u_j^f)_x(0,t), \text{ and } P(t) = -\int_0^t p(s)ds. \tag{25}$$

Here, p is independent of j by (12). We have that u^f is given by

$$u_1^f = \tilde{u}_1^p + v^f, \text{ and for } j \geq 2,\ u_j^f = \tilde{u}_j^p. \tag{26}$$

Note that v^f has already been explicitly determined in (23). Thus, by (21), we have an explicit solution for u^f if we can solve for p. We now prove the existence, uniqueness, and regularity of P.

By (21) and (26), we have for $j \geq 2$,

$$\begin{aligned} g_j(t) &= P(t) + \int_0^t w_j(0,s)P(t-s)ds \\ &\quad +2\sum_{n\geq 1}(-1)^n\big(P(t-2n\ell_j) + \int_{2n\ell_j}^t w_j(2n\ell_j,s)P(t-s)ds\big). \end{aligned} \tag{27}$$

For $j = 1$, we have by (21), (24), and (26) that

$$\begin{aligned} g_1(t) &= P(t) + \int_0^t w_1(0,s)P(t-s)ds \\ &+ 2\sum_{n\geq 1}(-1)^r\big(P(t-2n\ell_1) + \int_{2n\ell_1}^t w_1(2n\ell_1,s)P(t-s)ds\big) \\ &+ 2\sum_{n\geq 1}\big(F(t-(2n-1)\ell_1) + \int_{(2n-1)\ell_1}^t k_1((2n-1)\ell_1,s)F(t-s)\big). \end{aligned} \tag{28}$$

We remark that at the moment, we have not yet solved for either P or g_j for any j. Let

$$\alpha = \min\{\ell_j,\ j = 1, ..., N\}.$$

We solve for P with an iterative argument using steps of length 2α. The iterations are necessary because the upper limits in the sums in (27), (28) increase with time due to reflections of the wave at the various vertices. In what follows, we label by $G(t)$ various terms that we have already solved for, which by (24), includes $v^f(0,t)$. For $t \leq \ell_1$ we have by unit wave speed that $P(t) = 0$. Suppose now $t \in [\ell_1, \ell_1 + 2\alpha]$. Then,

$$t - 2\ell_j \leq \ell_1 + 2\alpha - 2\ell_j < \ell_1,$$

and hence $P(t-s) = 0$ for $s \geq 2n\ell_j$, for all j with $n \geq 1$. By (13), we have

$$\sum_1^N g_j(t) = 0,$$

and hence from (27) and (28) we get

$$NP(t) + \int_0^t \Big(\sum_{j=1}^N w_j(0,s)\Big)P(t-s)ds = G(t),\ t \in [\ell_1, \ell_1 + 2\alpha]. \tag{29}$$

It is easy to show that this is a Volterra equation of the second kind (VESK), and so admits a unique solution P with $||P||_{L^2(\ell_1,\ell_1+2\alpha)} \leq ||F||_{L^2(0,2\alpha)}$. Furthermore, by differentiating this equation we get $||p||_{L^2(\ell_1,\ell_1+2\alpha)} \leq ||f||_{L^2(0,2\alpha)}$.

Having solved for P on $[0, \ell_1 + 2\alpha]$, we now suppose $t \in [\ell_1 + 2\alpha, \ell_1 + 4\alpha]$. Thus for any j and any $n \geq 1$, we have $t - 2n\ell_j \leq \ell_1 + 2\alpha$, so all terms in (27), (28) involving $P(t-s)$, with $s \leq 2n\ell_j$ and $n \geq 1$, are known. Thus by (27), (28), and $\sum_1^N g_j(t) = 0$,

$$NP(t) + \int_0^t (\sum_{j=1}^N w_j(0,s))P(t-s)ds = G(T),\ t \in [\ell_1 + 2\alpha, \ell_1 + 4\alpha].$$

We can solve this VESK to determine uniquely $P(t)$ for $t \in [\ell_1 + 2\alpha, \ell_1 + 4\alpha]$, with the estimate $\|p\|_{L^2(\ell_1, \ell_1+4\alpha)} \leq \|f\|_{L^2(0,4\alpha)}$ holding. Iterating this process, we solve for the unique $P(t)$ for $t \in [0, T]$ as desired. The case for $f \in L^2(0, T)$ is then obtained by continuity. Part (a) of the Theorem follows easily from (21),(23), and (26). Part (b) of the theorem follows from Part (a) and (27) and (28). □

Define

$$R_1 f = u_1^f(\ell_1, t).$$

Proposition 1. *For R_1 one can determine q_1, ℓ_1, and N.*

Proof. Let $f(t) = \delta(t)$, so $F(t) = -H(t)$. From (24) and (29) one has, for $t < 3\ell_1$,

$$NP(t) + \int_0^t (\sum_{j=1}^N w_j(0,s))P(t-s)ds = 2H(t-\ell_1) + 2\int_{\ell_1}^t k_1(\ell_1, s)ds.$$

Thus, we have $P(t) = \frac{2}{N}H(t-\ell_1) + cont$, where *cont* denotes various continuous functions. We have by (22)

$$\begin{aligned} u_1^f(\ell_1, t) &= v^f(\ell_1, t) + \tilde{u}(\ell_1, t) \\ &= -H(t) - \int_0^t k_1(0,s)ds + \frac{2}{N}H(t - 2\ell_1) + cont. \end{aligned}$$

Clearly, the discontinuity at $t = 2\ell_1$ gives us ℓ_1 and N. That $R_{0,1}^{2\ell_1}$ determines q_1 is proven in [16]. □

Define the "reduced response operator" on e_j, with $j \geq 2$, by

$$(\tilde{R}_{0,j}p)(t) = \tilde{u}_j^p(0, t)$$

associated to the IBVP (17)–(20). From (21), we immediately obtain

Lemma 1. *For $j = 2, ..., N$, and any $h \in C_0^\infty(\mathbb{R}^+)$, we have*

$$(\tilde{R}_{0,j}h)(t) = \int_0^t \tilde{R}_{0,j}(s)h(t-s)ds,$$

with

$$\tilde{R}_{0,j}(s) = -1 - 2\sum_{n\geq 1}(-1)^n H(s - 2n\ell_j) - \tilde{r}_{0j}(s), \tag{30}$$

with $\tilde{r}_{0j}(0) = 0$. If T is finite, the sums above are finite.

Proof. Using (25), it is easy to see that

$$P(t - 2n\ell_j) = \int_0^t H(s - 2n\ell_j)p(t-s)ds.$$

The lemma now follows easily from (21). □

In what follows, we refer to $\tilde{R}_{0,j}(s)$ as the "reduced response function". For $f(t) = \delta(t)$, we denote the solution to the system (10)–(15) as u^δ. We also use the following.

Lemma 2. *Let $p(t) = (u_j^\delta)_x(0,t)$. For $j = 2, ..., N$, we have*

$$p(t) = \sum_{m\geq 1} \psi_m \delta(t - \zeta_m) + \theta_m H(t - \zeta_m) + a(t). \tag{31}$$

Here $a \in \mathcal{F}^1$, and $a(s) = 0$ for $s < \zeta_1, \zeta_1 = \beta_1 = \ell_1$ and $\psi_1 \neq 0$.

This result holds from the proof of Theorem 2, the unit speed of wave propagation, and the properties of wave reflections off $x = \ell_j$, see [33]. The details are left to the reader.

The following result follows from (22), (25), and Lemma 2. The details of the proof are left to the reader.

Corollary 1. *Let $g_j(t) = u_j^\delta(0,t)$. For $j = 2, ..., N$, we have*

$$g_j(t) = \sum_{k\geq 1} \phi_k H(t - \gamma_k) + A(t). \tag{32}$$

Here $A \in \mathcal{F}^1$, and $A(s) = 0$ for $s < \gamma_1, \gamma_1 = \ell_1$.

4.3. Solution of Inverse Problem

Here, we establish some notation. We recall the following notation: for v_k we list the incident edges by $\{e_{k,j} : j = 1, ..., Y_k\}$. Here, $e_{k,1}$ is chosen to be the edge lying on the path from γ_0 to v_k, and the remaining edges are labeled randomly.

Now let k_0 be some fixed interior vertex, and let j_0 satisfy $1 < j_0 \leq Y_{k_0}$. Denote by $\Omega_{k_0}^{j_0}$ the unique subtree of Ω having v_{k_0} as root with incident edge e_{k_0,j_0}, and by $V_{k_0}^{j_0}$ the set of its vertices, see Figure 4.

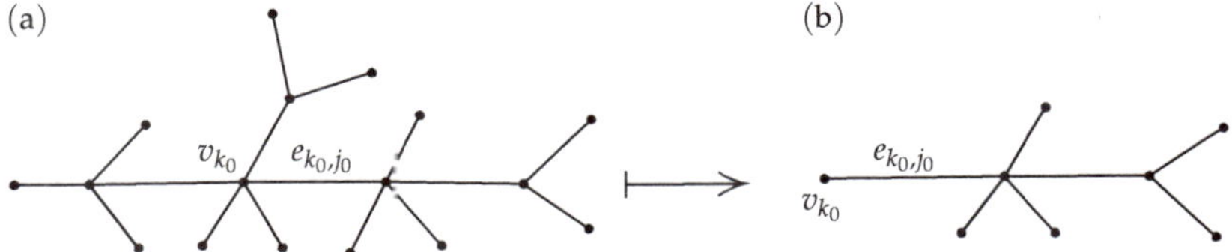

Figure 4. **(a)** Ω, **(b)** Subtree $\Omega_{k_0}^{j_0}$.

We define an associated response operator as follows. Let $\Gamma_{k_0}^{j_0} = \{v_{k_0}, \gamma_{N_0}, ..., \gamma_N\}$ be the boundary vertices on $\Omega_{k_0}^{j_0}$. Suppose $\varphi = \varphi^b$ solves the IBVP

$$\frac{\partial^2 \varphi}{\partial t^2} - \frac{\partial^2 \varphi}{\partial x^2} + q\varphi = 0, \ x \in \Omega_{k_0}^{j_0} \setminus V_{k_0}^{j_0}, \ t \in \times[0,T], \tag{33}$$

$$\varphi|_{t=0} = \varphi_t|_{t=0} = 0, \tag{34}$$

$$\partial\varphi(v_k,t) = \partial\varphi_j(v_k,t), \ j \in J(v_k), \ v_k \in V_{k_0}^{j_0} \setminus \Gamma_{k_0}^{j_0}, \ t \in [0,T], \tag{35}$$

$$\sum_{j\in J(v_k)} \varphi_j(v_k,t) = 0, \ v_k \in V_{k_0}^{j_0} \setminus \Gamma_{k_0}^{j_0}, \ t \in [0,T], \tag{36}$$

$$\partial\varphi(v_{k_0},t) = b(t), \ t \in [0,T], \tag{37}$$

$$\varphi(\gamma_l,t) = 0, \ l = N_0, ..., N, \ t \in [0,T]. \tag{38}$$

Then we define an associated reduced response operator

$$(\tilde{R}_{k_0,j_0}b)(t) = \varphi^b_{j_0}(v_{k_0}, t),$$

with associated response function $\tilde{R}_{k_0,j_0}(s)$.

Suppose we determined $\tilde{R}_{k_0,j_0}$. It would follow from Proposition 1 that one could recover the following data: ℓ_{j_0}, q_{j_0}, and $\Upsilon_{k'}$, where $v_{k'}$ is the vertex adjacent to v_{k_0} in $\Omega^{j_0}_{k_0}$. In this section we will present an iterative method to determine the operator $\tilde{R}_{k_0,j_0}$ from the $(|\Gamma|-1)$-tuple of operators, R^T, which we know by hypothesis for some $T > 2\ell$. An important ingredient is the following generalization to a tree of Corollary 1.

Lemma 3. *Let $T > 0$, and let $R^T_{k,j}$ be associated with (33)–(38), defined by (7) and (8). The response function for $R^T_{k,j}$ has the form*

$$R_{k,j}(s) = r_{k,j}(s) + \sum_{n \geq 1} \phi_n H(s - \gamma_n).$$

Here, $r_{k,j} \in \mathcal{F}^1$, and the sequence $\{\gamma_n\}$ is positive and strictly increasing. If T is finite then the sums are finite.

Proof. The proof follows from the proof of Corollary 1, together with the transmission and reflection properties of waves at interior vertices, and reflection properties at boundary vertices. □

Fix $T > 2\ell$. The rest of this section shows how to recover $\tilde{R}_{k_0,j_0}$ from R^T.

Step 1

For the first step, let v_{k_0} be the vertex adjacent to the root γ_0, with associated edge labeled e_1. By Proposition 1, we can use $R^T_{0,1}$ to recover Υ_{k_0}, ℓ_1, q_1.

Step 2

Consider $e_{k_0,2}$. In Step 2, we show how to solve for $\tilde{R}_{k_0,2}$, see Figure 5.

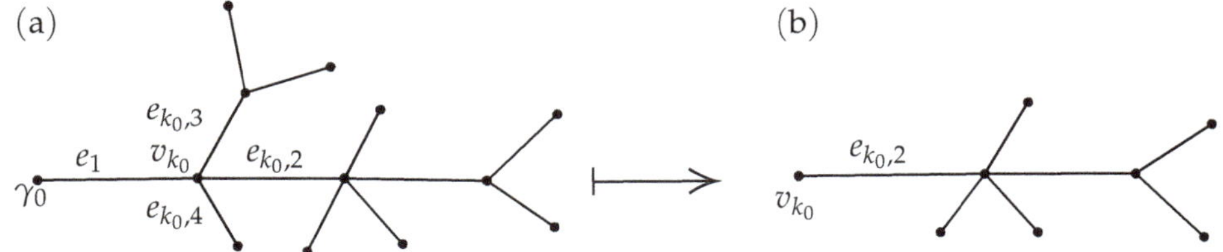

Figure 5. (**a**) Ω. (**b**) Subtree $\Omega^2_{k_0}$, with $e_1 = e_{k_0,1}$.

Since v_{k_0} is the root of $\Omega^2_{k_0}$, the following equation is essentially a restatement of Lemma 1 to trees; the details of its proof are left to the reader.

$$\tilde{R}_{k_0,2}(s) = \tilde{r}_{k_0,2}(s) + \sum_{m \geq 0} \alpha_m H(s - \xi_m). \tag{39}$$

Here, $0 = \xi_0 < \xi_1 < \ldots$, and $\tilde{r}_{k_0,2}(s) \in \mathcal{F}^1$. In what follows in Step 2, for readability, we rewrite $\tilde{r}_{k_0,2}$ as $\tilde{r}$.

Since we know ℓ_1 and q_1, we can solve the wave equation on e_1 with known boundary data. We identify e_1 as the interval $(0, \ell_1)$ with v_{k_0} corresponding to $x = 0$. Then u^f, restricted to e_1, solves the following Cauchy problem, where we view x as the "time" variable:

$$\begin{aligned} u_{tt} - u_{xx} + q_1 u &= 0,\ x \in (0, \ell_1),\ t > 0, \\ \partial u(\ell_1, t) &= f(t),\ t > 0, \\ u(\ell_1, t) &= (R_{0,1} f)(t),\ t > 0, \\ u(x, 0) &= 0,\ x \in (0, \ell_1). \end{aligned}$$

Since the function $R_{0,1}(s)$ is known, we can thus uniquely determine $u^f(0,t) = u^f(v_{k_0}, t)$ and $\partial u_1^f(v_{k_0}, t)$. Thus $p(t) = \partial u_2^\delta(v_{k_0}, t)$ is determined.

We now show how p and $u_2^\delta(v_{k_0}, t)$ can be used to determine $\tilde{R}_{k_0,2}(s)$. The following equation follows from the definition of the response operators for any $f \in L^2$:

$$\int_0^t \tilde{R}_{k_0,2}(s) p(t-s) ds = u_2^\delta(v_{k_0}, t) = \int_0^t R_{k_0,2}(s) \delta(t-s) ds. \tag{40}$$

In what follows, it is convenient to extend $f(t) \in L^2(0,T)$ as zero for $t < 0$. By Lemma 1 and by an adaptation of Lemma 2 to general trees, we have the following expansions:

$$R_{k_0,2}(s) = r_{k_0,2}(s) + \sum_{n \geq 1} \phi_n H(t - \gamma_n)\ r_{k_0,2}|_{s \in (0,\beta_1)} = 0,\ \beta_1 = \ell_1, \tag{41}$$

$$p(s) = a(s - \ell_1) + \sum_{l \geq 1} \psi_l \delta(s - \zeta_l) + \theta_l H(s - \zeta_l), \zeta_1 = \nu_1 = \ell_1, \psi_1 \neq 0,. \tag{42}$$

Here, $r_{k_0,2} \in \mathcal{F}^1$ and $a(s) \in \mathcal{F}^1$, and $\{\xi_k\}$ and $\{\beta_n\}$ are positive and increasing. Clearly $a(s), r_{k_0,2}(s), \{\psi_m\}, \{\theta_m\}, \{\phi_n\}, \{\gamma_n\}$, can all be determined by $R_{0,1}$ and $R_{k_0,2}$, whereas for now $\tilde{r}$ and the sets $\{\alpha_m\}, \{\xi_m\}$ are unknown. Inserting (39),(41) and (42) into (40), we get

$$\begin{aligned} r_{k_0,2}(t) + \sum_n \phi_n H(t - \gamma_n) = \int_C^t \tilde{r}(s) a(t - s - \ell_1) ds + \sum_l \psi_l \tilde{r}(t - \zeta_l) + \int_0^t \sum_l \theta_l H(t - s - \zeta_l) \tilde{r}(s) ds \\ + \sum_m \alpha_m \int_0^t a(t - s - \ell_1) H(s - \xi_m) ds + \sum_{m,l} \psi_l \alpha_m H(t - \zeta_l - \xi_m) + \sum_{m,l} \theta_l \alpha_m \int_0^t H(s - \xi_m) H(t - s - \zeta_l) ds. \end{aligned} \tag{43}$$

Here all sums have 1 as lower limit of summation.

Lemma 4. *The sets $\{\alpha_m\}, \{\xi_m\}$ can be determined by $R_{0,1}$ and $R_{k_0,2}$.*

Proof. We mimic an iterative argument in [26]. Differentiating (43) and then matching the delta singularities, we get

$$\sum_{n \geq 1} \phi_n \delta(t - \gamma_n) = \sum_{m \geq 1} \sum_{l \geq 1} \psi_l \alpha_m \delta(t - \zeta_l - \xi_m). \tag{44}$$

Since the sequences $\{\gamma_n\}, \{\zeta_l\}, \{\xi_m\}$ are all strictly increasing, clearly we have $\gamma_1 = \zeta_1 + \xi_1$, so that $\phi_1 = \alpha_1 \psi_1$, and so $\xi_1 = \gamma_1 - \zeta_1$ and $\alpha_1 = \phi_1 / \psi_1$. We represent that the set $\{\phi_1, \gamma_1\}$, $\{\zeta_1, \psi_1\}$ determines the set $\{\xi_1, \alpha_1\}$ by

$$\{\phi_1, \gamma_1\},\ \{\xi_1, \psi_1\} \implies \{\xi_1, \alpha_1\}.$$

We now match the term $\delta(t - \gamma_2)$ with its counterpart on the right hand side of (44). There are three possible cases.

Case 1: $\gamma_2 \neq \zeta_2 + \xi_1$.

In this case, we must have

$$\xi_2 = \gamma_2 - \zeta_1,\ \alpha_2 = \phi_2/\psi_1.$$

Case 2a: $\gamma_2 = \zeta_2 + \xi_1$ and $\phi_2 \neq \psi_2\alpha_1$. Note that the last inequality can be verified by an observer at this stage. Then $\gamma_2 = \zeta_1 + \xi_2$ and $\phi_2 = \psi_1\alpha_2 + \psi_2\alpha_1$. and hence

$$\xi_2 = \zeta_1 - \gamma_2,\ \alpha_2 = (\phi_2 - \psi_2\alpha_1)/\psi_1.$$

Case 2b: $\gamma_2 = \zeta_2 + \xi_1$ and $\phi_2 = \psi_2\alpha_1$. Then $\gamma_2 \neq \zeta_1 + \xi_2$. Note we have not yet solved for $\{\xi_2, \alpha_2\}$. In this case, we now repeat the matching coefficient argument just used with $\delta(t - \gamma_3)$.

Again there are three cases:

Case 2bi: $\gamma_3 \neq \zeta_3 + \xi_1$. Note all of these terms are known, so this inequality can be verified. In this case, $\gamma_3 = \zeta_1 + \xi_2$, so $\xi_2 = \gamma_3 - \zeta_1$ and $\alpha_2 = \phi_3/\psi_1$.

Case 2bii: $\gamma_3 = \zeta_3 + \xi_1$ and $\phi_3 \neq \alpha_1\psi_3$. Then $\gamma_3 = \zeta_1 + \xi_2$, and $\phi_3 = \alpha_1\psi_3 + \alpha_2\psi_1$. Thus $\xi_2 = \gamma_3 - \zeta_1$ and $\alpha_2 = (\phi_3 - \alpha_1\psi_3)/\psi_1$.

Case 2biii: $\gamma_3 = \zeta_3 + \xi_1$ and $\phi_3 = \alpha_1\psi_3$. Then $\gamma_3 < \zeta_1 + \xi_2$, and we need to continue our procedure with γ_4.

Repeating this procedure as necessary, say for a total of N_2 times, we solve for $\{\xi_2, \alpha_2\}$. We represent this process as

$$\{\phi_k, \gamma_k\}_{k=1}^{N_2} \implies \{\xi_k, \alpha_k\}_{k=1}^{2}.$$

We must have N_2 finite by (44) and the finiteness of the graph.

Iterating this procedure, suppose for $p \in \mathbb{N}$ we have

$$\{\phi_k, \gamma_k\}_{k=1}^{N_p} \implies \{\xi_k, \alpha_k\}_{k=1}^{p}.$$

Here N_p is chosen to be minimal, and so $\gamma_{N_p} = \zeta_1 + \xi_p$. We wish to solve for $\{\zeta_{p+1}, \phi_{p+1}\}$.

We can again distinguish three cases:

Case 1: $\gamma_{(N_p+1)} \neq \zeta_k + \xi_j,\ \forall j \leq p,\ \forall k$. Note that we know $\{\xi_j\}_1^p$ and $\{\zeta_k\}$, so these inequalities are verifiable. In this case, we must have $\gamma_{(N_p+1)} = \zeta_1 + \xi_{p+1}$ and $\psi_1\alpha_{p+1} = \phi_{(N_p+1)}$, so we have determined α_{p+1}, ξ_{p+1} in this case.

Case 2: There exists an integer Q and pairs $\{\zeta_{i_n}, \xi_{j_n}\}_{n=1}^{Q}$, with $j_n \leq p$, such that

$$\gamma_{(N_p+1)} = \zeta_{i_1} + \xi_{j_1} = \ldots = \zeta_{i_Q} + \xi_{j_Q}. \tag{45}$$

Note that all the numbers $\{\zeta_{i_n}, \xi_{j_n}\}$ have been determined, so these equations can be all verified. We can assume all pairs $\{\zeta_{i_n}, \xi_{j_n}\}$ satisfying (45) with $j_n \leq p$ are listed. In this case, we have either

Case 2i: $\phi_{(N_p+1)} \neq \alpha_{j_1}\psi_{i_1} + \ldots + \alpha_{j_Q}\psi_{i_Q}$. It follows then that $\gamma_{(N_p+1)} = \zeta_1 + \xi_{p+1}$, and

$$\phi_{(N_p+1)} = \alpha_{p+1}\psi_1 + \phi_{j_1}\psi_{i_1} + \ldots + \phi_{j_Q}\psi_{i_Q}.$$

We thus solve for ξ_{p+1}, α_{p+1}.

Case 2ii: $\gamma_{(N_p+1)} = \phi_{j_1}\psi_{i_1} + \ldots + \phi_{j_Q}\psi_{i_Q}$. It follows then that $\alpha_{(N_p+1)} \neq \zeta_1 + \xi_{p+1}$, and we have to repeat this process with $\gamma_{(N_p+2)}$.

Repeating the reasoning in Case 2ii as often as necessary, we eventually solve for $\{\xi_{p+1}, \alpha_{p+1}\}$. Thus,

$$\{\phi_k, \gamma_k\}_{k=1}^{N_{p+1}} \implies \{\xi_k, \alpha_k\}_{k=1}^{p+1}.$$

Hence, we can solve for $\{\xi_p : p \leq L\}, \{\alpha_p : p \leq L\}$ for any positive integer L given knowledge of $R_{0,1}^T, R_{k_0,2}^T$ for $T = T(L)$ sufficiently large. □

It remains to solve for $\tilde{r}$. In what follows, we set $\tilde{R}(s) = 0$ for $s < 0$. We use $G(t)$ to denote various functions that we have already established to be determined by $R_{0,1}$ and $R_{k_0,2}$. Having already solved for $\{\zeta_n, \alpha_n\}$, we can eliminate from (43) the Heavyside functions to get, recalling $\zeta_1 = \ell_1$,

$$G(t) = \sum_{l\geq 1} \psi_l \tilde{r}(t-\zeta_l) + \int_0^t \tilde{r}(s)\big(a(t-s-\zeta_1)\big) + \sum_{l\geq 1} \theta_l H(t-s-\zeta_l)ds. \tag{46}$$

We solve this with an iterative argument. Let $\alpha = \min_m\{\zeta_{l+1} - \zeta_l\}$. For $t < \zeta_1 + \alpha$, we have for $l > 1$ that $t - \zeta_l < 0$ so $r(t) = 0$. Hence

$$G(t) = \psi_1 \tilde{r}(t-\zeta_1) + \int_C^t (\theta_1 H(t-s-\zeta_1) + a(t-s-\zeta_1))\tilde{r}(s)ds, \; t < \zeta_1 + \alpha. \tag{47}$$

Letting $r = t - \zeta_1$, we get

$$\begin{aligned} G(r) &= \psi_1 \tilde{r}(r) + \int_0^{r+\zeta_1} (\theta_1 H(r-s) + a(r-s))\tilde{r}(s)ds, \\ &= \psi_1 \tilde{r}(r) + \int_0^{r} (\theta_1 H(r-s) + a(r-s))\tilde{r}(s)ds, \; r < \alpha. \end{aligned}$$

We solve this VESK to determine $\tilde{r}(s)$, $r < \alpha$. Now for $t < \zeta_1 + 2\alpha$, we have for $l > 1$ that $t - \zeta_l < \alpha$, and so those terms in (46) with $t - \zeta_l$ can be absorbed in G to again give

$$G(r) = \psi_1 \tilde{r}(r) + \int_0^{r} (\theta_1 H(r-s) + a(r-s))\tilde{r}(s)ds, \; r < 2\alpha.$$

We solve this VESK to determine $\tilde{r}(s)$, $r < 2\alpha$. Iterating this procedure, we solve for $\tilde{r}(s)$ for any finite s.

Step 3 Because $R_{k_0,j}$ are determined by assumption for $j = 2, ..., \Upsilon_{k_0} - 1$, the functions $u_j^f(v_{k_0}, t)$ are determined. In Step 2, we showed $u_1^f(v_{k_0}, t)$ is also determined. Hence by (4), $u_{\Upsilon_{k_0}}^f(v_{k_0}, t)$ is also determined. We can now carry out the argument in Step 2 on the remaining edges $e_{k_0,3}, ..., e_{k_0,\Upsilon_{k_0}}$ incident on v_{k_0} to determine $\tilde{R}_{k_0,j}$ for all j.

Step 4 For each $j = 2, ..., \Upsilon_{k_0}$, we use Proposition 1, to find the associated ℓ_j, q_j together with the valence of the vertex adjacent to v_{k_0}. Careful reading of Steps 2 and 3 shows that we can use $R_{0,1}^T$ and $R_{k_0,j}^T$ for any $T > 2(\ell_1 + \ell_j)$.

Step 5 Let $v_{k_1}, ...$ be the vertices adjacent to v_{k_0}, other than γ_0. We now iterate Steps 2–4 for the each of these vertices. Choose for instance v_{k_1}. If it were a boundary vertex, this fact would be determined in Step 4, and then this algorithm goes to the next vertex, which we, for convenience, still label v_{k_1}. We can thus assume v_{k_1} is an interior vertex. Let us label an incident edge (other than $e_2 := e_{k_0,2}$) as $e_3 := e_{k_1,3}$, see Figure 6.

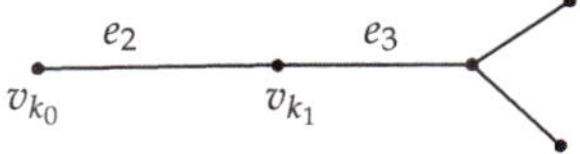

Figure 6. For Step 5: a subtree of $\Omega_{k_0}^2$.

We wish to determine $\tilde{R}_{k_1,3}$. Mimicking Step 2, let u^δ solve (1)–(6), let $b(t) = \partial u_3^\delta(v_{k_1}, t)$. We have the following formula holding by the definition of response operators:

$$\int_0^t \tilde{R}_{k_1,3}(s)\; b(t-s)ds = \int_0^t R_{k_1,3}(s)\delta(t-s)ds.$$

Of course $R_{k_1,3}(s)$ is assumed to be known. We determine b as follows. We have, from Step 2, that $p(t) = \partial u_1^\delta(v_{k_0}, t)$ is known. We identify e_2 as the interval $(0, \ell_2)$ with v_{k_1} corresponding to $x = 0$.

Then $b(t) = \partial u_2^f(v_{k_1}, t)$ arises as a solution to the following Cauchy problem on e_2, where we view x as the "time" variable:

$$\begin{aligned} y_{tt} - y_{xx} + q_2 y &= 0,\ x \in (0, \ell_2),\ t > 0 \\ y_x(\ell_2, t) &= p(t),\ t > 0 \\ y(\ell_2, t) &= (R_{k_0,2}\delta)(t),\ t > 0 \\ y(x, 0) &= 0,\ x \in (0, \ell_2). \end{aligned}$$

Since q_2, ℓ_2, and $R_{k_0,2}$ are all known, we can thus determine $b(t) = y_x(0, t)$.

The rest of the argument here is a straightforward adaptation of Steps 2–4 above. The details are left to the reader.

Step 6 Arguing as in Step 5, we determine $\tilde{R}_{k,j}$ for all other vertices adjacent to v_{k_0} and their associated edges. The details are left to the reader.

Steps above 6 Clearly this procedure can be iterated until all edges of our finite graph have been covered.

5. Conclusions

In this paper, we applied the ideas of the boundary control and leaf peeling methods to solve an inverse problem on a tree featuring non-standard, delta-prime vertex conditions on the interior. Our method required using only one boundary actuator and one boundary sensor, all other observations being internal. Using the Neumann-to-Dirichlet map (acting from one boundary vertex to one boundary and all internal vertices) we recovered the topology and geometry of the graph together with the coefficients q_j of the equations. It would be interesting to see a numerical implementation of our method. It would also be interesting to adapt our methods to quantum graphs with cycles.

Author Contributions: Conceptualization, S.A. and J.E.; methodology, S.A. and J.E.; formal analysis, S.A. and J.E.; writing—original draft preparation, S.A. and J.E.; writing—review and editing, S.A. and J.E. All authors have read and agreed to the published version of the manuscript.

Funding: The research of the first author was supported in part by the National Science Foundation, grant DMS 1909869.

Conflicts of Interest: The authors declare no conflict of interest. The funders had no role in the design of the study; in the collection, analyses, or interpretation of data; in the writing of the manuscript, or in the decision to publish the results.

Abbreviations

The following abbreviations are used in this manuscript:

VESK Volterra equation of the second kind

References

1. Lagnese, J.; Leugering, G.; Schmidt, E.J.P.G. *Modelling, Analysis, and Control of Dynamical Elastic Multilink Structures*; Birkhauser: Basel, Switzerland, 1994.
2. Gugat, M.; Leugering, G. Global boundary controllability of the Saint-Venant system for sloped canals with friction. *Ann. Inst. H. Poincaré Anal. Non Lineaire* **2009**, *26*, 257–270. [CrossRef]
3. Alì, G.; Bartel, A.; Günther, M. Parabolic Differential-Algebraic Models in Electrical Network Design. *Multiscale Model. Simul.* **2005**, *4*, 813–838. [CrossRef]
4. Bastin, G.; Coron, J.M; dÁndrèa Novel, B. Using hyperbolic systems of balance laws for modeling, control and stability analysis of physical networks. In *Proceedings of The Lecture Notes for the Pre-Congress Workshop on Complex Embedded and Networked Control Systems 17th IFAC World Congress, Seoul, Korea, 16–20 July 2008*; Elsevier: Amsterdam, The Netherlands, 2008.

5. Colombo, R.M.; Guerra, G.; Herty, M.; Schleper, V. Optimal control in networks of pipes and canals. *SIAM J. Control Optim.* **2009**, *48*, 2032–2050. [CrossRef]
6. Hurt, N.E. *Mathematical Physics of Quantum Wires and Devices*; Kluwer: Dordrecht, The Netherlands, 2000.
7. Joachim, C.; Roth, S. (Eds.) *Atomic and Molecular Wires*; NATO Science Series. Series E: Applied Sciences; Kluwer: Dordrecht, The Netherlands, 1997; p. 341.
8. Kostrykin, V.; Schrader, R. Kirchoff's rule for quantum wires. *J. Phys. A Math. Gen.* **1999**, *32*, 595–630. [CrossRef]
9. Kostrykin, V.; Schrader, R. Kirchoff's rule for quantum wires II: The inverse problem with possible applications to quantum computers. *Fortschritte Derphysik* **2000**, *48*, 703–716. [CrossRef]
10. Kottos, T.; Smilansky, U. Quantum chaos on graphs. *Phys. Rev. Lett.* **1997**, *79*, 4794–4797. [CrossRef]
11. Kottos, T.; Smilansky, U. Periodic orbit theory and spectral statistics for quantum graphs. *Ann. Phys.* **1999**, *274*, 76–124. [CrossRef]
12. Melnikov, Y.B.; Pavlov, B.S. Two-body scattering on a graph and application to simple nanoelectronic devices. *J. Math. Phys.* **1995**, *36*, 2813–2825. [CrossRef]
13. Adam, S.; Hwang, E.H.; Galits, V.M.; Das Sarma, S. A self-consistent theory for graphene transport. *Proc. Natl. Acad. Sci. USA* **2007**, *104*, 18392–18397. [CrossRef]
14. Peres, N.M.R. Scattering in one-dimensional heterostructures described by the Dirac equation. *J. Phys. Condens. Matter* **2009**, *21*, 095501. [CrossRef]
15. Peres, N.M.R.; Rodrigues, J.N.B.; Stauber, T.; Lopes dos Santos, J.M.B. Dirac electrons in graphene-based quantum wires and quantum dots. *J. Phys. Condens. Matter* **2009**, *21*, 344202. [CrossRef] [PubMed]
16. Avdonin, S.A.; Bell, J. Determining a distributed conductance parameter for a neuronal cable model defined on a tree graph. *Inverse Probl. Imaging* **2015**, *9*, 645–659. [CrossRef]
17. Bell, J.; Craciun, G. A distributed parameter identification problem in neuronal cable theory models. *Math. Biosci.* **2005**, *194*, 1–19. [CrossRef] [PubMed]
18. Rall, W. Core conductor theory and cable properties of neurons. In *Handbook of Physiology, The Nervous System*; American Physiological Society: Rockville, MD, USA, 1977; pp. 39–97.
19. Berkolaiko, G.; Kuchment, P. *Introduction to Quantum Graphs, (Mathematical Surveys and Monographs*; American Mathematical Society: Providence, RI, USA, 2013; Volume 186.
20. Exner, P. Vertex couplings in quantum graphs: Approximations by scaled Schrödinger operators. In *Mathematics in Science and Technology*; World Sci. Publ.: Hackensack, NJ, USA, 2011; pp. 71–92.
21. Avdonin, S.A.; Kurasov, P. Inverse problems for quantum trees. *Inverse Probl. Imaging* **2008**, *2*, 1–21. [CrossRef]
22. Al-Musallam, F.; Avdonin, S.A.; Avdonina, N.; Edward, J. Control and inverse problems for networks of vibrating strings with attached masses. *Nanosyst. Phys. Chem. Math.* **2016**, *7*, 835–841. [CrossRef]
23. Avdonin, S.A.; Avdonina, N.; Edward, J. Boundary inverse problems for networks of vibrating strings with attached masses. In Proceedings of the Dynamic Systems and Applications, Volume 7, Dynamic, Atlanta, GA, USA, 27–30 May 2015, pp. 41–44.
24. Avdonin, S.A.; Mikhaylov, V.S.; Nurtazina, K.B. On inverse dynamical and spectral problems for the wave and Schrödinger equations on finite trees. The leaf peeling method. *J. Math. Sci.* **2017**, *224*, 1–15. [CrossRef]
25. Avdonin, S.A.; Zhao, Y. Leaf peeling method for the wave equation on metric tree graphs. *Inverse Probl. Imaging* **2020**. [CrossRef]
26. Avdonin, S.A.; Edward, J. An inverse problem for quantum trees. **2020**, submitted. [CrossRef]
27. Avdonin, S.A.; Zhao, Y. Exact controllability of the 1-d wave equation on finite metric tree graphs. *Appl. Math. Optim.* **2020**. [CrossRef]
28. Avdonin, S.A. Control problems on quantum graphs. In *Analysis on Graphs and Its Applications (Proceedings of Symposia in Pure Mathematics)*; AMS: Pawtucket, RI, USA, 2008; Volume 77, pp. 507–521.
29. Avdonin, S.A.; Nicaise, S. Source identification problems for the wave equation on graphs. *Inverse Probl.* **2015**, *31*, 095007. [CrossRef]
30. Belishev, M.I.; Vakulenko, A.F. Inverse problems on graphs: Recovering the tree of strings by the BC-method. *J. Inverse Ill-Posed Probl.* **2006**, *14*, 29–46. [CrossRef]
31. Dager, R.; Zuazua, E. *Wave Propagation, Observation and Control in 1-d Flexible Multi-Structures, (Mathematiques and Applications)*; Springer: Berlin/Heidelberg, Germany, 2006; Volume 50.

32. Avdonin, S.A.; Kurasov, P.; Nowaczyk, M. Inverse problems for quantum trees II: On the reconstruction of boundary conditions for star graphs. *Inverse Probl. Imaging* **2010**, *4*, 579–598. [CrossRef]
33. Avdonin, S.A.; Edward, J. Controllability for string with attached masses and Riesz bases for asymmetric spaces. *Math. Control Relat. Fields* **2019**, *9*, 453–494. [CrossRef]

MDPI

Article

Data-Informed Decomposition for Localized Uncertainty Quantification of Dynamical Systems

Waad Subber *, Sayan Ghosh, Piyush Pandita, Yiming Zhang and Liping Wang

Probabilistic Design and Optimization Group, GE Research, 1 Research Circle, Niskayuna, NY 12309, USA; sayan.ghosh1@ge.com (S.G.); piyush.pandita@ge.com (P.P.); yiming.zhang@ge.com (Y.Z.); wangli@ge.com (L.W.)
* Correspondence: Waad.Subber@ge.com

Abstract: Industrial dynamical systems often exhibit multi-scale responses due to material heterogeneity and complex operation conditions. The smallest length-scale of the systems dynamics controls the numerical resolution required to resolve the embedded physics. In practice however, high numerical resolution is only required in a confined region of the domain where fast dynamics or localized material variability is exhibited, whereas a coarser discretization can be sufficient in the rest majority of the domain. Partitioning the complex dynamical system into smaller easier-to-solve problems based on the localized dynamics and material variability can reduce the overall computational cost. The region of interest can be specified based on the localized features of the solution, user interest, and correlation length of the material properties. For problems where a region of interest is not evident, Bayesian inference can provide a feasible solution. In this work, we employ a Bayesian framework to update the prior knowledge of the localized region of interest using measurements of the system response. Once, the region of interest is identified, the localized uncertainty is propagate forward through the computational domain. We demonstrate our framework using numerical experiments on a three-dimensional elastodynamic problem.

Keywords: Bayesian inference; uncertainty quantification; dynamical systems; inverse problem; machine learning; system Identification; Gaussian process; polynomial chaos

Citation: Subber, W.; Ghosh, S.; Pandita, P.; Zhang, Y.; Wang, L. Data-Informed Decomposition for Localized Uncertainty Quantification of Dynamical Systems. *Vibration* **2021**, *4*, 49–63. https://doi.org/10.3390/vibration4010004

Received: 19 November 2020
Accepted: 16 December 2020
Published: 31 December 2020

Publisher's Note: MDPI stays neutral with regard to jurisdictional claims in published maps and institutional affiliations.

1. Introduction

With the increase in demand for high-performance and highly-efficient systems, the complexity of industrial design and manufacturing processes are increasing proportionally, exposing many opportunities for novel technologies as well as many associated technical challenges. For example, advancement in the design of composite structures allows us to reduce weight, advancement in additive manufacturing enables us to reduce cost. Introducing a new technology typically happens at the lowest level of the systems hierarchy (e.g., at the parts or sub-component levels). Extending new technologies to the system level requires rigorous testing for many years. For example, in the eighties, composite material was only used for limited components of an aircraft (i.e., the wing and tail [1]). Recently, however, about 50% of the material used in the Boeing 787 Dreamliner are composite material [2]. In the industrial setting, the process of adaptation of a new technology can be accelerated by proper assessment of uncertainty at various aspects of the products life cycle. For example, at the design stage of an aircraft wing rib, it is crucial to consider the effect of uncertainty in the material and operation conditions on the aeroelastic dynamics of the wing [3]. At the manufacturing stage, it is important to consider the impact of material uncertainty on the quality control [4,5]. The maintenance stage requires a holistic assessment of the effect of measurement uncertainty on the static and dynamic responses of the wing during structural health monitoring [6].

Quantifying uncertainty at the system level often requires a physics-based computational model for the entire structure. However, in structures such as an aircraft wing, traditional computational models may become too complex and costly for simulating the multi-scale dynamical response due to material heterogeneity at the sub-component level. The effect of the sub-component on the entire structure depends on the size, location and loading conditions of the part. It is therefore, necessary to consider a different level of fidelity for the analysis of the sub-components in order to reduce the cost and complexity of uncertainty quantification. To this end, the concept of localized uncertainty propagation for dynamical systems having multi spatio-temporal scales can be utilized to address such issues [7–9].

In this work, we consider assessing the effect of localized uncertainty in a region of interest within the entire structure. The framework is based on two uncoupled steps: (1) identification of the region of interest, (2) quantifying the effect of localized uncertainty. For structures composed of distinct parts, the localized region of interest for uncertainty propagation can be easily identified. Alternatively, measurement data of the system response can be used to identify the localized region of interest. The Bayesian paradigm integrates computational models and observational data in one framework to update the current state of knowledge [10,11]. Estimating the posterior probability density function in the Bayesian method requires solving the forward model many times, which may become challenging for limited computational budget. This issue is often addressed by building a surrogate model such as a Gaussian Process (GP) regression model [12]. The GP models are non-parametric and Bayesian in nature, and they provide uncertainty bounds on their predictions. Once the region of interest is identified, a Polynomial Chaos (PC) expansion [13,14] is used to propagate the uncertain material properties of the localized region through the entire domain. In contrast to References [7,8] the contributions of this work—(1) The partitioning of the domain is inferred from measurement of the system response, (2) Gaussian process model is used as a surrogate in the Bayesian framework, (3) non-intrusive polynomial chaos approach is used for localized uncertainty propagation. The rest of this work is organized as follows—in Section 2, we provide the problem statement and the associated mathematical formulations. Our numerical demonstrations are provided in Section 3. We provide the conclusions of the current work in Section 4.

2. Methodology

In this section, we present the mathematical framework of our approach for data-driven partitioning scheme for localized uncertainty quantification. In particular, in Section 2.1, we introduce the problem statement in the Bayesian setting. For problems where the localized region of interest is not defined explicitly, we rely on measurement data of the system response to infer the localized region of interest using Bayesian framework. The Bayesian framework requires a computational model (the forward problem) to estimate the response of the system for a given set of the input parameters. Consequently, in Section 2.2, we discuss the stochastic elastodynamic problem and its finite element discretization. Estimating the localized region of interest in the Bayesian setting necessitates many solutions to the stochastic elastodynamic problem which can become computationally demanding. A surrogate model for the system response can be used to reduce the computational cost of the Bayesian framework as will be presented in Section 2.3. Once the localized region of interest is estimated, uncertainty representation of the material properties within the region of interest can be performed. The localized uncertainty is propagated forward through the entire computational domain in order to estimate the effect on the material variability on the response. For this task, we use the polynomial chose expansion for efficient assessment of uncertainty with less computational cost. The polynomial chose expansion is reviewed in Section 2.4.

2.1. Bayesian Inference

In the Bayesian inference, the prior knowledge is updated to posterior using noisy measurements and the response of a physical model [10,11]. The update is based on the Bayes' rule defined as

$$p(\theta|\mathbf{d}) = \frac{p(\theta)p(\mathbf{d}|\theta)}{p(\mathbf{d})}, \tag{1}$$

where θ is the unknown parameter to be estimated, $\mathbf{d}$ is the measurement of an observable quantity, $p(\theta|\mathbf{d})$ is the posterior probability density function, $p(\theta)$ is the prior probability density function and $p(\mathbf{d}|\theta)$ denotes the likelihood of the observations given the parameter. We assume that the measured data, $\mathbf{d}$, is generated from a statistical model represented as

$$\mathbf{d} = \mathbb{M}(\theta) + \epsilon, \tag{2}$$

where $\mathbb{M}(\theta)$ denotes a physical model and ϵ is a measurement noise represented as a Gaussian random variable with unknown variance $\epsilon \sim \mathcal{N}(0, \sigma_n^2)$. For a Gaussian noise, the likelihood function becomes

$$p(\mathbf{d}|\theta) = \frac{1}{(2\pi\sigma_n^2)^{-N/2}} \exp\left(-\sum_i^N \frac{[d_i - \mathbb{M}(\theta_i)]^2}{2\sigma_n^2} \right). \tag{3}$$

The task in hand is to utilize the measurement $\mathbf{d}$ and the physical model $\mathbb{M}(\theta)$ to estimate the system parameters θ. The process requires many executions to the physical model $\mathbb{M}(\theta)$, which can be computationally expensive. The computational model is often approximated by a simpler easy to evaluate model as:

$$\mathbb{M}(\theta) \simeq \mathcal{M}(\theta), \tag{4}$$

where $\mathcal{M}(\theta)$ denotes the surrogate model that is constructed using limited runs of the physical model $\mathbb{M}(\theta)$. In our work, we represent $\mathcal{M}(\theta)$ as the Gaussian process surrogate model [15]. Once we construct the surrogate model, the localized features parameterized by θ is estimated using Markov Chain Monte Carlo (MCMC) sampling technique [16,17]. Having identified the region of interest, a localized uncertainty quantification of the material properties can be performed efficiently using polynomial chaos expansion [13].

2.2. The Forward Problem

We consider an arbitrary physical domain $\Omega \in \mathbb{R}^d$ with $\partial\Omega$ being its boundary as shown in Figure 1a, and define the following problem:

Find a random function $\mathbf{u}(\mathbf{x}, t, \boldsymbol{\xi}) : \Omega \times [0, T_f] \times \Xi \rightarrow \mathbb{R}$, such that the following equations hold

$$\begin{aligned}
\rho(\xi)\ddot{\mathbf{u}}(\mathbf{x}, t, \boldsymbol{\xi}) &= \nabla \cdot \sigma + \mathbf{b} && \text{in } \Omega \times [0, T_f] \times \Xi, \\
\mathbf{u}(\mathbf{x}, t, \boldsymbol{\xi}) &= \bar{\mathbf{u}} && \text{on } \partial\Omega_u \times [0, T_f] \times \Xi, \\
\sigma \cdot \mathbf{n} &= \bar{\mathbf{t}} && \text{on } \partial\Omega_t \times [0, T_f] \times \Xi, \\
\mathbf{u}(\mathbf{x}, 0, \boldsymbol{\xi}) &= \mathbf{u}_0 && \text{in } \Omega \times \Xi, \\
\dot{\mathbf{u}}(\mathbf{x}, 0, \boldsymbol{\xi}) &= \dot{\mathbf{u}}_0 && \text{in } \Omega \times \Xi,
\end{aligned} \tag{5}$$

where $\rho(\xi)$ is the mass density, σ is the stress tensor, $\mathbf{u}$ is the displacement field, $\mathbf{b}$ is the body force per unit volume, $\bar{\mathbf{u}}$ is the prescribed displacement on $\partial\Omega_u$, $\bar{\mathbf{t}}$ is the prescribed traction on $\partial\Omega_t$, $\mathbf{n}$ is a unit normal to the surface, and $\mathbf{u}_0$ and $\dot{\mathbf{u}}_0$ are the initial displacement and velocity, respectively. Here, we define the stochastic space by (Θ, Σ, P), where Θ denoting the sample space, Σ being the σ-algebra of Θ, and P representing an appropriate probability measure. The stochastic space is parameterized by a finite set of standardized identically distributed random variables $\boldsymbol{\xi} = \{\xi_i(\theta)\}_{i=1}^M$, where $\theta \in \Theta$. The support of the random variables is defined as $\Xi = \Xi_1 \times \Xi_2 \times \cdots \Xi_M \in \mathbb{R}^M$ with a joint probability density function given as $p(\boldsymbol{\xi}) = p_1(\xi_1) \cdot p_2(\xi_2) \cdots p_M(\xi_M)$.

For linear isotropic elastic martial, the constitutive relation between the stress and strain tensors is given by:

$$\sigma = \lambda(\boldsymbol{\xi})\mathrm{tr}(\boldsymbol{\varepsilon})\mathbf{I} + 2\mu(\boldsymbol{\xi})\boldsymbol{\varepsilon}, \tag{6}$$

where $\lambda(\boldsymbol{\xi})$ and $\mu(\boldsymbol{\xi})$ are the Lamé's parameters, $\mathbf{I}$ is an identity tensor and $\boldsymbol{\varepsilon}$ is the symmetric strain tensor defined as

$$\boldsymbol{\varepsilon} = \frac{1}{2}\left(\nabla\mathbf{u} + \nabla\mathbf{u}^T\right). \tag{7}$$

For a random Young's modulus $E(\mathbf{x}, \boldsymbol{\xi})$ and deterministic Poisson's ratio ν , the Lamé's parameters can be expressed as

$$\lambda(\boldsymbol{\xi}) = \frac{E(\mathbf{x}, \boldsymbol{\xi})\nu}{(1+\nu)(1-2\nu)}, \quad \mu(\boldsymbol{\xi}) = \frac{E(\mathbf{x}, \boldsymbol{\xi})}{2(1+\nu)}. \tag{8}$$

We consider the case that uncertainty stems from a localized variability in a confined region within the physical domain. For example as shown in Figure 1b, the variability in the quantity of interest can be attributed to random material properties within the subdomain Ω_2. The artificial martial boundaries shown in Figure 1b for subdomain Ω_2 is estimated using Bayesian inference. Localizing random variability in the neighborhood of the quantity of interest reduces the computational cost of uncertainty propagation in problems where a region of interest can be specified. Depending on the interest in the region, each subdomain can have its local uncertainty representation and the corresponding mesh and time resolution. As a result, the Asynchronous Space-Time Domain Decomposition Method with Localized Uncertainty Quantification (PASTA-DDM-UQ) [7–9] can be utilized. In PASTA-DDM-UQ, spatial, temporal and material decomposition are considered. In this work however, we only consider material decomposition and apply non-intrusive approach for uncertainty propagation.

Consequently, let the physical domain Ω be partitioned based on the martial variability into n_s non-overlapping subdomains $\Omega_s, 1 \leqslant s \leqslant n_s$ as shown in Figure 1b and such that:

$$\Omega = \bigcup_{s=1}^{n_s} \Omega_s, \quad \Omega_s \bigcap \Omega_r = \varnothing \text{ for } s \neq r, \quad \Gamma = \bigcup_{s=1}^{n_s} \Gamma_s, \quad \Gamma_s = \partial\Omega_s \backslash \partial\Omega. \tag{9}$$

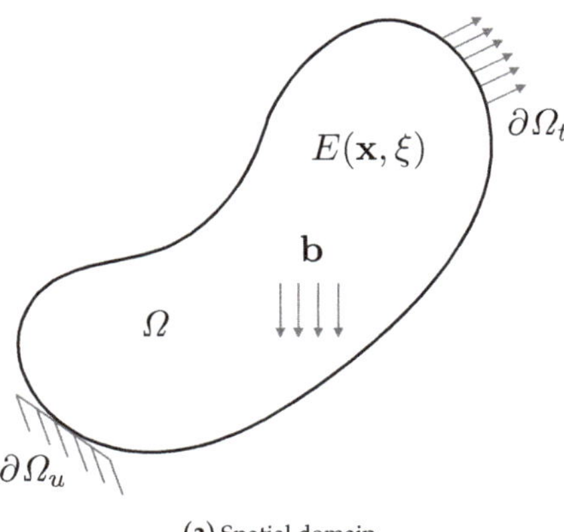

(a) Spatial domain **(b)** Domain decomposition.

Figure 1. An arbitrary computational domain Ω with a random material property (i.e., $E(\mathbf{x}, \boldsymbol{\xi})$) and its partitioning into non-overlapping subdomains. The partitioning is based on material variability.

Note that the partitioning boundaries are not set *a priori* as it will be estimated using noisy measurement of the system response. According to the decomposition in Equation (9), the stochastic dynamical problem in Equation (5) can be transformed into the following minimization problem:

Find a random function $\mathbf{u}(\mathbf{x}, t, \boldsymbol{\xi}) : \Omega \times [0, T_f] \times \Xi \to \mathbb{R}$, such that

$$\mathcal{L}(\mathbf{u}, \dot{\mathbf{u}}) = \sum_{s=1}^{n_s} (\mathcal{T}_s(\dot{\mathbf{u}}) - \mathcal{V}_s(\mathbf{u})) \to \min, \quad s = 1, \cdots, n_s, \tag{10}$$

where $\mathcal{L}(\mathbf{u}, \dot{\mathbf{u}})$ is the Lagrangian of the system, $\mathcal{T}_s(\dot{\mathbf{u}})$ denotes the subdomain kinetic energy and $\mathcal{V}_s(\mathbf{u})$ is the subdomain potential energy defined as:

$$\mathcal{T}_s(\dot{\mathbf{u}}) = \int_\Xi \int_{\Omega_s} \frac{1}{2} \rho_s(\xi) \dot{\mathbf{u}} \cdot \dot{\mathbf{u}} \, \mathrm{d}\Omega \mathrm{d}\Xi, \tag{11}$$

$$\mathcal{V}_s(\mathbf{u}) = \int_\Xi \left(\int_{\Omega_s} \frac{1}{2} \boldsymbol{\varepsilon} : \boldsymbol{\sigma}_s \, \mathrm{d}\Omega + \int_{\Omega_s} \mathbf{u} \cdot \mathbf{b}_s \, \mathrm{d}\Omega + \int_{\partial\Omega_t} \mathbf{u} \cdot \bar{\mathbf{t}}_s \, \mathrm{d}\Gamma \right) \mathrm{d}\Xi, \tag{12}$$

where $\mathbf{b}_s$ and $\bar{\mathbf{t}}_s$ are the subdomain body force and the prescribed traction, respectively. The Hamilton's principle with a dissipation term reads

$$\int_0^{T_f} \left(\delta\mathcal{L} - \frac{\partial \mathcal{Q}}{\partial \dot{\boldsymbol{\varepsilon}}} : \delta\boldsymbol{\varepsilon} \right) \mathrm{d}t = 0, \tag{13}$$

where $\delta\mathcal{L}$ is the first variation of the augmented Lagrangian defined as

$$\begin{aligned} \delta\mathcal{L} = \sum_{s=1}^{n_s} \int_\Xi \Bigg(\int_{\Omega_s} \rho_s(\xi) \delta\dot{\mathbf{u}} \cdot \dot{\mathbf{u}} \, \mathrm{d}\Omega - \int_{\Omega_s} \delta\boldsymbol{\varepsilon} : \mathbf{D}_s(\boldsymbol{\xi}) : \boldsymbol{\varepsilon} \, \mathrm{d}\Omega + \\ \textstyle\int_{\Omega_s} \delta\mathbf{u} \cdot \mathbf{b}_s \, \mathrm{d}\Omega + \int_{\partial\Omega_t} \delta\mathbf{u} \cdot \bar{\mathbf{t}}_s \, \mathrm{d}\Gamma \Bigg) \mathrm{d}\Xi, \end{aligned} \tag{14}$$

here we define $\mathbf{D}_s(\boldsymbol{\xi})$ as the uncertain linear elasticity tensor. The dissipation function $\mathcal{Q}(\dot{\mathbf{u}})$ in the Hamilton is defined as

$$\mathcal{Q}(\dot{\mathbf{u}}) = \sum_{s=1}^{n_s} \frac{1}{2} \int_\Xi \int_{\Omega_s} \dot{\boldsymbol{\varepsilon}} : \widehat{\mathbf{D}}_s : \dot{\boldsymbol{\varepsilon}} \, \mathrm{d}\Omega \mathrm{d}\Xi, \quad s = 1, \cdots, n_s, \tag{15}$$

where $\widehat{\mathbf{D}}_s$ is the damping tensor assumed to be deterministic. Substituting Equations (15) and (14) into the Hamilton's principle Equation (13) gives the following stochastic equation of motion for a typical subdomain Ω_s

$$\begin{aligned} \textstyle\int_\Xi \int_{\Omega_s} \rho_s(\xi) \ddot{\mathbf{u}} \cdot \delta\mathbf{u} \, \mathrm{d}\Omega \, \mathrm{d}\Xi + \int_\Xi \int_{\Omega_s} \dot{\boldsymbol{\varepsilon}} : \widehat{\mathbf{D}}_s : \delta\boldsymbol{\varepsilon} \, \mathrm{d}\Omega \, \mathrm{d}\Xi + \int_\Xi \int_{\Omega_s} \boldsymbol{\varepsilon} : \mathbf{D}_s(\boldsymbol{\xi}) : \delta\boldsymbol{\varepsilon} \, \mathrm{d}\Omega \, \mathrm{d}\Xi \\ = \textstyle\int_\Xi \int_{\Omega_s} \delta\mathbf{u} \cdot \mathbf{b}_s \, \mathrm{d}\Omega \, \mathrm{d}\Xi + \int_\Xi \int_{\partial\Omega_t} \delta\mathbf{u} \cdot \bar{\mathbf{t}}_s \, \mathrm{d}\Gamma \, \mathrm{d}\Xi. \end{aligned} \tag{16}$$

In the next section, we describe the finite element discretization of the weak form defined in Equation (16).

Spatial and Temporal Discretizations

Let the spatial domain Ω be triangulated with finite elements of size h and let the associated finite element subspace be defined as $\mathcal{X}_h \subset H_0^1(\Omega)$, the spatial component of the solution to the stochastic problem is then sought in the tensor product function space $W = H_0^1(\Omega) \otimes L^2(\Theta)$ defined as [14,18]

$$W = \{w(\mathbf{x}, \theta) : \Omega \times \Theta \to \mathbb{R} \mid \|w\|_W^2 < \infty\}, \subset H_0^1(\Omega) \otimes L^2(\Theta), \tag{17}$$

where the energy norm $\|\cdot\|_W^2$ is defined as

$$\|w(\mathbf{x}, \theta)\|_W^2 = \int_\Theta \left(\int_\Omega \kappa(\mathbf{x}, \theta) |\nabla w(\mathbf{x}, \theta)|^2 \mathrm{d}\mathbf{x} \right) \mathrm{dP}(\theta). \tag{18}$$

The tensor product space W can be viewed as a stochastic space consists of random functions satisfying the Dirichlet boundary condition and having a finite second order moment. For a given realization of the underlying random variables of the stochastic space, an approximate finite element solution to the deterministic part can be expressed as

$$\mathbf{u}^h = \sum_i^{n_i} \mathbf{N}_i(\mathbf{x})\tilde{\mathbf{u}}^i(t), \tag{19}$$

where $\mathbf{N}_i(\mathbf{x})$ are traditional spatial finite element basis functions and $\tilde{\mathbf{u}}^i(t)$ are the nodal values of the solution as a function of time [19]. Substituting the discrete field, Equation (19) in the weak form Equation (16) gives the following semi-discretized stochastic equation of motion :

$$\int_{\Xi} (\mathbf{M}\ddot{\mathbf{u}}(t) + \mathbf{C}\dot{\mathbf{u}}(t) + \mathbf{K}\mathbf{u}(t))\mathrm{d}\Xi = \int_{\Xi} \mathbf{F}(t)\mathrm{d}\Xi. \tag{20}$$

We drop the nodal finite element marks (tilde) for brevity of the representation and define the following matrices:

$$\mathbf{M} = \sum_{s=1}^{n_s} \int_{\Omega_s} \rho_s \mathbf{N}^T \mathbf{N} \mathrm{d}\Omega, \quad \mathbf{C} = \sum_{s=1}^{n_s} \int_{\Omega_s} \mathbf{B}^T \widehat{\mathbf{D}}_s \mathbf{B} \mathrm{d}\Omega,$$
$$\mathbf{K} = \sum_{s=1}^{n_s} \int_{\Omega_s} \mathbf{B}^T \mathbf{D}_s^i \mathbf{B} \mathrm{d}\Omega, \quad \mathbf{F}(t) = \sum_{s=1}^{n_s} \left(\int_{\Omega_s} \mathbf{b}_s^T \mathbf{N} \mathrm{d}\Omega + \int_{\partial\Omega_s} \bar{\mathbf{t}}_s^T \mathbf{N} \mathrm{d}\Gamma \right).$$

Here, $\mathbf{B}$ is the displacement-strain matrix. For time discretization, we use the Newmark time integration scheme to advance the stochastic system one time step as

$$\dot{\mathbf{u}}^{k+1} = \dot{\mathbf{u}}^k + (1-\gamma)\Delta t \ddot{\mathbf{u}}^k + \gamma \Delta t \ddot{\mathbf{u}}^{k+1}, \tag{21}$$

$$\mathbf{u}^{k+1} = \mathbf{u}^k + \Delta t \dot{\mathbf{u}}^k + \left(\frac{1}{2} - \beta\right)\Delta t^2 \ddot{\mathbf{u}}^k + \beta \Delta t^2 \ddot{\mathbf{u}}^{k+1}, \tag{22}$$

where γ and β are the integration parameters, and $\Delta t = \frac{T_f - T_0}{n_t}$. Substituting he Newmark scheme into the semi-discretized stochastic equation of motion Equation (20), gives the following fully discretized linear system for a give realization of the random vector $\boldsymbol{\xi}$:

$$\mathbf{A}(\boldsymbol{\xi})\mathbf{U}^{k+1}(\boldsymbol{\xi}) = \mathbf{F}^{k+1} - \mathbf{G}\mathbf{U}^k(\boldsymbol{\xi}), \tag{23}$$

where for compact representation, we define

$$\mathbf{A}(\boldsymbol{\xi}) = \begin{bmatrix} \mathbf{M}(\boldsymbol{\xi}) & \mathbf{C} & \mathbf{K}(\boldsymbol{\xi}) \\ -\gamma\Delta T\mathbf{I} & \mathbf{I} & \mathbf{0} \\ -\beta\Delta T^2\mathbf{I} & \mathbf{0} & \mathbf{I} \end{bmatrix}, \quad \mathbf{G} = \begin{bmatrix} \mathbf{0} & \mathbf{0} & \mathbf{0} \\ -(1-\gamma)\Delta T\mathbf{I} & -\mathbf{I} & \mathbf{0} \\ -(\frac{1}{2}-\beta)\Delta T^2\mathbf{I} & -\Delta T\mathbf{I} & -\mathbf{I} \end{bmatrix},$$
$$\mathbf{U}(\boldsymbol{\xi}) = \begin{Bmatrix} \ddot{\mathbf{u}}(\boldsymbol{\xi}) \\ \dot{\mathbf{u}}(\boldsymbol{\xi}) \\ \mathbf{u}(\boldsymbol{\xi}) \end{Bmatrix}, \quad \mathbf{F} = \begin{Bmatrix} \mathbf{f} \\ \mathbf{0} \\ \mathbf{0} \end{Bmatrix}.$$

For the data-driven decomposition approach, many solutions to the forward problem Equation (23) are required to estimate the appropriate decomposition for localized uncertainty propagation. To mitigate the computational cost involved with identifying the underlying localized region of interest, a Gaussian Process (GP) surrogate model is utilized as explained in the next section.

2.3. Surrogate Modeling

The Gaussian Process (GP) surrogate model is widely used for engineering problems as a cost-effective alternative to costly computer simulator [20,21]. In GP for dynamical systems, we consider $\mathcal{D} = \{(\mathbf{x}_i, \mathbf{y}_i) \mid i = 1, 2, \cdots, N\}$ to be a set of training data consists of N samples, where $\mathbf{x}_i \in \mathbb{R}^d$ represents the input sample i, and $\mathbf{y}_i$ is the corresponding output vector of size n_T. For time-series data, the output is observed at a sequence of time steps $t_j \in [t_1, t_2, \cdots, t_{n_T}]$. We concatenate all the input and output into the design matrix $\mathcal{X}$ and the corresponding observation matrix $\mathcal{Y}$, respectively as:

$$\mathcal{X} = \begin{bmatrix} t_1 & \mathbf{x}_1 \\ \vdots & \vdots \\ t_{n_T} & \mathbf{x}_1 \\ \vdots & \vdots \\ t_1 & \mathbf{x}_N \\ \vdots & \vdots \\ t_{n_T} & \mathbf{x}_N \end{bmatrix}, \qquad \mathcal{Y} = \begin{bmatrix} y_1^1 \\ \vdots \\ y_{n_T}^1 \\ \vdots \\ y_1^N \\ \vdots \\ y_{n_T}^N \end{bmatrix}, \tag{24}$$

where y_j^i is the response at time t_j for the input parameters $\mathbf{x}_i$. The sizes of the design matrix $\mathcal{X}$ and the observation matrix $\mathcal{Y}$ are $(N \times n_T) \times (d+1)$ and $(N \times n_T) \times 1$, respectively. In compact form, the training data set $(\mathcal{X}, Y)$ can be rewritten as:

$$\mathcal{X} = \begin{bmatrix} \mathbf{1}_N \otimes \mathbf{T} & \mathbf{X} \otimes \mathbf{1}_{n_T} \end{bmatrix}, \qquad \mathcal{Y} = \text{vec}(\mathbf{Y}), \tag{25}$$

where $\mathbf{1}_N$ is an identity vector of size N, $\mathbf{X} = [\mathbf{x}_1, \cdots, \mathbf{x}_N]^T$, $\mathbf{T} = [t_1, \cdots t_{n_T}]^T$, $\mathbf{1}_{n_T}$ is an identity vector of size n_T, $\mathbf{Y} = \begin{bmatrix} \mathbf{y}^1 & \cdots & \mathbf{y}^N \end{bmatrix}$ and $\mathbf{y}^i = [y_1^i, \cdots y_{n_T}^i]^T$. Here the symbols $\otimes$ and vec($\bullet$) represent Kronecker product and vectorization operators, respectively. Consequently, a general regression model for time-dependent data can be expressed as a function $f(\mathcal{X})$ that maps the input $\mathcal{X}$ to time-series observation $\mathcal{Y}$. In GP regression, the goal is to infer the function $f(\mathcal{X})$ from noisy observation of the output $\mathcal{Y}$. To this end, the function $f(\mathcal{X})$ is viewed as a random realization of a GP as $f(\mathcal{X}) \sim \mathcal{GP}(\mu(\mathcal{X}), \mathbf{K}(\mathcal{X}, \mathcal{X}'))$, where $\mu(\mathcal{X})$ and $\mathbf{K}(\mathcal{X}, \mathcal{X}')$ are the mean vector and covariance matrix of the process, respectively. Training the GP model can be performed by finding the optimal values to the covariance parameters. Systematically, this is done by maximizing the evidence or the marginal likelihood with respect to the hyperparameter parameters of the kernel. For a zero mean and $\mathbf{K}(\mathcal{X}, \mathcal{X}')$ covariance matrix, the prediction of the GP based on noisy observation for a new input $\mathbf{x}_*$ is a Gaussian process with the following posterior mean and covariance [12]

$$\mu(\mathbf{x}_*) = \mathbf{k}(\mathbf{x}_*, \mathcal{X})[\mathbf{K}(\mathcal{X}, \mathcal{X}') + \sigma_n^2 \mathbf{I}]^{-1} \mathcal{Y}, \tag{26}$$

$$\sigma^2(\mathbf{x}_*) = \mathbf{k}(\mathbf{x}_*, \mathbf{x}_*) - \mathbf{k}(\mathbf{x}_*, \mathcal{X})[\mathbf{K}(\mathcal{X}, \mathcal{X}') + \sigma_n^2 \mathbf{I}]^{-1} \mathbf{k}(\mathcal{X}, \mathbf{x}_*). \tag{27}$$

The covariance function in the GP framework encodes the smoothness and it measures the similarity of the process between two points. The covariance function also encodes the prior belief over the regression function to model the measurements. The prior belief can be on the level of the function smoothness, or behavior and trend such as periodicity. Selecting the right covariance kernel can be challenging for time-dependent data and may require a composition of several covariance functions together to model the right behavior of the data. On the other hand, for problems where the training data is given in the form as in Equation (24), the size of the data may grow exponentially demanding large computational budged. In such a scenario, a scaleable framework for the GP regression of large data set can be exploited to efficiently address the computational cost [22].

In this work, the ultimate goal of the GP model is to serve as a surrogate for the costly simulation code. Thus, we follow a simplified approach to reduce the computational cost of building the surrogate [23]. For the case when the time index of measurement is set *a priori* and prediction at intermediate time instant is not required, the inter correlation between the time steps can be relaxed. Specifically, the prediction of the model in this case is always set at the location of the measured data, and the model only considers the correlation between the input variables $\mathbf{x}_i$. Thus the GP can be constructed on the subset of the data $(\mathbf{X}, \mathbf{Y})$ instead of $(\mathcal{X}, \mathcal{Y})$ as $\mathrm{GP}(\mu(\mathbf{X}), \mathbf{K}(\mathbf{X}, \mathbf{X}'))$, where

$$\mathbf{X} = \begin{bmatrix} \mathbf{x}_1 \\ \vdots \\ \mathbf{x}_N \end{bmatrix}, \qquad \mathbf{Y} = \begin{bmatrix} y_1^1, & \dots, & y_{n_T}^1 \\ & \vdots & \\ y_N^1, & \dots, & y_{n_T}^N. \end{bmatrix} \tag{28}$$

2.4. Polynomial Chaos

The Polynomial Chaos (PC) expansion is based on the decomposition of a stochastic process into deterministic coefficients scaling random functions. In particular, the PC approximates a stochastic process as a linear combination of stochastic orthogonal basis functions as

$$\mathbf{u}(t, \xi) = \sum_{j=0}^{N} \mathbf{\Psi}_j(\xi)\mathbf{u}_j(t), \tag{29}$$

where $\mathbf{\Psi}_j(\xi)$ are a set of multivariate orthogonal random polynomials and $\mathbf{u}_j(t)$ are the deterministic projection coefficients. The PC coefficients can be estimated non-intrusively as

$$\mathbf{u}_j(t) = \frac{\int_{\Xi} \mathbf{u}(t, \xi)\mathbf{\Psi}_j(\xi)\mathrm{d}\Xi}{\int_{\Xi} \mathbf{\Psi}_j^2(\xi)\mathrm{d}\Xi}, \tag{30}$$

where $\int_{\Xi}(\bullet)\,\mathrm{d}\Xi$ denotes the expectation operator with respect to the probability density function of the underlying random variables. The expectation integral can be estimated using random sampling or deterministic quadrature rule [24].

3. Numerical Example

For the numerical demonstration, we consider the problem of detecting the desired geometry (e.g., localized features) for a given specimen from noisy measurements of its dynamical response. We parameterize the geometry by the dimensions of the inner section (the inner length l_i and radius r_i) as shown in Figure 2. The inner dimensions are inferred from noisy measurement of the beam deflection at the mid-span. Once the dimensions are estimated, we perform localized uncertainty propagation of the material parameters of the inner core. For the verification of the framework, a modular approach is considered, where each sub-component of the framework is verified as a standalone unit. This approach is justified by the sequential nature of the coupling mechanism between the building blocks of the framework (i.e., features identification and then uncertainty propagation). This approach quantifies the error in each step and prevents the error measures from being overly influenced by one component compared to another.

3.1. The Forward Problem

We consider a 3-D Aluminum beam with mean elastic properties of $E = 70$ GPa, $\nu = 0.3$ and $\rho = 26.25\ \mathrm{kN/m^3}$. For the damping representation, we consider Rayleigh damping as $\mathbf{C} = \eta_m\mathbf{M} + \eta_k\mathbf{K}$ with the constants $\eta_m = 0$ and $\eta_k = 0.001$. The stiffness $\mathbf{K}$ in the Rayleigh damping is based on the mean properties. We utilize FEniCS [25] for the forward finite element simulations. Figure 2 shows a 2D projection of the beam geometry, where we parameterize the inner cylinder by length l_i and radius r_i, and the outer cylinder by length l_o and radius r_o. For the reference case, the inner and outer dimensions are $l_i = 0.45$ m, $r_i = 0.025$ m and $r_o = 0.05$ m and, $l_o = 1.0$ m, respectively. The beam is subjected to an impact force defined as: $F(t, \mathbf{x}) = [0, 0, F_0 t/t_c \delta(t - tc)]^T$, where $F_0 = -5.0$ GN and the ramp time is $t_c = 0.5$ ms. The beam is fixed at both ends and subjected to zero initial displacement and velocity.

Figure 2. Schematic showing a 2D projection of a typical beam. For the reference case the inner and outer dimensions are ($l_i = 0.45$ m, $r_i = 0.025$ m) and ($r_o = 0.05$ m and, $l_o = 1.0$ m), respectively.

We consider the vertical deflection at the mid-span to be the quantity of interest (QoI) in identifying the underlying beam geometry. Figure 3 shows the mid-span displacement and velocity for a the reference case.

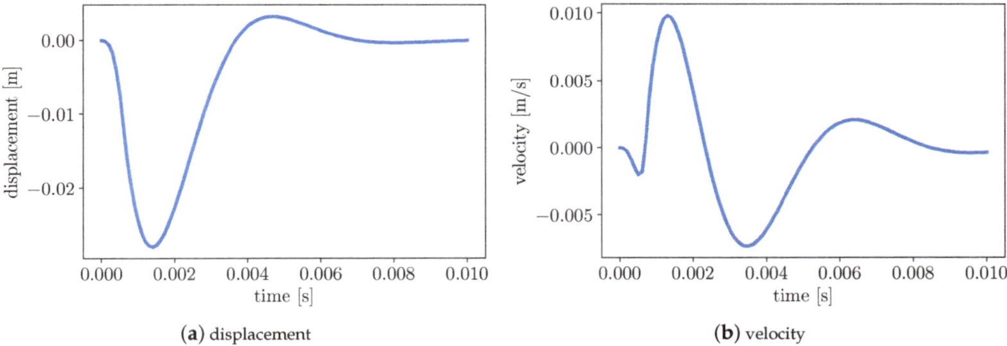

(**a**) displacement (**b**) velocity

Figure 3. The displacement and velocity at the mid-span of the reference case for the mean material properties $E = 70$ GPa, $\nu = 0.3$ and $\rho = 26.25$ kN/m^3.

3.2. The Surrogate Model

In order to infer the beam geometry from measurement of the QoI, many runs of the forward model, the 3D finite element dynamical code, are required. A surrogate model can overcome this issue by utilizing a limited number of a prespecified runs. The design of computer experiments concept can be used to optimally select the required runs [26–28]. For multi-fidelity simulations, where a high-cost high-accuracy and a low-cost low-accuracy simulators are available, a balance between the computational cost and accuracy can be achieved in designing the numerical simulations experiments [29].

The surrogate model is constructed based on samples that can represent the variability in the beam geometry due to different values of the inner dimensions. We define the variability of the inner dimensions by assigning a uniform random distribution with a specified bounds as $l_i \sim U(0.25, 0.75)$m and $r_i \sim U(0.01, 0.05)$m. Using Latin hypercube sampling technique, we generate 50 independent samples for the inner dimensions. Using these samples, we generate the geometry of the beam followed by constructing the corresponding finite element mesh, and executing the forward model to calculate the mid-span deflection (QoI). Samples of the training geometries are shown in Figure 4. Clearly, the samples span a wide range of the probable geometries of the beam. The corresponding scatter of the mid-span vertical displacement of the 50 samples is shown in Figure 5. The variability of the inner dimensions not only affect the geometry, but also the location and magnitude of the bouncing deflection at around time $t = 0.002$ s and $t = 0.005$ s.

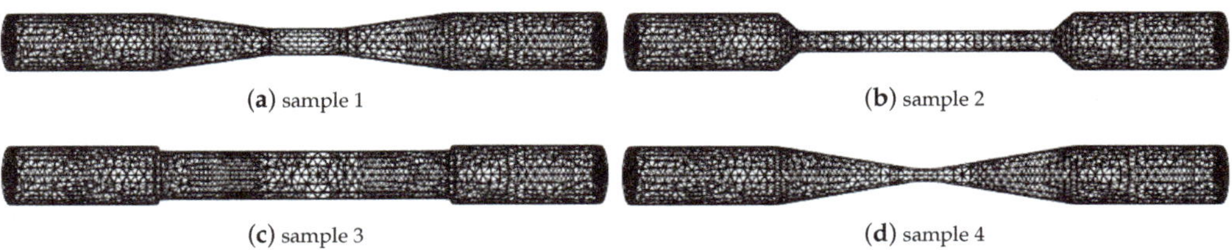

Figure 4. Four samples showing the variability in the beam geometry due to different values of the inner dimensions (l_i, r_i).

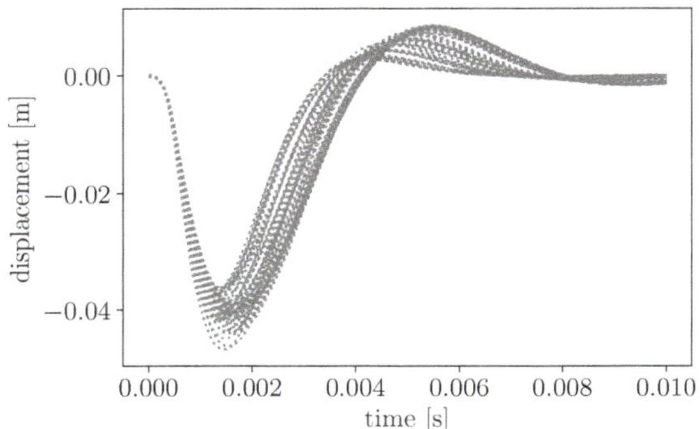

Figure 5. The mid-span vertical displacement of the 50 samples.

We randomly split the 50 samples into two groups as follows—40 samples for training and 10 for testing. For numerical implementation and to mimic the real world, we add a Gaussian random noise of strength (e.g., $10^{-3} \times \max(u)$) to the deflection measurements. Figure 6 shows samples of observed and predicted responses with confidence bounds for different values of the inner dimensions. The errorbars (based on two standard deviation) are indistinguishable within the scale of the graph. The maximum and minimum values of the mean squared error between the prediction and the observed response are 2.10×10^{-7} and 5.35×10^{-9}, respectively. Given the fact that the testing samples are not seen by the model during the training phase, the GP model can predict the unseen data within the given accuracy.

To summarize the quality of the prediction, in Figure 7, we show the L_2-norm of the observed and predicted QoI. The observed/predicted validation plot indicates that the coefficient of determination between the prediction and observation is $r^2 = 0.98$, and the corresponding mean squared error is 2.53×10^{-6}. These statistical metrics indicate that the GP model can estimate the unseen geometry from a noisy measurement of the QoI within a given accuracy.

Once the GP model is validated, it can be deployed as a low-cost surrogate for the 3D finite element analysis code. The prediction of GP model takes only a fraction of the time that is needed by the finite element code to estimate the QoI with a fair accuracy.

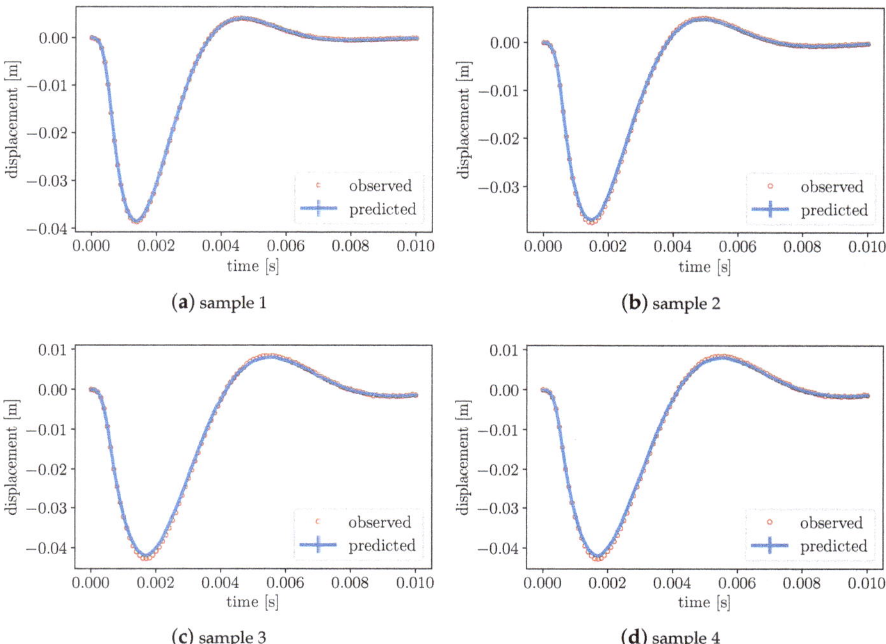

Figure 6. Observed and predicted quantity of interest (QoI) for different testing samples. The test samples are not part of the training set. The errorbars are indistinguishable within the scale of the graph.

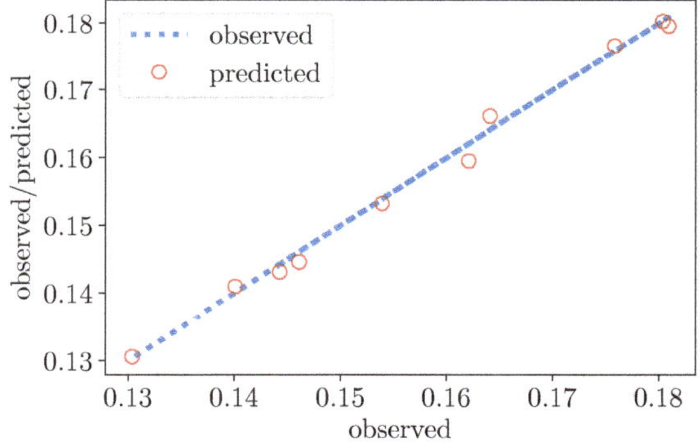

Figure 7. The observed/predicted validation plot showing the norm of the observed (test data) and the corresponding model predictions.

3.3. The Backward Problem

In the backward problem, we try to estimate the inner dimensions (l_i, r_i) of the beam from noisy measurements of the QoI. We assume that a noisy measurement for the QoI is available as shown in Figure 8. The synthetic data is generated using inner dimension $l_i = 0.313$ m and $r_i = 0.055$ m plus ($\sigma_n = 0.1 \times \max(u)$) Gaussian noise to mimic a real experiment setting.

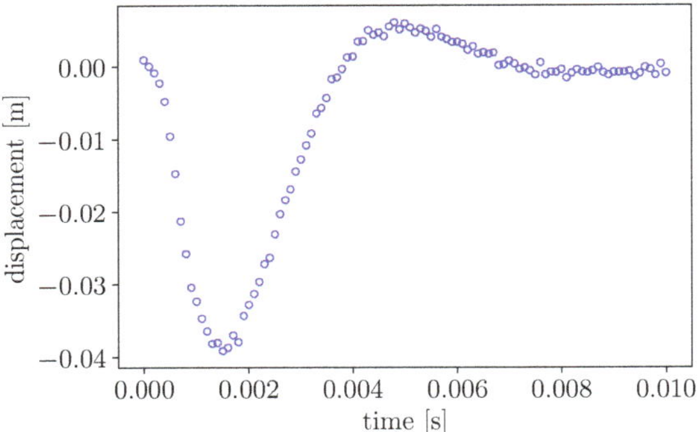

Figure 8. Noisy measurement of the QoI.

For the Bayesian calculation, we use non-informative prior for both the parameters $\theta = [l_i, r_i]$ to assess the robustness of the inversion process. An adaptive MCMC method (DRAM) [16,17] is utilized to estimate the posterior density. In Figure 9, we show the estimated posterior density of the parameters $\theta = [l_i, r_i]$. We also show the prior density and the true value of the parameters. Note that the true parameters are not part of either the training nor the testing data sets. This highlights the robustness of the framework. The mean of the estimated values are $l_i = 0.310 \pm 0.048$ m and $r_i = 0.054 \pm 0.004$ m (the confidence bounds are based on two standard deviation).

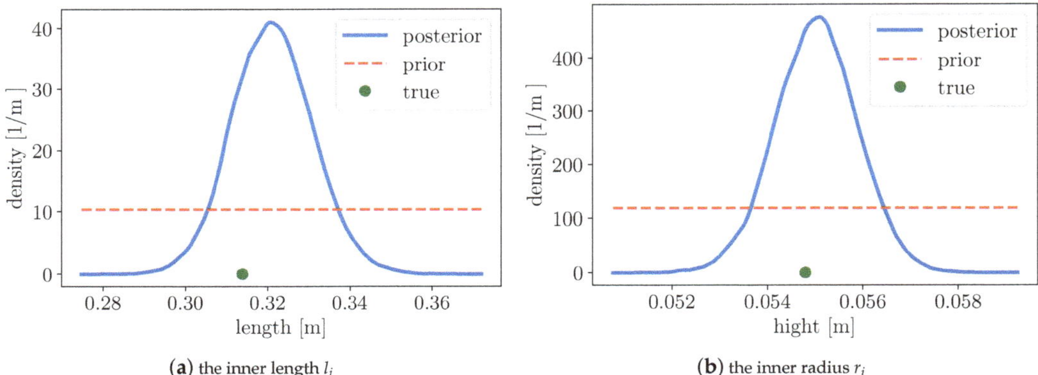

Figure 9. The estimated posterior density function of the inner dimensions $\theta = [l_i, r_i]$. The sold line is the posterior PDF, the dotted line is the prior PDF and the bullet dot represents the true value $l_i = 0.313$ m and $r_i = 0.055$ m.

Next, the uncertainty in the parameter estimation represented by the posterior density in Figure 9 is propagated forward through the surrogate model to estimate a confidence bounds on the prediction of the QoI. In Figure 10, we show the model prediction and the 95% confidence interval as well as the true measured response. The L_2-norm of the discrepancy between the mean model prediction and the measured data is 0.005 m. This conforms that the response due to the estimated parameters uncertainty agrees reasonably well with the true response. Note that, in the estimation of the localized region of interest, the material properties are assumed deterministic. For uncertainty propagation, the Maximum A Posteriori (MAP) estimation is used for the inner dimensions, while assuming random material properties in the region of interest. The relativity small errorbars (within the scale of the graph) indicates that the single point estimation MAP can be used to set the inner dimension sufficiently.

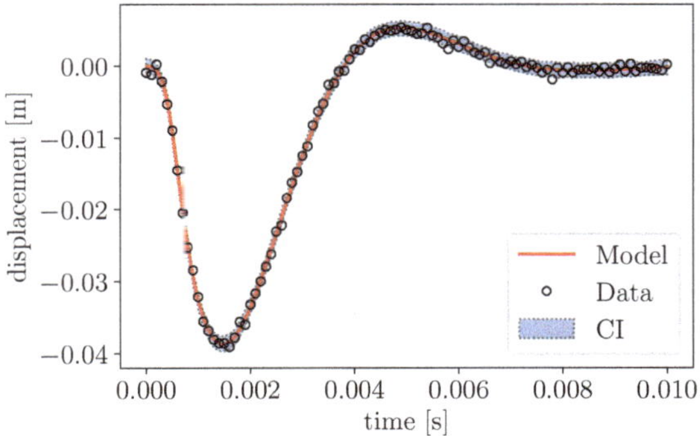

Figure 10. The prediction of the surrogate model and its confidence interval due to uncertainty propagation of the variability in the estimated inner dimensions.

3.4. Localized Uncertainty Propagation

The QoI is confined within the core cylinder defined by inner dimensions $\theta = [l_i, r_i]$. Once these dimensions are available, the effect of the random variability in the material properties of the inner subdomain can be estimated using PC expansion. Without loss of generality, here we assume that for the inner cylinder, the Young's modulus and material density are random quantities, while Poisson's ratio is deterministic as

$$E(\mathbf{x}, \xi_1) = \begin{cases} E_0(1 + \sigma_E \xi_1), & \text{for } \mathbf{x} \in \Omega_2 \\ E_0, & \text{otherwise} \end{cases} \tag{31}$$

and

$$\rho(\mathbf{x}, \xi_2) = \begin{cases} \rho_0(1 + \sigma_\rho\, \xi_2), & \text{for } \mathbf{x} \in \Omega_2 \\ \rho_0, & \text{otherwise} \end{cases} \tag{32}$$

where the artificial boundary for Ω_2 are defined by MAP estimation of the inner dimensions $\theta = [l_i, r_i]$, $E_0 = 70$ GPa, $\rho_0 = 26.25$ kN/m^3, $\sigma_E = 0.25$ and $\sigma_\rho = 0.15$ and ξ_1, ξ_2 are standard normal random variables. Note that, not only the solution over Ω_2 is stochastic, but also over all the whole domain since the spatial finite element and stochastic basis functions are continuous across the domains interfaces. We use second order PC expansion to propagate the localized uncertainty due to the random Young's modulus and material density as shown in Figure 11.

To verify the PCE order, Figure 12 shows the error between the predictions of both the displacement and velocity using second and third order expansion. The error measure is defined as error$(\bullet) = (\bullet)_{3rd} - (\bullet)_{2nd}$. The relatively small values of the error confirm that the second order expansion is sufficient for uncertainty propagation for this problem.

The uncertainty bounds follow the trend of the response, with a higher value near the shock location. Although not explored here, high spatio-temporal resolution solver can be directed toward the region of interest, while a less resolution alternative can be assigned to the regions away from the QoI. As demonstrated in References [7–9], PASTA-DDM-UQ approach leads to a customized solver for localized uncertainty propagation with less computational cost.

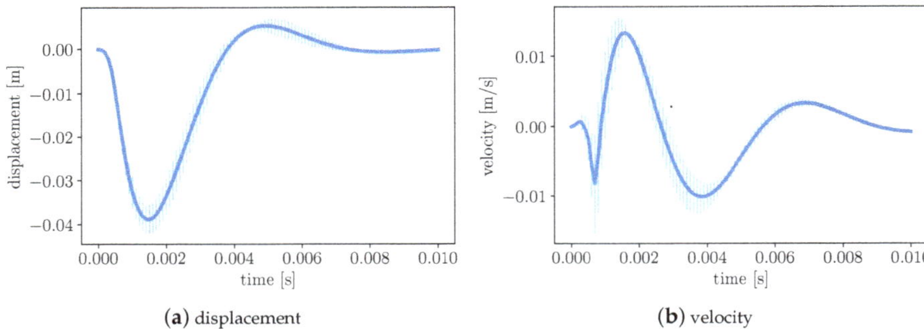

(**a**) displacement (**b**) velocity

Figure 11. The Polynomial Chaos (PC) prediction of the displacement and velocity at the mid-span. The uncertainty bounds represent two standard deviation.

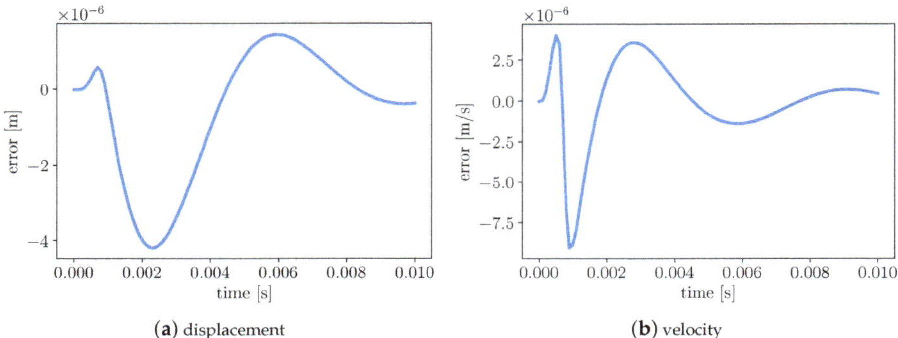

(**a**) displacement (**b**) velocity

Figure 12. The error between prediction of the 2nd and 3rd PC order for the displacement and velocity at the mid-span. The error measure is defined as $\text{error}(\bullet) = (\bullet)_{3rd} - (\bullet)_{2nd}$.

4. Conclusions

We present a data-based partitioning scheme for localized uncertainty quantification in elastodynamic system. The localized region of interest is identified using Bayesian inference framework. Measurement of the system response at one location in conjunction with a physics-based computational model is used to infer the localized features of the region of interested. A data-based surrogate model for the physics-based simulator is constructed using Gaussian process regression in order to reduce the computational cost of the Bayesian inversion. Material uncertainty in the region of interest is propagated through the system using polynomial chaos. We exercise our framework on a three-dimensional beam with localized feature and subjected to an impact load. The presented framework can facilitate quantifying the effect of the confined uncertainty in a localized region of interest within the global computational domain. Proper assessment of uncertainty at various level can accelerate the adaptation process of a new component introduced to an existing system.

Author Contributions: Conceptualization, W.S.; methodology, W.S.; software, W.S.; validation, W.S.; writing—original draft preparation, W.S., S.G., P.P. and Y.Z.; writing—review and editing, W.S., S.G., P.P.; supervision, L.W. All authors have read and agreed to the published version of the manuscript.

Funding: This research received no external funding.

Conflicts of Interest: The authors declare no conflict of interest.

References

1. Dutton, S.; Kelly, D.; Baker, A. *Composite Materials for Aircraft Structures*; American Institute of Aeronautics and Astronautics: Reston, VA, USA, 2004.
2. Mrazova, M. Advanced composite materials of the future in aerospace industry. *Incas Bull.* **2013**, *5*, 139.
3. Pettit, C.L. Uncertainty quantification in aeroelasticity: Recent results and research challenges. *J. Aircr.* **2004**, *41*, 1217–1229. [CrossRef]
4. Munk, C.L.; Nelson, P.E.; Strand, D.E. Determinant Wing Assembly. U.S. Patent 6,808,143, 26 November 2004.
5. Katunin, A.; Dragan, K.; Dziendzikowski, M. Damage identification in aircraft composite structures: A case study using various non-destructive testing techniques. *Compos. Struct.* **2015**, *127*, 1–9. [CrossRef]
6. Diamanti, K.; Soutis, C. Structural health monitoring techniques for aircraft composite structures. *Prog. Aerosp. Sci.* **2010**, *46*, 342–352. [CrossRef]
7. Subber, W.; Matouš, K. Asynchronous space-time domain decomposition method with localized uncertainty quantification. *Comput. Methods Appl. Mech. Eng.* **2017**, *325*, 369–394. [CrossRef]
8. Subber, W.; Salvadori, A.; Lee, S.; Matouš, K. Uncertainty quantification of the reverse Taylor impact test and localized asynchronous space-time algorithm. In *AIP Conference Proceedings, St. Louis, MO, USA, 9–14 July 2017*; AIP Publishing LLC: Melville, NY, USA, 2018; Volume 1979, p. 140005.
9. Subber, W.; Matouš, K. Asynchronous space-time algorithm based on a domain decomposition method for structural dynamics problems on non-matching meshes. *Comput. Mech.* **2016**, *57*, 211–235. [CrossRef]
10. Smith, R.C. *Uncertainty Quantification: Theory, Implementation, and Applications*; SIAM: Philadelphia, PA, USA, 2013; Volume 12.
11. Gelman, A.; Carlin, J.B.; Stern, H.S.; Rubin, D.B. *Bayesian Data Analysis*; Chapman & Hall/CRC: Boca Raton, FL, USA, 2014; Volume 2.
12. Williams, C.K.; Rasmussen, C.E. *Gaussian Processes for Machine Learning*; MIT Press: Cambridge, MA, USA, 2006; Volume 2.
13. Ghanem, R.G.; Spanos, P.D. *Stochastic Finite Elements: A Spectral Approach*; Courier Corporation: North Chelmsford, MA, USA, 2003.
14. Le Maître, O.; Knio, O.M. *Spectral Methods for Uncertainty Quantification: With Applications to Computational Fluid Dynamics*; Springer Science & Business Media: Berlin/Heidelberg, Germany, 2010.
15. Rasmussen, C.E. Gaussian processes in machine learning. In *Summer School on Machine Learning*; Springer: Berlin/Heidelberg, Germany, 2003; pp. 63–71.
16. Haario, H.; Laine, M.; Mira, A.; Saksman, E. DRAM: Efficient adaptive MCMC. *Stat. Comput.* **2006**, *16*, 339–354. [CrossRef]
17. Miles, P.R. pymcmcstat: A Python Package for Bayesian Inference Using Delayed Rejection Adaptive Metropolis. *J. Open Source Softw.* **2019**, *4*, 1417. [CrossRef]
18. Babuška, I.; Nobile, F.; Tempone, R. A stochastic collocation method for elliptic partial differential equations with random input data. *SIAM J. Numer. Anal.* **2007**, *45*, 1005–1034. [CrossRef]
19. Hughes, T.J. *The Finite Element Method: Linear Static and Dynamic Finite Element Analysis*; Courier Corporation: North Chelmsford, MA, USA, 2012.
20. Ghosh, S.; Pandita, P.; Subber, W.; Zhang, Y.; Wang, L. Efficient bayesian inverse method using robust gaussian processes for design under uncertainty. In Proceedings of the AIAA Scitech 2020 Forum, Orlando, FL, USA, 6–10 January 2020; p. 1877.
21. Ghosh, S.; Pandita, P.; Atkinson, S.; Subber, W.; Zhang, Y.; Kumar, N.C.; Chakrabarti, S.; Wang, L. Advances in Bayesian Probabilistic Modeling for Industrial Applications. *ASCE-ASME J. Risk Uncert. Eng. Syst. Part B Mech. Eng.* **2020**, *6*, doi:10.1115/1.4046747. [CrossRef]
22. Zhang, Y.; Ghosh, S.; Pandita, P.; Subber, W.; Khan, G.; Wang, L. Remarks for scaling up a general gaussian process to model large dataset with sub-models. In Proceedings of the AIAA Scitech 2020 Forum, Orlando, FL, USA, 6–10 January 2020; p. 0678.
23. Shabouei, M.; Subber, W.; Williams, C.W.; Matouš, K.; Powers, J.M. Chemo-thermal model and gaussian process emulator for combustion synthesis of ni/al composites. *Combust. Flame* **2019**, *207*, 153–170. [CrossRef]
24. Feinberg, J.; Langtangen, H.P. Chaospy: An open source tool for designing methods of uncertainty quantification. *J. Comput. Sci.* **2015**, *11*, 46–57. [CrossRef]
25. Logg, A.; Mardal, K.A.; Wells, G. *Automated Solution of Differential Equations by the Finite Element Method: The FEniCS Book*; Springer Science & Business Media: Berlin/Heidelberg, Germany, 2012; Volume 84.
26. Kristensen, J.; Subber, W.; Zhang, Y.; Ghosh, S.; Kumar, N.C.; Khan, G.; Wang, L. Industrial applications of intelligent adaptive sampling methods for multi-objective optimization. In *Design Engineering and Manufacturing*; IntechOpen: London, UK, 2019.
27. Kristensen, J.; Bilionis, I.; Zabaras, N. Adaptive simulation selection for the discovery of the ground state line of binary alloys with a limited computational budget. In *Recent Progress and Modern Challenges in Applied Mathematics, Modeling and Computational Science*; Springer: Berlin/Heidelberg, Germany, 2017; pp. 185–211.
28. Pronzato, L.; Müller, W.G. Design of computer experiments: Space filling and beyond. *Stat. Comput.* **2012**, *22*, 681–701. [CrossRef]
29. Ghosh, S.; Kristensen, J.; Zhang, Y.; Subber, W.; Wang, L. A Strategy for Adaptive Sampling of Multi-Fidelity Gaussian Processes to Reduce Predictive Uncertainty. In Proceedings of the International Design Engineering Technical Conferences and Computers and Information in Engineering Conference, Anaheim, CA, USA, 18–21 August 2019; American Society of Mechanical Engineers: New York, NY, USA, 2019; Volume 59193, p. V02BT03A024.

Article

A Shake Table Frequency-Time Control Method Based on Inverse Model Identification and Servoactuator Feedback-Linearization

José Ramírez Senent *, Jaime H. García-Palacios and Iván M. Díaz

Escuela Técnica Superior de Ingenieros de Caminos, Canales y Puertos, Universidad Politécnica de Madrid, Calle del Profesor Aranguren, 3, 28040 Madrid, Spain; jaime.garcia.palacios@upm.es (J.H.G.-P.); ivan.munoz@upm.es (I.M.D.)

* Correspondence: jose.ramirez.senent@alumnos.upm.es

Received: 29 September 2020; Accepted: 2 November 2020; Published: 3 November 2020

Abstract: Shake tables are one of the most widespread means to perform vibration testing due to their ability to capture structural dynamic behavior. The shake table acceleration control problem represents a challenging task due to the inherent non-linearities associated to hydraulic servoactuators, their low hydraulic resonance frequencies and the high frequency content of the target signals, among other factors. In this work, a new shake table control method is presented. The procedure relies on identifying the Frequency Response Function between the time derivative of pressure force exerted on the actuator's piston rod and the resultant acceleration at the control point. Then, the Impedance Function is calculated, and the required pressure force time variation is estimated by multiplying the impedance by the target acceleration profile in frequency domain. The pressure force time derivative profile can be directly imposed on an actuator's piston by means of a feedback linearization scheme, which approximately cancels out the actuator's non-linearities leaving only those related to structure under test present in the control loop. The previous architecture is completed with a parallel Three Variable Controller to deal with disturbances. The effectiveness of the proposed method is demonstrated via simulations carried over a non-linear model of a one degree of freedom shake table, both in electrical noise free and contaminated scenarios. Numerical experiments results show an accurate tracking of the target acceleration profile and better performance than traditional control approaches, thus confirming the potential of the proposed method for its implementation in actual systems.

Keywords: shake table control; vibration testing; system identification; inverse dynamics; feedback linearization; servohydraulics

1. Introduction

Shake table testing constitutes a widespread method of laboratory vibration testing due to its intrinsic ability of capturing dynamic behavior of the structure under test (SuT) [1]. Despite the fact that this structural testing approach originated within the Earthquake Engineering field, it is commonly employed nowadays in the Automotive, Railway and Aerospace industries, on a complete system or component basis, both for homologation and research purposes [2,3].

These testing facilities reproduce a controlled motion in a very stiff platform, onto which the SuT is installed, in one or more degrees of freedom (DoF), depending on the particular geometric configuration of the actuators that drive the table. Target motion is frequently defined in terms of acceleration time histories. These systems are very often powered by hydraulic servoactuators owing to their high performances in terms of stroke, velocity, specific force and frequency range. Actuators' rod kinematics is governed by high performance servovalves. Normally, advanced features, such as hydraulic rod bearings, close-coupled accumulators and adjustable backlash swivels, are equipped in actuators to enhance their performance and controllability.

Nevertheless, the use of hydraulic actuation systems leads to a challenging associated motion control problem. This fact is due to (i) the inherent non-linearity associated to hydraulic components, (ii) the low resonant frequency associated to oil column [4], usually falling within the operation frequency range, (iii) the high frequency range of the target acceleration time histories due to scaling issues [5], and (iv) the tight tolerances allowed for acceleration tracking.

Control approaches employed in shake table testing fall within two main groups, i.e., time domain methods and frequency domain methods. Useful reviews of shake table control algorithms can be found in [4,6]. Time domain methods range from Proportional Integral Derivative (PID) controllers, with constant or variable control gains [7], to sophisticated Model Based Control (MBC) architectures, which make use of a feedforward term which models the (inverse) dynamics between the control order received by the servovalve and the resultant controlled kinematic variable [8–13]. Repetitive control approaches are also used in shake table control when simple oscillatory waveforms are to be reproduced [14,15]. Other remarkable examples of time domain methods are the Three Variable Control (TVC) algorithm which includes feedback and feedforward control loops for displacement, velocity and acceleration [16,17] and the Minimal Control Synthesis (MCS) algorithm, which aims at matching actual system response with that of a reference system [18,19]. Figure 1 shows a block diagram describing a generic time domain control architecture for shake tables; the output of the feedforward controller u_{FF}, which is calculated from the desired acceleration, a_{ref}, is added to the output of the feedback controller, u_{FB}, to yield the total voltage to be injected into the servovalve u_{SV}. The feedback controller in this example calculates its control order accounting for the actual values of table displacement, x_t, and acceleration, $\ddot{x}_t$.

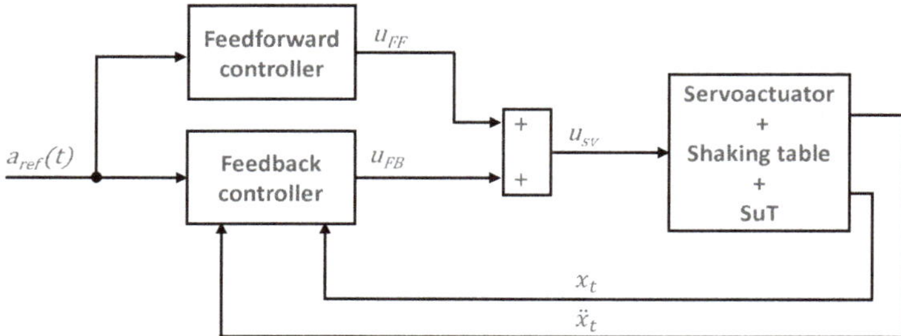

Figure 1. Generic time domain control architecture for shake table.

Frequency domain methods, on the other hand, are iterative in nature and constitute the industry standard for vibration tests [4]. This approach relies on identifying, at a first stage, the Frequency Response Function (FRF) which relates the resultant acceleration measured at the control point to the control order sent to the servovalve. For this purpose, several blocks of excitation signal are output by the controller while simultaneously acquiring system response. Excitation and output blocks are transformed into frequency domain by means of Fast Fourier Transform (FFT) and the FRF is estimated through an averaging process. Later on, this FRF is inverted to obtain the Impedance Function (IF), which is multiplied by the required output of the system, transformed into frequency domain, therefore yielding an initial estimate of the drive signal, d. The worked out drive must, of course, be transformed back into time domain prior to being injected into the system; this is accomplished by means of an Inverse Fast Fourier Transform (IFFT) process. This initially obtained drive block is refined, usually at a low level testing stage, by an iterative scheme, which accounts for error in prior iteration, e, and may update the IF, until a satisfactory control order is found [4,20]. When frequency domain methods are used in servohydraulic testing systems, the identification and iterative schemes are implemented in an Outer Control Loop (OCL) while a faster Inner Control Loop (ICL) directly commands actuator servovalve. This ILC is usually based on a displacement PID but may also include advanced features such as TVC or differential pressure, ΔP, feedback [4,6]. The FRF and IF identification procedure employed in this family of methods does not specifically account for non-linearities present in hydraulic actuation system and therefore obtains a FRF corresponding to a linearization around a working point. This circumstance may lead to a high number of iterations to obtain a system response within allowed limits. Figure 2 shows a generic block diagram describing frequency domain methods for shake tables.

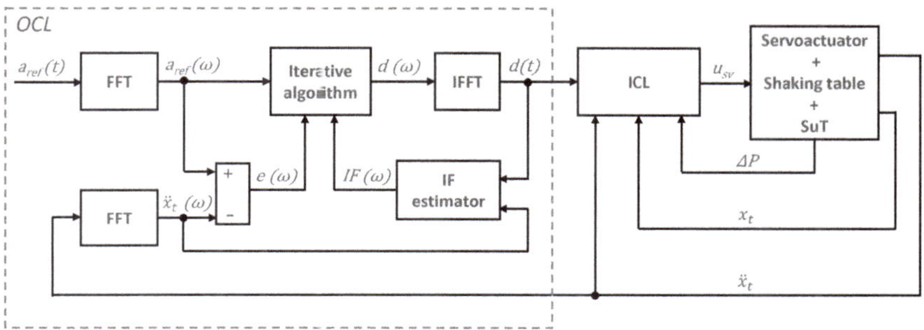

Figure 2. Generic frequency domain control architecture for shake table.

In this paper, a mixed frequency-time approach for a one horizontal DoF shake table is presented. The suggested methodology is based on the identification of an FRF-IF pair relating the time derivative of the pressure force exerted on actuator's rod to the acceleration measured at control point, is presented. The IF obtained in this way, allows for the synthesis of a target pressure force time derivative drive, that can be directly imposed on cylinder piston rod thanks to a feedback linearization scheme, which approximately cancels out non-linearities present in hydraulic actuation system. The presented procedure requires an initial system's FRF-IF identification stage; however, iterations in test mode are not needed and the non-linearities associated to hydraulic system are excluded from the control loop, having to deal only with those associated to SuT behavior. System usability and tracking performance are thus improved with respect to those of traditional iterative methods. A parallel TVC controller, which accounts for model imperfections and external disturbances, completes the abovementioned architecture and represents the time domain component of the suggested control method. The effectiveness of the proposed procedure has been assessed by means of numerical simulations carried out in electrical noise free and contaminated scenarios and compared to that of the classical iterative schemes traditionally used for shake table control.

The remainder of this paper is organized as follows. Section 2 describes the non-linear model implemented to assess the potential performance of the proposed methodology. Section 3 covers in detail the suggested methodology explaining the implemented feedback linearization and servovalve dynamics inversion algorithms (Section 3.1), the IF and hydraulic parameters identification processes (Section 3.2), the drive calculation procedure (Section 3.3) and the TVC controller (Section 3.4). Section 4 presents the simulation results obtained for a random acceleration target waveform in both noise-free and noise-contaminated cases and a performance comparison between the proposed and the classical iterative control approaches. Finally, Section 5 outlines the main conclusions drawn from this research.

2. Shake Table System Modeling

This work is focused on a one horizontal DoF shake table system (see Figure 3). Its main components are the table where the SuT is installed, the linear guidance system (based on low friction roller bearings and linear rails), the hydraulic servoactuator (equipped with hydrostatic bearings and adjustable backlash swivels), the servovalve installed on actuator's manifold and the servoactuator reaction structure. A shear building with two identical stories was selected as the SuT selected for the numerical experiments.

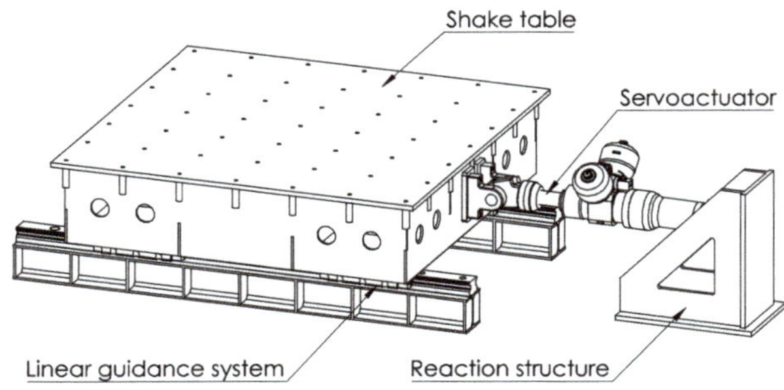

Figure 3. Shake table system. Courtesy of Vzero Engineering Solutions, SL.

A model of the previously mentioned elements has been implemented to assess the goodness of the proposed control methodology, in what follows, this model is described in detail. Figure 4 shows a scheme of the components of the system which have been modelled, along with the sign criteria adopted.

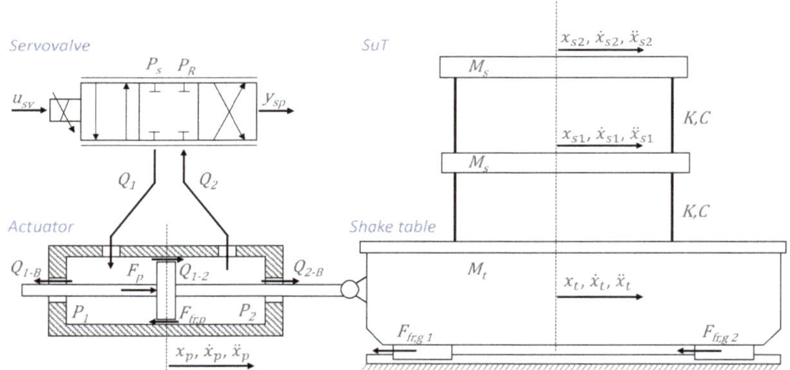

Figure 4. Scheme of the shake table system modeled elements.

The motion of the spool of servo-valve's main stage has been modelled according to a first order linear system [21]:

$$C_{sp}u_{sv} = \tau_{sp}\dot{y}_{sp} + y_{sp}, \tag{1}$$

where u_{sv} is the voltage injected into the servovalve, y_{sp} is servovalve's main stage spool motion, τ_{sp} is the time constant of the system and C_{sp} is the spool gain.

Flow through servovalve ports has been computed assuming a critically lapped spool with symmetrical and matched orifices [22] and a linear characteristic [23] as shown in the next equations:

$$Q_1 = \begin{cases} C_d K_{sp} y_{sp} sgn(P_s - P_1)\sqrt{2|P_s - P_1|/\rho}\,; & y_{sp} \geq 0 \\ C_d K_{sp} y_{sp} sgn(P_1 - P_s)\sqrt{2|P_1 - P_R|/\rho}\,; & y_{sp} < 0 \end{cases}, \tag{2}$$

$$Q_2 = \begin{cases} C_d K_{sp} y_{sp} sgn(P_2 - P_R)\sqrt{2|P_2 - P_R|/\rho}; & y_{sp} \geq 0 \\ C_d K_{sp} y_{sp} sgn(P_S - P_2)\sqrt{2|P_S - P_2|/\rho}\,; & y_{sp} < 0 \end{cases}, \tag{3}$$

where Q_1 and Q_2 are the volumetric flow rates across ports 1 and 2 of servovalve, P_1 and P_2 are the pressures at chambers 1 and 2 of the servoactuator, P_S and P_R are supply and return pressures at servoactuator's manifold, C_d is the discharge coefficient of inlet orifices to chambers, K_{sp} is the passage area to spool displacement ratio, ρ is hydraulic fluid density and sgn represents the sign function.

The evolution of pressures at actuator's chambers has been modelled making use of the Continuity Equation, defining an average mass density per chamber and utilizing the Bulk modulus definition [22]:

$$\left(v_{01} + A_w x_p\right)\dot{P}_1/\beta_1 + A_w \dot{x}_p = Q_1 - Q_{1-2} - Q_{1B}, \tag{4}$$

$$\left(v_{02} - A_w x_p\right)\dot{P}_2/\beta_2 - A_w \dot{x}_p = -Q_2 + Q_{1-2} - Q_{2B}, \tag{5}$$

where x_p is rod displacement, Q_{1-2} is the leakage flow between chambers through piston-sleeve annular passage area, Q_{1B} and Q_{2B} are leakage flows between each chamber and its respective hydrostatic bearing, A_w is actuator's effective area, v_{02} and v_{01} are the initial volumes of chambers and β_1 and β_2 are the equivalent Bulk moduli of each compartment. Overdot notation has been employed to denote time differentiation. Leakage flows are normally assumed to be laminar and their corresponding flow rate is therefore modeled using expressions proportional to the difference of pressures seen by the fluid:

$$Q_{1-2} = C_{l12}(P_1 - P_2), \tag{6}$$

$$Q_{iB} = C_{iB}(P_i - P_{Bi}), \tag{7}$$

where C_{l12} and C_{iB} represent, respectively, the across-chambers and chamber-bearing leakage coefficients, P_{Bi} is the operating pressure of each chamber bearing and i stands for the related actuator chamber. Nevertheless, due to their reduced values, all leakage flows have been neglected in the ensuing analysis.

The resultant force, F_t, exerted on the shake table (including the piston rod in it) can be expressed as:

$$F_t = (P_1 - P_2)A_w - F_{fr,p}, \tag{8}$$

in which $F_{fr,p}$ represents friction force between piston and cylinder sleeve and rod and bearings. The term $(P_1 - P_2)A_w$ constitutes the pressure force. Its time derivative, $(\dot{P}_1 - \dot{P}_2)A_w = A_w\Delta\dot{P}$, will be later exhaustively referred to. Friction force has been considered viscous and equal to $C_p\dot{x}_p$, where x_p represents actuator's rod displacement and C_p its damping coefficient. This is a common practice when modelling low friction, high performance servoactuators.

Finally, the motion of the shake table and SuT has been evaluated by:

$$\begin{bmatrix} M_t + m_p & 0 & 0 \\ 0 & M_s & 0 \\ 0 & 0 & M_s \end{bmatrix} \begin{Bmatrix} \ddot{x}_t \\ \ddot{x}_{s1} \\ \ddot{x}_{s2} \end{Bmatrix} + \begin{bmatrix} C_{11} & C_{12} & C_{13} \\ C_{21} & C_{22} & C_{23} \\ C_{31} & C_{32} & C_{33} \end{bmatrix} \begin{Bmatrix} \dot{x}_t \\ \dot{x}_{s1} \\ \dot{x}_{s2} \end{Bmatrix} + \begin{bmatrix} K & -K & 0 \\ -K & 2K & -K \\ 0 & -K & K \end{bmatrix} \begin{Bmatrix} x_t \\ x_{s1} \\ x_{s2} \end{Bmatrix} = \begin{Bmatrix} F_t - F_{fr,g} \\ 0 \\ 0 \end{Bmatrix} \tag{9}$$

where x_t is table displacement, considered throughout the subsequent analysis identical to rod displacement, x_p, x_{s1} and x_{s2} are the displacements of shear building stories, $F_{fr,g}$ is the friction force between linear bearings and rails, m_p is piston rod mass, M_s is the mass of each of the stories and K is the stiffness of the pillars of each story. The components of the damping matrix, C_{ij}, have been calculated starting from a modal damping matrix in which a damping ratio $\zeta = 5\%$ has been considered for all the flexible vibration modes. Later on, the damping matrix expressed in problem coordinates has been calculated making use of the change of coordinates matrix formed by the mass-normalized eigenvectors of the system.

The electrical noise affecting sensor signals and servovalve input has been modelled by means of gaussian waveforms characterized by their rms voltage value, $u_{n,rms} = 2.8 \times 10^{-3}$ V rms, which leads to a noise voltage peak value of $u_{n,peak} = 0.01$ V (see Section 3.2.1 for considerations on the noise peak value). In order to transform electrical noise into physical quantities influencing model behavior, the value of the noise voltage has been multiplied by the appropriate sensor gains: g_{dis}, g_{acc}, g_{press}, and g_{sp} for the acceleration, displacement, chamber pressures and servovalve spool position sensors respectively and g_{sv} for the servovalve input voltage.

Delays in sensor readings have been neglected throughout this paper due to the fact that the frequency range of the sensors commonly used in shake table facilities is sufficiently broader than the frequency range of interest, which in the case under study is up to 100 Hz.

A fixed step solver has been used to perform simulations. A time step, Δt, of 1.0×10^{-4} s has been used for all the simulations in this work. This time step has been selected taking into consideration that it is a loop rate achievable with commercial-off-the-shelf real-time controllers based on Field Programmable Gate Array (FPGA) technology. Finally, Table 1 lists the values of the employed in numerical simulations.

Table 1. Values of parameters used in the considered model.

Parameter	Value	Parameter	Value
A_w (m^2)	5.9000×10^{-3}	K_{sp}(m)	6.6797×10^{-2}
β_i (MPa) [1]	1.5000×10^{3}	K (N/m)	3.9478×10^{6}
C_p (Ns/m)	1.0000×10^{3}	m_p (kg)	8.0000×10^{1}
C_d (-)	6.1100×10^{-1}	M_t (kg)	3.0000×10^{3}
C_{sp} (m/V)	1.8000×10^{-4}	M_s (kg)	1.0000×10^{3}
Δt (s)	1.0000×10^{-4}	P_s (MPa)	2.8000×10^{1}
g_{acc} (m/s^2/V)	9.8100×10^{0}	P_R (MPa)	0
g_{dis} (m/V)	1.5000×10^{-2}	ρ (kg/m^3)	8.5000×10^{2}
g_{press} (Pa/V)	4.0000×10^{6}	τ_{sp} (s)	1.0000×10^{-2}
g_{sp} (%/V)	1.0000×10^{1}	$u_{n,peak}$ (V)	1.0000×10^{-2}
g_{sv} (V/V)	1.0000×10^{0}	$u_{n,rms}$ (V)	2.8000×10^{-3}
ζ (-)	5.0000×10^{-2}	v_{0i} (m^3) [1]	8.8600×10^{-4}

[1] i stands for servoactuator chamber number.

3. Description of the Proposed Control Methodology

The proposed control methodology comprises the following blocks:

- Feedback linearization. The purpose of this block is to cancel out, at least approximately, the non-linearities inherent to the servovalve-actuator system, leading to a control scheme where the time derivative of the pressure force exerted on the servoactuator's piston rod can be directly imposed.
- System identification. This module operates when the system is in identification mode, prior to the test itself. It is in charge of: (i) estimating and inverting the Accelerance Function (AF), which later is transformed into a more suitable IF representing the inverse model of the shake table-SuT system, and (ii) obtaining approximations for the values of the hydraulic parameters required by the feedback linearization scheme. It can also be implemented to operate, on a signal block basis, refining identification of IF and system parameters between one signal block and the following, as the test proceeds.
- Drive calculation. This algorithm operates when the system is in test mode, on a signal block basis. It calculates the necessary pressure force time derivative to be applied on servoactuator's rod by multiplying the IF from the system identification module by the desired acceleration output, in frequency domain, and transforming the result back into time domain.
- TVC controller. This feedback controller is necessary to compensate for the unavoidable imperfections present in the identified inverse model and to ensure overall system stability. It is implemented in parallel with the abovementioned architecture and accounts for errors in displacement, velocity and acceleration tracking in real-time.

Figure 5 shows a block diagram of the proposed control procedure, illustrating the interconnections between the previously enumerated modules. The proposed control methodology requires measuring the following variables: table (rod) displacement x_t, rod acceleration $\ddot{x}_t$, spool position y_{sp}, pressure at actuator's chambers P_1 and P_2, pressure and return pressures P_S and P_R at the servovalve's manifold and estimating the value of rod velocity, $\dot{x}_t$ [24,25].

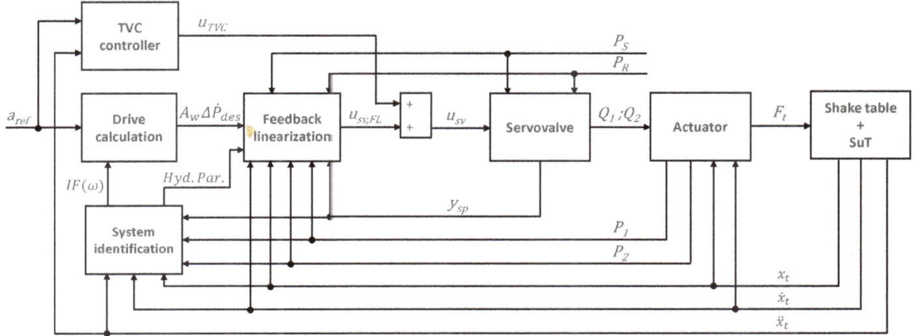

Figure 5. Control architecture block diagram.

3.1. Feedback Linearization

Given a state-space representation of a multiple-input-multiple-output non-linear system: $\dot{x} = f(x,u)$; $y = g(x)$, the aim of the feedback linearization scheme is to find a state transformation, $z = z(x)$ and an input transformation, $u = u(x,v)$, such that the non-linear system is transformed into an equivalent linear system of the form $\dot{z} = Az + Bv$ [26,27].

In the case under study, the non-linearities are present in servovalve flow-pressure (Equations (2) and (3)) and in chambers pressure time evolution expressions (Equations (4) and (5)). In this work, a direct approach has been employed to work out feedback linearization transformation.

By rearranging Equations (4) and (5), leaving the pressure derivatives on the left hand side, assuming $\beta_1 = \beta_2 = \beta$, and subtracting them, the time derivative of the pressure force acting on rod can be casted as:

$$\begin{aligned}\left(\dot{P}_1 - \dot{P}_2\right)A_w = \; & \beta A_w\left(\frac{Q_1}{v_{01}+A_w x_p} + \frac{Q_2}{v_{02}-A_w x_p}\right) \\ & -\beta A_w^2 \dot{x}_p\left(\frac{1}{v_{01}+A_w x_p} + \frac{1}{v_{02}-A_w x_p}\right) \\ & -\beta A_w C_{l12}(P_1 - P_2)\left(\frac{1}{v_{01}+A_w x_p} + \frac{1}{v_{02}-A_w x_p}\right) \\ & -\beta A_w\left(\frac{C_{l1B}(P_1-P_{B1})}{v_{01}+A_w x_p} - \frac{C_{l2B}(P_2-P_{B2})}{v_{02}-A_w x_p}\right)\end{aligned} \quad (10)$$

Now, for the case $y_{sp} \geq 0$, if Equations (2) and (3) are substituted in Equation (10), the following expression is obtained:

$$\left(\dot{P}_1 - \dot{P}_2\right)A_w = y_{sp} \cdot F_1^+\left(P_1, P_2, P_S, P_R, x_p\right) + F_2^+\left(x_p, \dot{x}_p\right) + F_3^+\left(P_1, P_2, x_p\right), \quad (11)$$

where:

$$F_1^+\left(P_1, P_2, P_S, P_R, x_p\right) = \beta A_w C_d K_{sv}\left(\frac{sgn(P_s - P_1)\sqrt{2|P_s - P_1|/\rho}}{v_{01} + A_w x_p} + \frac{sgn(P_2 - P_R)\sqrt{2|P_2 - P_R|/\rho}}{v_{02} - A_w x_p}\right), \quad (12)$$

$$F_2^+\left(x_p, \dot{x}_p\right) = -\beta A_w^2 \dot{x}_p\left(\frac{1}{v_{01} + A_w x_p} + \frac{1}{v_{02} - A_w x_p}\right), \quad (13)$$

and

$$F_3^+(P_1, P_2, x_p) = -\beta A_w \left[C_{l12}(P_1 - P_2)\left(\frac{1}{v_{01} + A_w x_p} + \frac{1}{v_{02} - A_w x_p}\right) + \left(\frac{C_{l1B}(P_1 - P_{B1})}{v_{01} + A_w x_p} - \frac{C_{l2B}(P_2 - P_{B2})}{v_{02} - A_w x_p}\right)\right]. \tag{14}$$

When $y_{sp} < 0$, Equation (3) is substituted in Equation (10) and the time derivative of pressure force results in:

$$\left(\dot{P}_1 - \dot{P}_2\right)A_w = y_{sp} \cdot F_1^-\left(P_1, P_2, P_S, P_R, x_p\right) + F_2^-\left(x_p, \dot{x}_p\right) + F_3^-\left(P_1, P_2, x_p\right), \tag{15}$$

where:

$$F_1^-\left(P_1, P_2, P_S, P_R, x_p\right) = \beta A_w C_d K_{sv}\left(\frac{sgn(P_1 - P_R)\sqrt{2|P_1 - P_R|/\rho}}{v_{01} + A_w x_p} + \frac{sgn(P_S - P_2)\sqrt{2|P_S - P_2|/\rho}}{v_{02} - A_w x_p}\right), \tag{16}$$

$$F_2^-\left(x_p, \dot{x}_p\right) = F_2^+\left(x_p, \dot{x}_p\right), \tag{17}$$

and

$$F_3^-\left(P_1, P_2, x_p\right) = F_3^+\left(P_1, P_2, x_p\right). \tag{18}$$

Let us now define a desired change in pressure force time derivative, $A_w \Delta \dot{P}_{des}$. Then, if the spool position y_{sp} could be forced to instantaneously take the values defined by:

$$y_{sp} = \begin{cases} \dfrac{A_w \Delta \dot{P}_{des} - F_2^+\left(x_p, \dot{x}_p\right) - F_3^+\left(P_1, P_2, x_p\right)}{F_1^+\left(P_1, P_2, P_S, P_R, x_p\right)} & ;\ y_{sp,prev} \geq 0 \\ \dfrac{A_w \Delta \dot{P}_{des} - F_2^-\left(x_p, \dot{x}_p\right) - F_3^-\left(P_1, P_2, x_p\right)}{F_1^-\left(P_1, P_2, P_S, P_R, x_p\right)} & ;\ y_{sp,prev} < 0 \end{cases} \tag{19}$$

$(\dot{P}_1 - \dot{P}_2)A_w = A_w \Delta \dot{P}_{des}$ would hold, and an arbitrary time variation of pressure force could be imposed on the servoactuator. Above, $y_{sp,prev}$ denotes the value of the spool position at the previous iteration of the real-time control system, in which feedback linearization scheme is implemented.

This is the core idea to the control procedure presented in this work: finding the required time derivative of pressure force on the actuator's rod, so that the acceleration reference profile is fulfilled. Due to the feedback linearization transformation found (Equation (19)), this value of $A_w \Delta \dot{P}_{des}$ will be effectively imposed on servoactuator's rod.

In order to force the spool position to accurately track the value determined by Equation (19), a servovalve spool dynamics inversion algorithm must be implemented. Assuming that spool motion is governed by the first order system in Equation (1), the dynamics inversion can be expressed as:

$$C_{sp} u_{sv,FL} = \tau_{sp} \dot{y}_{sp,des} + y_{sp,des}, \tag{20}$$

in which $u_{sv,FL}$ is the voltage output of the spool dynamics inversion algorithm and $y_{sp,des}$ is the desired spool position obtained from Equation (19). The calculation of the required voltage input to the servovalve therefore implies calculating the time derivative of $y_{sp,des}$. For that purpose, a fourth order backward differentiation scheme has been utilized here [28]:

$$\dot{y}_n = \frac{3y_{n-4} - 16y_{n-3} + 36y_{n-2} - 48y_{n-1} + 25y_n}{12\Delta t} \tag{21}$$

where n denotes the actual time step, y_n the function evaluated at that time step and Δt the time step which equals 1×10^{-4} s (see Table 1). Figure 6 illustrates the block diagram of the described feedback linearization scheme.

3.2. System Identification

As mentioned before, the aim of the system identification module is twofold: (i) to work out an estimate of system's IF and (ii) to identify hydraulic system parameters which are required to carry out feedback linearization. These procedures are dealt with in next two subsubsections.

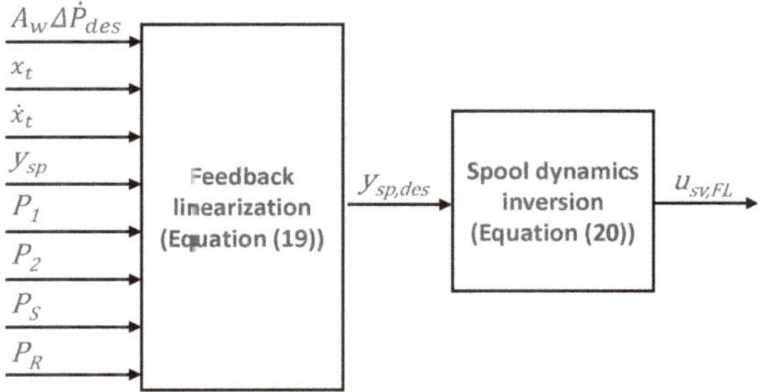

Figure 6. Feedback linearization scheme.

3.2.1. Impedance Function Identification Procedure

The first step taken in finding a suitable approximation of the IF of the system has been to work out an estimate of its AF to later derive an appropriate IF. Two scenarios for IF identification are considered in what follows: (i) a noise-free environment and (ii) a more realistic situation in which noise contaminates servovalve voltage input and force and acceleration measurements. The former is presented for theoretical validation purposes, while the latter constitutes a robustness check of the IF identification procedure necessary to correctly assess the potential of the proposed methodology.

The followed procedure has been essentially the same for both cases and consists in feeding the servovalve with several (voltage) blocks of random stimuli and recording simultaneously the force on the table and the acceleration at the control point. These constitute, respectively, the input and the output of the shake table-SuT system (see Figure 5). Later on, the AF has been estimated making use of classical FRF estimation algorithms [29]. In the noise-free environment the H_1 estimator has been employed. With this approach the AF is expressed as $AF(\omega) = G_{AF}(\omega)/G_{FF}(\omega)$, where $G_{AF}(\omega)$ is the cross-spectrum between force and acceleration and $G_{FF}(\omega)$ is force autospectrum. For the noise-contaminated scenario, the H_2 estimator, which minimizes the effect of the noise at both the input and the output of the system, has been employed.

The input to the servovalve has been selected to be a gaussian random waveform with a duration of 20 s, a flat frequency content between 0.1 Hz and 100 Hz and a maximum amplitude of 20 mV. This signal has been windowed with a unit square signal with a duty cycle of 50%. In this way, excitation is only effective during the first half of the block, allowing for system response (acceleration) to attenuate towards the end of the block, therefore minimizing leakage errors. Figure 7 shows signals obtained in one iteration of the identification stage in time and frequency domains.

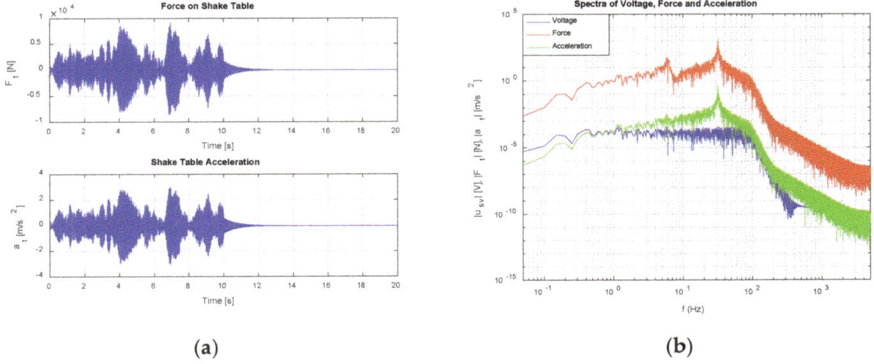

(a) **(b)**

Figure 7. Identification signals (one block): (**a**) Force on table and table acceleration in time domain; (**b**) Servovalve voltage, Force on table and table acceleration in frequency domain.

The AF approximation has been computed by linearly averaging the results obtained with sixteen input-output blocks. Figure 8 shows the achieved estimate for AF and its theoretical shape calculated analytically by transforming into frequency domain Equation (9) and performing the appropriate manipulations.

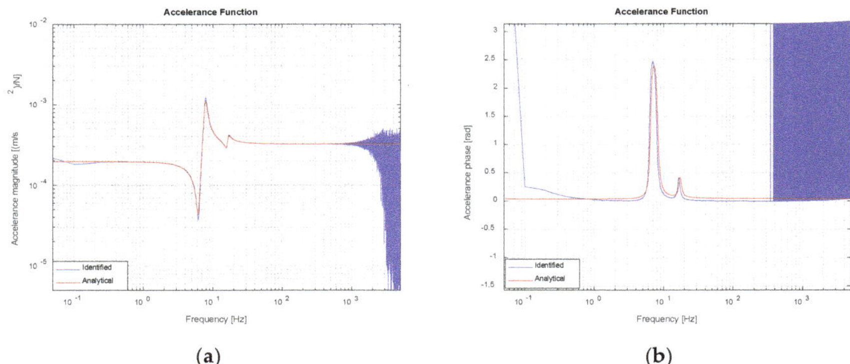

Figure 8. Identified and analytical AF (noise-free scenario): (**a**) Magnitude; (**b**) Phase.

As it was explained in Section 3.1, the feedback linearization scheme, in theory, allows for the imposition of an arbitrary time derivative of pressure force exerted on actuator's rod. Consequently, the IF sought must relate table acceleration to the time derivative of pressure force. This frequency function can be easily obtained by differentiating in frequency domain, without resorting to perform numerical derivatives on the desired pressure force obtained in time domain. Figures 9 and 10 show the FRF (acceleration over pressure force time derivative) and the IF (pressure force time derivative over acceleration) finally used by the drive calculation module, along with their theoretical value. The quality of the identified inverse model is quite acceptable within the complete frequency range of interest, except for the very low frequencies and in the neighborhood of the first modal frequency of the system under control, where small differences can be observed.

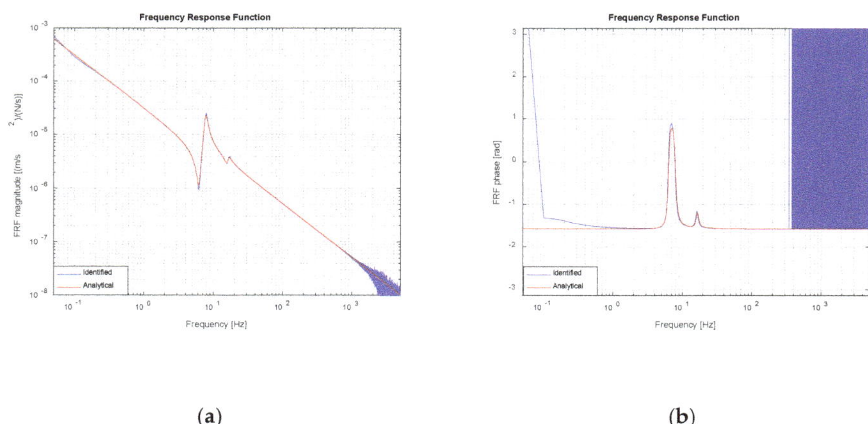

(**a**) (**b**)

Figure 9. Identified and analytical FRF (noise-free scenario): (**a**) Magnitude; (**b**) Phase.

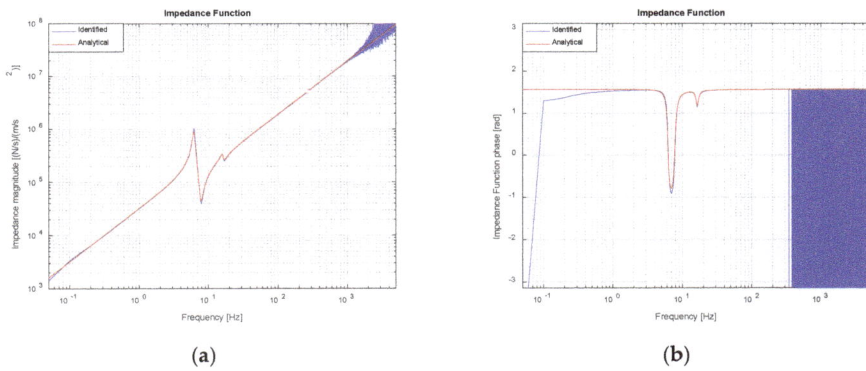

(a) (b)

Figure 10. Identified and analytical IF (noise-free scenario): (**a**) Magnitude; (**b**) Phase.

In order to correctly assess the potential of the proposed methodology in a more realistic scenario, in what follows, the outcomes of an IF identification procedure in which servovalve voltage, force and acceleration signals have been influenced by electrical noise is presented.

The noise in measurements has been simulated by adding gaussian noise to servovalve input voltage and force and acceleration signals. The peak magnitude of the noise has been set to $u_{n,peak} = 0.01$ V (see Section 2), which is an attainable value, when good industrial practices for low distance voltage signals wiring and shielding are observed. Later, the value of the noise affecting the physical quantities has been calculated by multiplying the electrical noise by each sensor's gain as explained in Section 2. Figures 11 and 12 show, in the presence of noise, the same information as Figures 9 and 10. Clearly, the quality of the estimates of FRF and IF decreases; nevertheless, identification error remains within reasonable limits and the obtained estimates are sufficiently good in the whole frequency range of interest. So as to better compare the FRF estimates, coherence functions associated to FRF identification, both in the noise-free and noise-contaminated cases are shown in Figure 13. The coherence function is defined as $\gamma^2 = |G_{FA}(\omega)|^2/[G_{FF}(\omega)G_{AA}(\omega)]$ and measures the degree of linear relationship between two signals. Even though coherence in the noise-contaminated case is evidently worse than its noise-free counterpart, its values remain quite close to unity in the frequency range of interest, consequently confirming the validity of the FRF and IF estimates in the presence of noise of a reasonable magnitude.

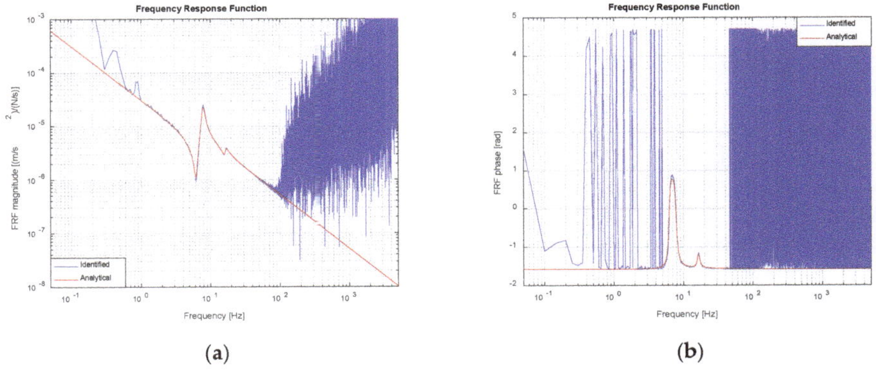

(a) (b)

Figure 11. Identified and analytical FRF (noise-contaminated scenario): (**a**) Magnitude; (**b**) Phase.

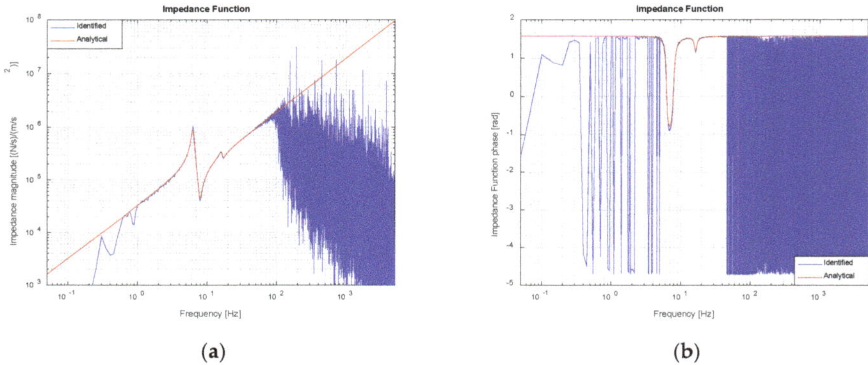

(a) (b)

Figure 12. Identified and analytical IF (noise-contaminated scenario): (**a**) Magnitude; (**b**) Phase.

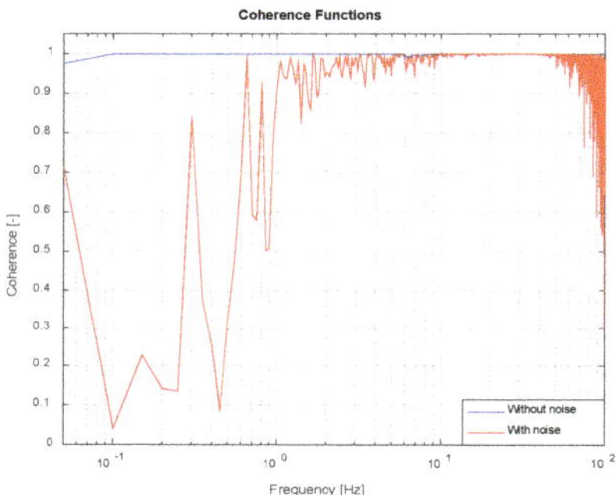

Figure 13. Coherence functions comparison.

3.2.2. Hydraulic Parameters Identification Procedure

The implementation of the feedback linearization scheme implies knowing accurate estimates of hydraulic parameters. In order to identify the sought values, and taking advance of the collection of data available from IF identification stage, a linear state-space model of the servoactuator has been identified. The inputs to this state-space model are, on the one hand, the voltage input to the servovalve, u_{sv}, and the force exerted on piston rod, F_p, by the shake table, on the other. The latter can be calculated by means of:

$$F_p = m_p\ddot{x}_t - (P_1 - P_2)A_w + F_{fr,p}, \tag{22}$$

which would correspond to the reading of a load cell installed between rod tip and shake table.

The state variables of the model have been selected as the velocity of the piston rod, which, as mentioned before, is identical to table velocity, $\dot{x}_t$, the difference of pressures across chambers denoted by ΔP and the servovalve's main spool position, y_{sp}. It has been assumed that Bulk moduli of each compartment are identical and are denoted by β. As mentioned in Section 2, the leakage flows between chambers and from chambers to bearings have been neglected. Accounting for these assumptions and linearizing Equations (1)–(5) around the mid-stroke operating point of the hydraulic cylinder, leads to the following analytical form of the state equations of the servoactuator system:

$$\begin{Bmatrix} \ddot{x}_t \\ \Delta\dot{P} \\ \dot{y}_{sp} \end{Bmatrix} = \begin{bmatrix} -C_p/m_p & A_w/m_p & 0 \\ -2A_w\beta/v_0 & 0 & 2C_dK_{sv}\beta\sqrt{P_s/\rho}/v_0 \\ 0 & 0 & -1/\tau_{sp} \end{bmatrix} \begin{Bmatrix} \dot{x}_t \\ \Delta P \\ y_{sp} \end{Bmatrix} + \begin{bmatrix} 0 & 1/m_p \\ 0 & 0 \\ C_{sp}/\tau_{sp} & 0 \end{bmatrix} \begin{Bmatrix} u_{sv} \\ F_p \end{Bmatrix} \tag{23}$$

By means of a least squares procedure, the components of the matrices in Equation (23) have been identified. This process implies approximating the values of the rod velocity $\dot{x}_t$ and time derivatives of state variables ΔP and y_{sp} (see Equation (21)).

Once the estimates of matrices components are available, it is possible to estimate directly the values of A_w, β/v_0, $C_d K_{sv}\sqrt{2/\rho}$, τ_{sp} and C_{sp} used in the feedback linearization scheme and also the values of m_p and C_p. In this work, it has been considered that the initial volumes of actuators chambers are known, and therefore, β can be estimated from β/v_0 value.

A check of the robustness against noise of the hydraulic parameters estimation process has been performed in the same way as for the IF estimation case. Table 2 shows the nominal and identified values of the hydraulic parameters and the identification relative error both for the noise-free and noise-contaminated identification cases. Despite the fact that parameters estimation quality decreases when a noisy environment is considered, the obtained values still represent with reasonable accuracy system actual parameters.

Table 2. Identified hydraulic parameters.

Parameter	Model Value (S.I. Units)	Identified Value without Noise (S.I. Units)	Relative Error without Noise (%)	Identified Value with Noise (S.I. Units)	Relative Error with Noise (%)
A_w	5.9000×10^{-3}	5.9000×10^{-3}	1.5458×10^{-6}	5.9000×10^{-3}	2.0000×10^{-3}
β	1.5000×10^{9}	1.5000×10^{9}	4.3000×10^{-2}	1.4926×10^{9}	-4.9010×10^{-1}
$C_d K_{sv}\sqrt{2/\rho}$	3.5635×10^{-6}	3.5602000×10^{-6}	-9.2000×10^{-2}	3.5607×10^{-6}	-7.9700×10^{-1}
C_p	1.0000×10^{3}	1.0000×10^{3}	1.2000×10^{-3}	9.9632×10^{2}	-3.6830×10^{-1}
C_{sp}	1.8000×10^{-4}	1.8036×10^{-4}	2.0090×10^{-1}	1.8744×10^{-4}	4.1317×10^{0}
τ_{sp}	1.0000×10^{-2}	1.0000×10^{-2}	2.0170×10^{-1}	1.0500×10^{-2}	4.6080×10^{0}

3.3. Drive Calculation Module

The drive calculation module computes the desired pressure force time derivative to be injected to the feedback linearization module. Firstly, the reference acceleration profile is transformed into frequency domain by means of an FFT process. Then, the drive signal is calculated multiplying the identified IF by the transformed acceleration profile. Finally, the result is transformed back into time domain by means of an IFFT process. Figure 14 schematizes the described process.

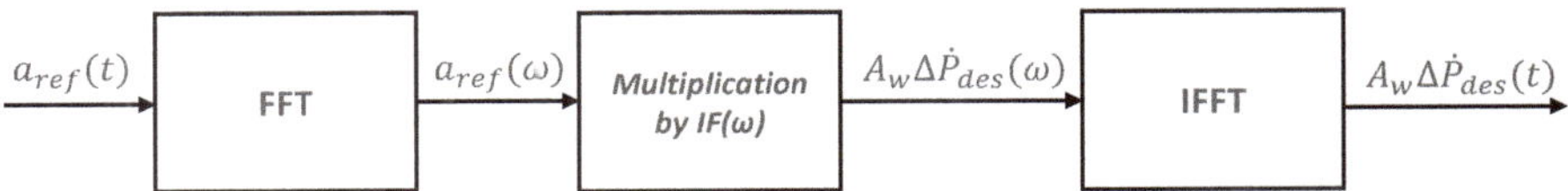

Figure 14. Drive calculation process.

3.4. Three Variable Controller

A feedback controller has been implemented, in parallel with the previously described architecture, with the aim to cope with the unavoidable errors occurring within AF and hydraulic parameters identification processes. TVC philosophy has been adopted so as to provide real-time simultaneous corrections to rod displacement, velocity and acceleration errors. The control law of the TVC is defined as follows:

$$u_{TVC} = K_d\left(d_{ref} - x_p\right) + K_v\left(v_{ref} - \dot{x}_p\right) + K_a\left(a_{ref} - \ddot{x}_p\right), \tag{24}$$

where u_{TVC} is the control voltage output by TVC, d_{ref}, v_{ref}, a_{ref} are respectively the displacement, velocity and acceleration reference waveforms and K_d, K_v and K_a are the control gains for displacement, velocity and acceleration errors. As it can be noticed, the implementation of this controller requires calculating, by integration, the reference displacements and velocities from the given reference acceleration.

4. Numerical Simulations Results and Control Methods Comparison

In this section, the numerical results obtained with the model described in Section 2 are presented. Section 4.1. shows and discusses simulation results for the new proposed control method in scenarios with and without noise present in measurements from sensors. Section 4.2 compares the performance of the suggested control procedure to that of the classical iterative control approach illustrating the main differences.

4.1. Numerical Simulations Results

The chosen acceleration reference in all the presented cases is a random gaussian waveform with a duration of 20 s and a flat frequency content between 1 Hz and 80 Hz (see Figure 17). A Hanning window has been applied to the reference profile to ensure null values at the block ends. The approximate peak displacement, velocity and acceleration values are 30 mm, 0.4 m/s and 25 m/s^2. A fixed step solver and a time step of 1×10^{-4} s has been used for all the simulations in this section.

A first set of simulations has been carried out with the TVC feature disabled in a noise-free environment. As explained in Section 3.1, the feedback linearization module calculates an instantaneous spool position, which is attempted to be imposed on servovalve's main stage by means of a spool dynamics inversion algorithm (see Equation (20)). Figure 15 shows the reference and achieved servovalve spool position, both in time and frequency domain. It can be concluded that the tracking achieved by the dynamic inversion algorithm is accurate within the whole frequency range of interest.

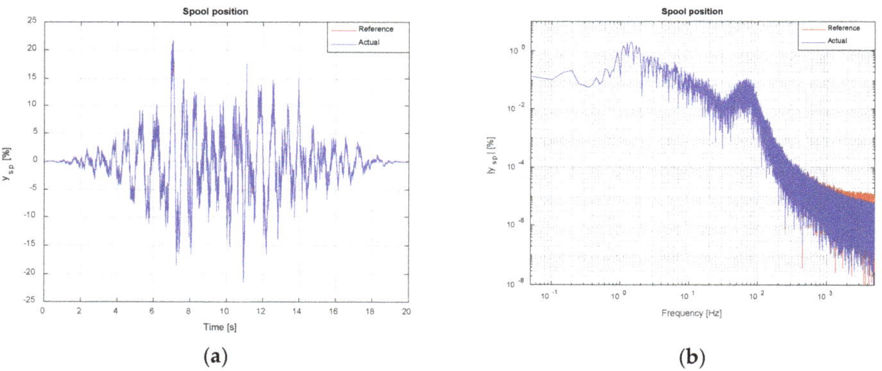

Figure 15. Spool position tracking (noise-free scenario). TVC disabled: (**a**) Time domain; (**b**) Frequency domain.

The time derivative of pressure force on actuator's rod, synthesized by the drive calculation module explained in Section 3.3, and the actual one effectively imposed on servoactuator owing to the feedback linearization process, are shown in Figure 16. Figure 16a has been zoomed around the area where maximum error takes place. Tracking is acceptable in the entire frequency range of interest with larger errors at low frequencies and around the first modal frequency of the table-SuT system, due to the poorer IF estimate obtained at those frequency values (see Figure 10).

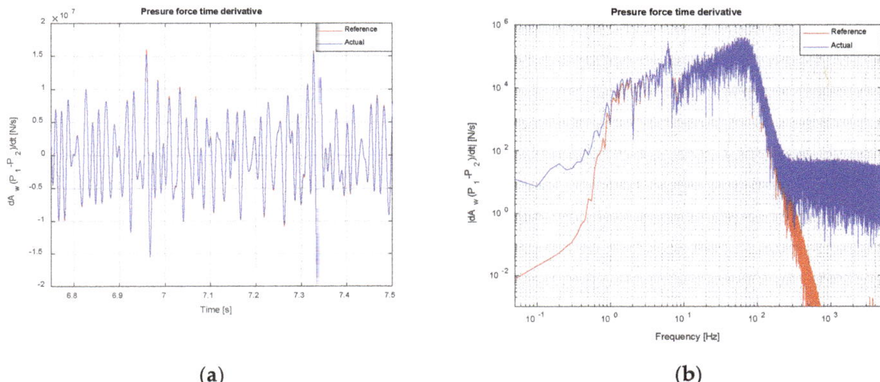

(a) (b)

Figure 16. Pressure force time derivative tracking (noise-free scenario). TVC disabled: (**a**) Time domain; (**b**) Frequency domain.

Figure 17 shows the target acceleration together with the one attained in numerical simulations. A tracking error of approximately 0.55 m/s^2 rms has been achieved. As in the previous case, and due to the same reasons stated there, reference acceleration tracking is reasonably satisfactory, except for the low frequencies and at the neighborhood of the first modal frequency of the system. Acceleration tracking error is shown in more detail in Figure 18. Despite the fact that tracking can be deemed acceptable, accumulation of errors within the low frequency region lead to increased velocity errors and displacement drifts which may hinder successful test execution due to limited servovalve flow rate capacity and actuator stroke. Therefore, it seems mandatory to enhance the control architecture with a parallel controller able to keep, simultaneously, all kinematic variables tracking errors within reasonable limits.

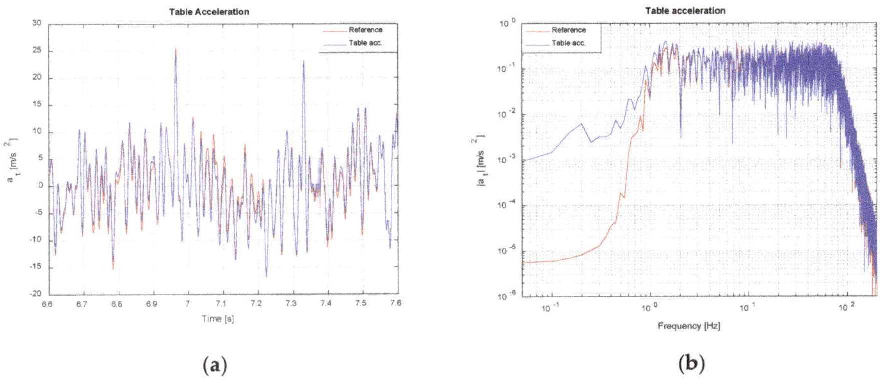

(a) (b)

Figure 17. Shake table acceleration tracking (noise-free scenario). TVC disabled: (**a**) Time domain; (**b**) Frequency domain.

A second round of simulations has been carried out with the TVC feature enabled in a noise-free environment. Figures 19–21 show the same information as that offered in Figures 16–18. The employed values of TVC control gains have been $K_d = 1$, $K_v = 0.5$ and $K_a = 0.25$. Figures below show a drastic decrease in acceleration tracking error of from 0.55 m/s^2 rms in the previous case to 0.087 m/s^2. This reduction especially marked in the low frequency range, down to 0.6 Hz, confirming the effectiveness of the TVC feedback controller.

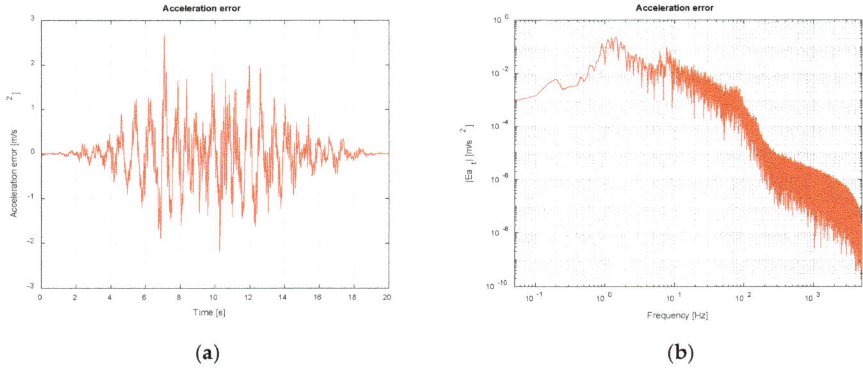

(**a**) (**b**)

Figure 18. Shake table acceleration error (noise-free scenario). TVC disabled: (**a**) Time domain; (**b**) Frequency domain.

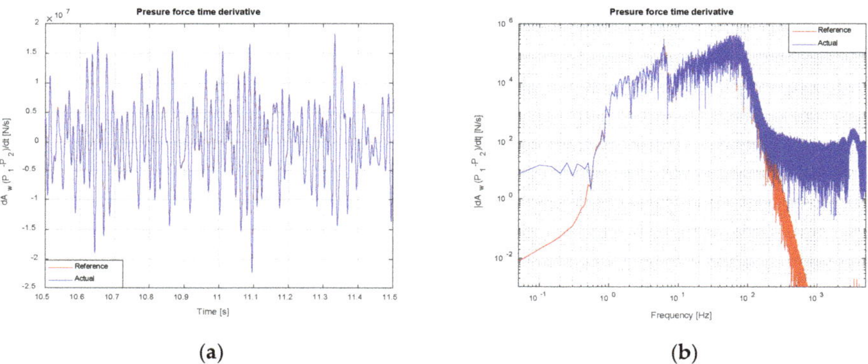

(**a**) (**b**)

Figure 19. Pressure force time derivative tracking (noise-free scenario). TVC Enabled: (**a**) Time domain; (**b**) Frequency domain.

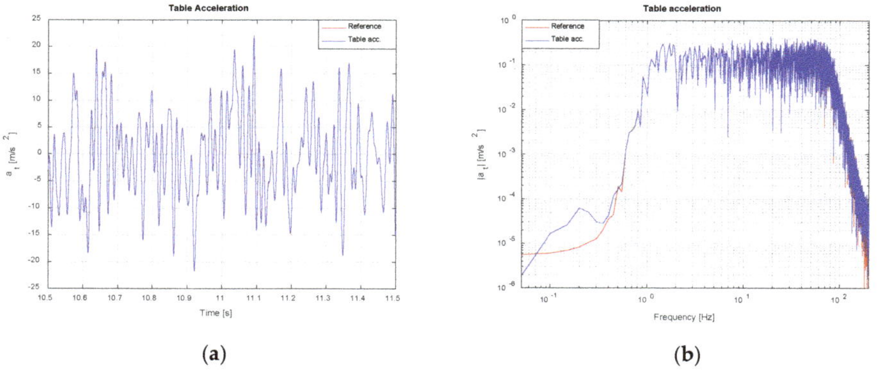

(**a**) (**b**)

Figure 20. Shake table acceleration tracking (noise-free scenario). TVC Enabled: (**a**) Time domain; (**b**) Frequency domain.

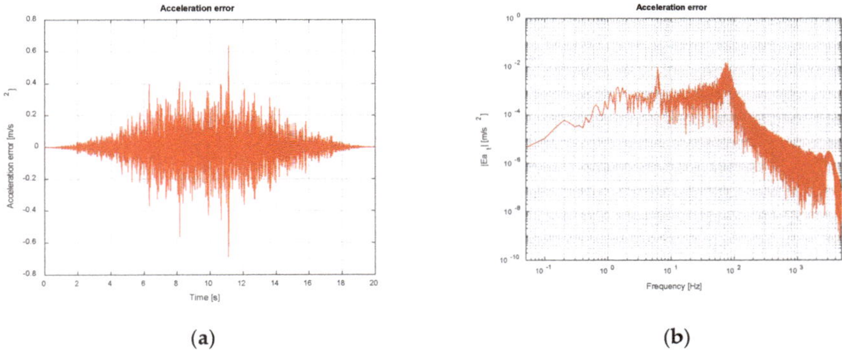

(**a**) (**b**)

Figure 21. Shake table acceleration error (noise-free scenario). TVC Enabled: (**a**) Time domain; (**b**) Frequency domain.

A third set of simulations has been carried out to assess the suggested control method in a scenario where noise is present in sensors measurements. The electrical noise has been modelled as a gaussian waveform with a peak value of 10 mV according to considerations made in Section 3.2.1. Sources of noise of this magnitude have been added to all the transducers present in the model (displacement, acceleration, chamber pressures and spool position) and to servovalve input voltage. Noise in voltage has been multiplied by the appropriate gains to translate it into the physical quantities affecting the model (see Section 2). The drive estimate has been synthesized making use of the IF obtained in a noisy environment (see Section 3.2.1) and the hydraulic parameters utilized by feedback linearization scheme have been those estimated in presence of noise. The TVC feature has been enabled and the control gains used have been the same as in the previous case, that is, $K_d = 1$, $K_v = 0.5$ and $K_a = 0.25$.

Figures 22–24 show the same information as that in Figures 19–21. An overall tracking error of 0.403 m/s^2 rms has been achieved. Tracking error has increased in the whole frequency range and specially around the higher frequency limit of the reference profile. The effect of noise is obviously more accused at lower target acceleration values due to the reduced signal to noise ratio at those sections. Nevertheless, despite the fact that electrical noise clearly affects negatively tracking quality, performance is still reasonably good and the stability of the system is maintained, therefore confirming the robustness of the proposed method when electrical noise of a reasonable magnitude contaminates sensors measurements.

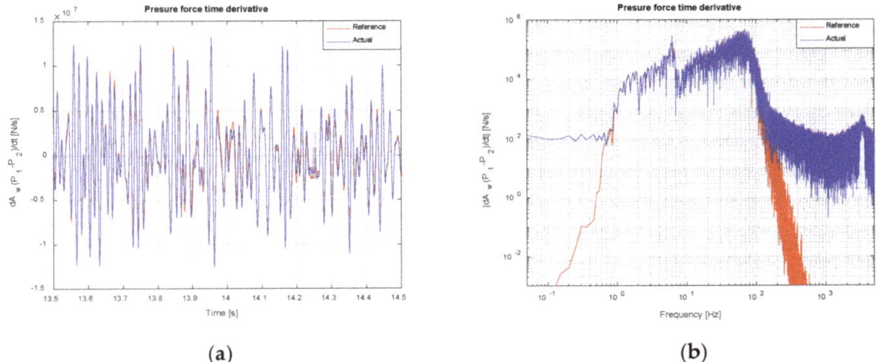

(**a**) (**b**)

Figure 22. Pressure force time derivative tracking (noise-contaminated scenario). TVC Enabled: (**a**) Time domain; (**b**) Frequency domain.

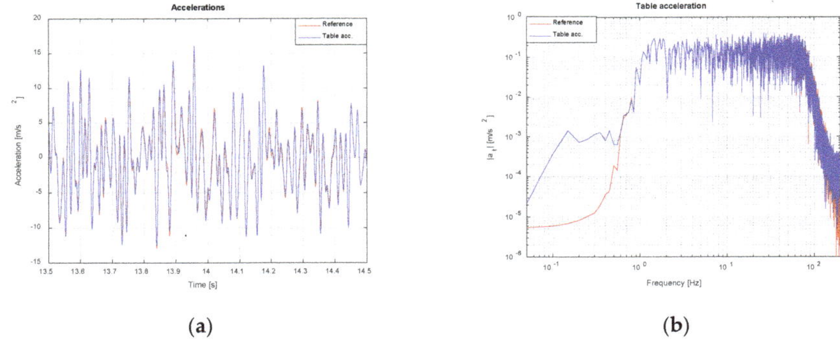

(a) (b)

Figure 23. Shake table acceleration tracking (noise-contaminated scenario). TVC Enabled: (**a**) Time domain; (**b**) Frequency domain.

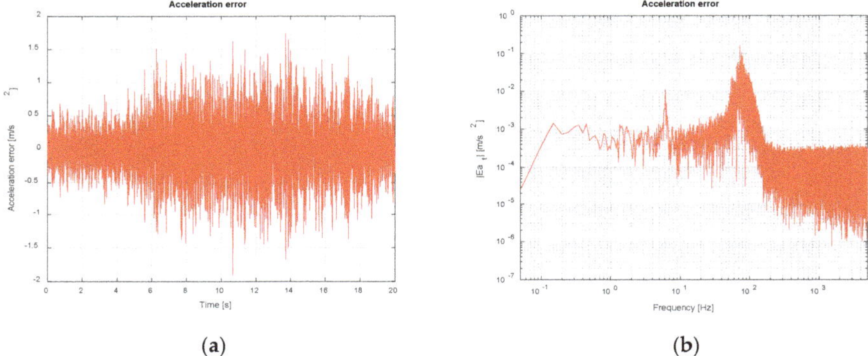

(a) (b)

Figure 24. Shake table acceleration error (noise-contaminated scenario). TVC Enabled: (**a**) Time domain; (**b**) Frequency domain.

Finally, a fourth round of simulations has been conducted to explore the trend of system behavior in a noisy environment, when the values of the control gains of the TVC are increased while keeping the rest of the parameters unaltered. The values of the control gains employed have been $K_d = 3$, $K_v = 2$ and $K_a = 1$. Figures 25 and 26 show, respectively, acceleration and acceleration error both in time and frequency domain. With the employed control parameters, the influence of the noise in the system is remarkably reduced yielding similar results as in the second set of simulations. In particular, a tracking error of 0.1154 m/s^2 rms has been attained.

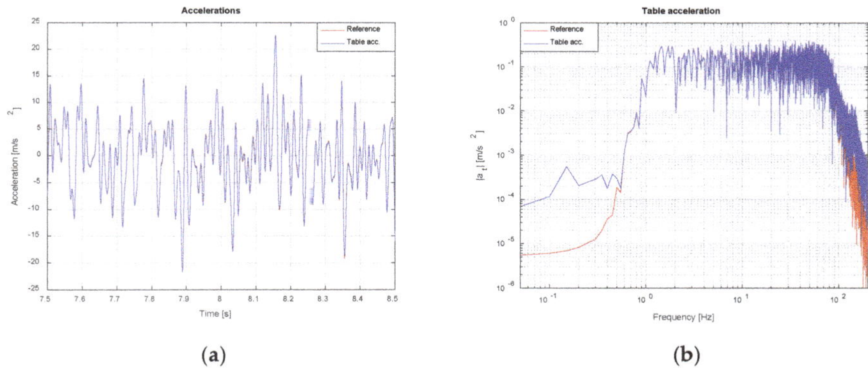

(**a**) (**b**)

Figure 25. Shake table acceleration tracking (noise-contaminated scenario). TVC Enabled with improved parameters: (**a**) Time domain; (**b**) Frequency domain.

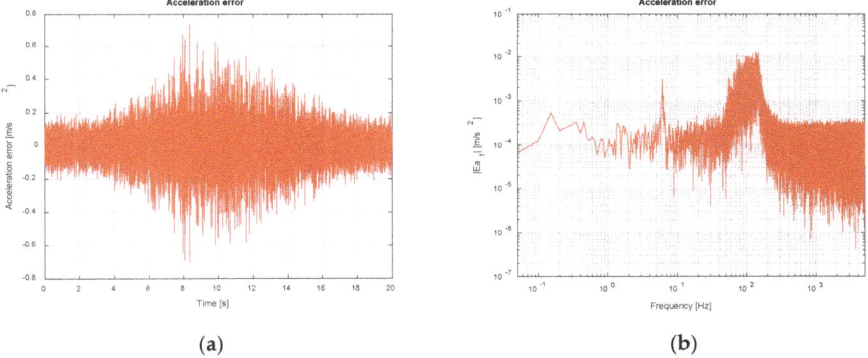

(**a**) (**b**)

Figure 26. Shake table acceleration error (noise-contaminated scenario). TVC Enabled with improved parameters: (**a**) Time domain; (**b**) Frequency domain.

4.2. Comparison between Control Methods

In this subsection, a comparison between the classical iterative control approach, which constitutes the industry standard for shake table testing, and the new proposed method is presented. A generic iterative scheme (see Figure 2), in a noise-free scenario, has been employed to obtain qualitative results representative of the classical method performance. First off, the FRF of the system has been identified making use of the H_1 estimator, according to traditional approach (acceleration over voltage). Then, it has been inverted to obtain the IF (voltage over acceleration). An estimate of the initial drive to be fed to the system has been calculated by means of: $d^0(\omega) = IF(\omega)a_{ref}(\omega)$, where $d^0(\omega)$ is the initial drive in frequency domain. The result has been transformed into time domain by means of an IFFT and has been injected into the servovalve. After the initial iteration, the drive signal would be updated in successive runs by an iterative scheme of the form: $d^{n+1}(\omega) = d^n(\omega) + kIF(\omega)\left[\ddot{x}_t{}^n(\omega) - a_{ref}(\omega)\right]$, formulated in frequency domain, where n denotes iteration number and k is the correction gain. However, only the acceleration results obtained at the first iteration have been considered, with the aim of evaluating the control methods in comparable operation conditions. In these simulations, the TVC feature in the new control procedure has been enabled and the values of control gains used have been: $K_d = 1$, $K_v = 0.5$ and $K_a = 0.25$.

Figures 27 and 28 show acceleration response and tracking error for both methods. The first iteration of classical approach reaches a tracking error of 1.484 m/s^2 rms as opposed to the 0.087 m/s^2 rms featured by the new implementation. The proposed method shows much better behavior than the first iteration of the classical approach over the complete frequency range. Nevertheless, this difference in performances is likely to decrease if a certain number of control iterations were carried out. According to Figure 28a, the error of the classical approach increases with the magnitude of the target acceleration. This tracking error rise is caused by the fact that this method relies on a linearization of a non-linear system around an operating point, which may no longer be valid when the target acceleration profile implies reaching large values of forces and displacements. In opposition, the new suggested procedure performs well even at high accelerations, in part, due to the fact that the implemented feedback linearization scheme excludes the non-linearities associated to hydraulic system from the control loop.

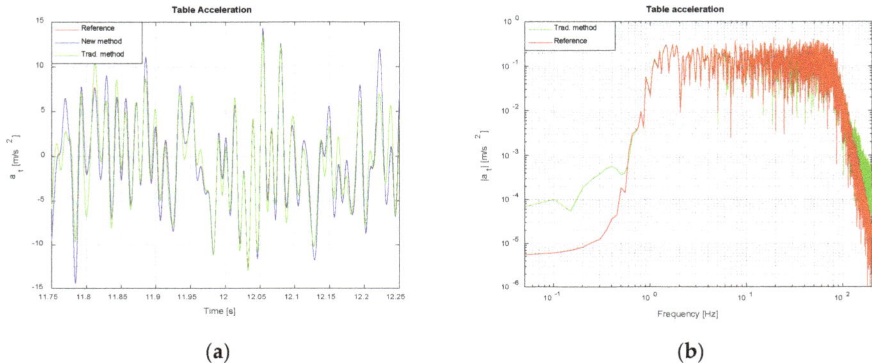

Figure 27. Comparison between classical and new control methods. Acceleration tracking: (**a**) Time domain; (**b**) Frequency domain.

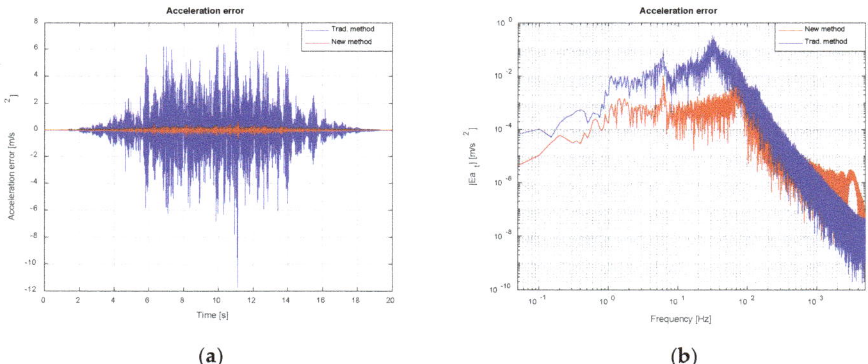

Figure 28. Comparison between classical and new control methods. Acceleration error: (**a**) Time domain; (**b**) Frequency domain.

5. Conclusions

This paper presents a novel mixed time-frequency acceleration control method for shake table systems. The suggested procedure includes identifying an FRF-IF pair which relates table acceleration to the time derivative of the pressure force acting on servoactuator's piston rod prior to the test. This approach allows for the calculation of a time variation of pressure force waveform which can be directly imposed by means of a feedback linearization scheme, which approximately cancels out non-linearities associated hydraulic actuation system leaving only those related to SuT present in the control loop. Modeling errors, feedback linearization imperfections and external perturbations are dealt with by a TVC implemented in parallel with the previously outlined architecture. Consequently, this method includes features belonging to time and frequency domain methods usually employed in shake table control systems.

The potential effectiveness of the methodology was assessed by means of numerical simulations carried over a model of the shake table loaded with a two stories shear building. Four groups of numerical simulations were performed:

1. without the parallel TVC feature enabled in an electrical noise free environment;
2. with the parallel TVC feature enabled in a noise-free environment;
3. with the parallel TVC feature enabled in an electrical noise contaminated environment;
4. with the same conditions as in 3. but with a better tuning of TVC parameters.

Results corresponding to the first group show quite acceptable acceleration tracking; however, errors at low frequencies may lead to undesired table drifts. When the TVC feature is enabled and electrical noise is not considered (group 2), tracking errors are drastically reduced, leading to almost-perfect acceleration tracking with a low control burden placed on the TVC controller. Noise affects negatively the performance in the whole frequency range, as it was demonstrated by the third group of simulations; however, tracking error remains within acceptable limits and the stability of the system is preserved, thus confirming the robustness of the proposed control procedure when electrical noise of a reasonable magnitude contaminates sensors measurements. Finally, the fourth group of simulations demonstrates that, by a proper tuning of the TVC control parameters, almost-perfect tracking is possible even in the presence of noise. The performance of the new proposed method is better than that of the classical iterative approaches when these operate on a single iteration basis.

The presented method thus appears quite promising for its implementation in real systems and features the following advantages over traditional iterative methods: (i) no iterations are required in test execution stage, (ii) non-linearities associated to hydraulic actuation are excluded from the control loop, improving tracking characteristics and (iii) the method is less sensitive to uncertainties in IF identification than the traditional control approaches due to the parallel TVC feature.

The proposed methodology requires, however, measuring or estimating rod displacement, velocity and acceleration, pressures at both actuator's chambers, pressures at supply and return ports of actuator manifolds and position of servovalve's main stage spool position. Therefore, its implementation implies increased instrumentation needs with respect to that used in traditional control methods.

Current works are focused on the implementation of the proposed architecture in an actual shake table system, paying special attention to the following points:

- Non-linearities present in the purely mechanical system.
- Rigorous studies on the uncertainty in IF and hydraulic parameters estimation and on the effect of noise present in sensors measurements.
- Development of differentiation schemes robust against noise in signals.
- Assessment of the effects of the delay due to control loop and sensors and their effect on feedback linearization scheme.

Author Contributions: Conceptualization, J.R.S., J.H.G.-P. and I.M.D.; methodology, J.R.S.; software, J.R.S.; validation, J.R.S.; formal analysis, J.R.S.; investigation, J.R.S. and I.M.D.; resources, J.R.S.; data curation, J.R.S.; writing—original draft preparation, J.R.S.; writing—review and editing, J.R.S., J.H.G.-P. and I.M.D.; visualization, J.R.S.; supervision, J.H.G.-P. and I.M.D.; project administration, J.H.G.-P. and I.M.D.; funding acquisition, J.H.G.-P. and I.M.D. All authors have read and agreed to the published version of the manuscript.

Funding: This research was funded by Spanish Ministry of Science, Innovation and Universities through the project SEED-SD (RTI2018-099639-B-I00).

Acknowledgments: The authors would like to express their gratitude to Vzero Engineering Solutions, S.L. for providing several pictures.

Conflicts of Interest: The authors declare no conflict of interest.

References

1. Williams, M.S.; Blakeborough, A. Laboratory testing of structures under dynamic loads: An introductory review. *Philos. Trans. R. Soc. Lond. Ser. A* **2001**, *359*, 1651–1669. [CrossRef]
2. Füllekrug, U. Utilization of multi-axial shaking tables for the modal identification of structures. *Philos. Trans. R. Soc. Lond. Ser. A* **2001**, *359*, 1753–1770. [CrossRef]
3. Severn, R. The development of shaking tables—A historical note. *Earthq. Eng. Struct. Dyn.* **2011**, *40*, 195–213. [CrossRef]
4. Plummer, A.R. Control techniques for structural testing: A review. *Proc. Inst. Mech. Eng. Part I J. Syst. Control Eng.* **2007**, *221*, 139–169. [CrossRef]
5. Bairrão, R.; Vaz, C.T. Shaking Table Testing of Civil Engineering Structures-The LNEC 3D Simulator Experience. In Proceedings of the 12th World Conference on Earthquake Engineering, Auckland, New Zealand, 30 January–4 February 2000.
6. Yao, J.; Dietz, M.; Xiao, R.; Yu, H.; Wang, T.; Yue, D. An overview of control schemes for hydraulic shaking tables. *J. Vib. Control* **2014**, *22*, 2807–2823. [CrossRef]
7. Hanh, D.L.; Ahn, K.K.; Kha, N.B.; Jo, W.K. Trajectory control of electro-hydraulic excavator using fuzzy self tuning algorithm with neural network. *J. Mech. Sci. Technol.* **2009**, *23*, 149–160. [CrossRef]
8. Phillips, B.M.; Wierschem, N.E.; Spencer, B.F. Model based multi-metric control of uniaxial shake tables. *Earthq. Eng. Struct. Dyn.* **2014**, *43*, 681–699. [CrossRef]
9. Najafi, A.; Spencer, B.F. Modified model-based control of shake tables for online acceleration tracking. *Earthq. Eng. Struct. Dyn.* **2020**. [CrossRef]
10. Plummer, A. Model-based motion control for multi-axis servohydraulic shaking tables. *Control Eng. Pract.* **2016**, *53*, 109–122. [CrossRef]
11. Shen, G.; Zheng, S.T.; Ye, Z.M.; Yang, Z.D.; Zhao, Y.; Han, J.W. Tracking control of an electro-hydraulic shaking table system using a combined feedforward inverse model and adaptive inverse control for real-time testing. *Proc. Inst. Mech. Eng. Part I J. Syst. Control Eng.* **2011**, *225*, 647–666. [CrossRef]
12. Nakata, N. Acceleration trajectory tracking control for earthquake simulators. *Eng. Struct.* **2014**, *32*, 2229–2236. [CrossRef]
13. Seki, K.; Iwasaki, M.; Kawafuku, M.; Hirai, H.; Yasuda, K. Improvement of control performance in shaking-tables by feedback compensation for reaction force. In Proceedings of the 34th Annual Conference of the IEEE Industrial Electronics Society, Orlando, FL, USA, 10–13 November 2008; IEEE: New York, NY, USA, 2008; pp. 2551–2556. [CrossRef]
14. Yao, J.; Di, D.T.; Jiang, G.L.; Gao, S. Acceleration amplitude-phase regulation for electro-hydraulic servo shaking table based on LMS adaptive filtering algorithm. *Int. J. Control* **2012**, *85*, 1581–1592. [CrossRef]
15. Yao, J.; Hu, S.H.; Fu, W.; Han, J.W. Harmonic cancellation for electro-hydraulic servo shaking table based on LMS adaptive algorithm. *J. Vib. Control* **2011**, *17*, 1862–1868. [CrossRef]
16. Nowak, R.F.; Kusner, D.A.; Larson, R.L.; Thoen, B.K. Utilizing modern digital signal processing for improvement of large scale shaking table performance. In Proceedings of the 12th World Conference on Earthquake Engineering, Auckland, New Zealand, 30 January–4 February 2000; pp. 2035–2042.
17. Tagawa, Y.; Kajiwara, K. Controller development for the E-Defense shaking table. *Proc. Inst. Mech. Eng. Part I J. Syst. Control Eng.* **2007**, *221*, 171–181. [CrossRef]
18. Stoten, D.P.; Gómez, E.G. Adaptive control of shaking tables using the minimal control synthesis algorithm. *Philos. Trans. R. Soc. Lond. Ser. A* **2001**, *359*, 1697–1723. [CrossRef]
19. Stoten, D.P.; Shimizu, N. The feedforward minimal control synthesis algorithm and its application to the control of shaking-tables. *Proc. Inst. Mech. Eng. Part I J. Syst. Control Eng.* **2007**, *221*, 423–444. [CrossRef]
20. De Cuyper, J.; Coppens, D.; Liefooghe, C.; Debille, J. Advanced system identification methods for improved service load simulation on multi-axial test rigs. *Eur. J. Mech. Environ. Eng.* **1999**, *44*, 27–39.

21. Zhao, J.; Catharine, C.S.; Posbergh, T. Nonlinear system modeling and velocity feedback compensation for effective force testing. *J. Eng. Mech.* **2005**, *131*, 244–253. [CrossRef]
22. Merrit, H.E. *Hydraulic Control. Systems*, 1st ed.; John Wiley & Sons: New York, NY, USA, 1991; pp. 76–100.
23. Servovalves with Integrated Electronics D791 and D792 Series. Available online: www.moog.com/content/dam/moog/literature/ICD/Moog-Valves-D791_D792-Catalog-en.pdf (accessed on 24 September 2020).
24. Stoten, D.P. Fusion of kinetic data using composite filters. *Proc. Inst. Mech. Eng. Part. I J. Syst. Control. Eng.* **2001**, *215*, 483–497. [CrossRef]
25. Plummer, A.R. Optimal complementary filters and their application in motion measurement. *Proc. Inst. Mech. Eng. Part. I J. Syst. Control. Eng.* **2006**, *6*, 489–507. [CrossRef]
26. Kwatny, H.G.; Blankenship, G.L. *Nonlinear Control and Analytical Mechanics: A Computational Approach*, 1st ed.; Birkenhäuser: Boston, MA, USA, 2000; pp. 185–201.
27. Slotine, J.E.; Li, W. *Applied Nonlinear Control.*, 1st ed.; Prentice-Hall, Inc.: Englewood Cliffs, NJ, USA, 1991; pp. 207–265.
28. Press, W.H.; Teukolsky, S.A.; Vetterling, W.T.; Flannery, B.P. Numerical Recipes in Fortran. In *The Art of Scientific Computing*, 2nd ed.; Cambridge University Press: New York, NY, USA, 1992; pp. 180–184.
29. Leclere, Q.; Roozen, B.; Sandier, C. On the use of the Hs estimator for the experimental assessment of transmissibility matrices. *Mech. Syst. Signal. Process.* **2014**, *43*, 237–245. [CrossRef]

Article

Dynamic Behaviour of High Performance of Sand Surfaces Used in the Sports Industry

Hasti Hayati [1,*], David Eager [1], Christian Peham [2] and Yujie Qi [3]

[1] School of Mechanical and Mechatronic Engineering, University of Technology Sydney, P.O. Box 123, Broadway, Australia; David.Eager@uts.edu.au

[2] Equine Clinics/Movement Science Group, University of Veterinary Medicine Vienna, Veterinärplatz 1, 1210 Wien, Austria; Christian.Peham@vetmeduni.ac.at

[3] Transport Research Centre, School of Civil and Environmental Engineering, University of Technology Sydney, P.O. Box 123, Broadway, Australia; Yujie.Qi@uts.edu.au

* Correspondence: Hasti.Hayati@uts.edu.au

Received: 15 September 2020; Accepted: 26 October 2020; Published: 29 October 2020

Abstract: The sand surface is considered a critical injury and performance contributing factor in different sports, from beach volleyball to greyhound racing. However, there is still a significant gap in understanding the dynamic behaviour of sport sand surfaces, particularly their vibration behaviour under impact loads. The purpose of this research was to introduce different measurement techniques to the study of sports sand surface dynamic behaviour. This study utilised an experimental drop test, accelerometry, in-situ moisture content and firmness data, to investigate the possible correlation between the sand surface and injuries. The analysis is underpinned by data gathered from greyhound racing and discussed where relevant.

Keywords: sports surfacing; sand surface; dynamic behaviour; impact tests; accelerometry; greyhound racing; equine racing

1. Introduction

Sand surfacing is seen on different sports such as, beach volleyball [1], equine racing [2,3] and greyhound racing [4–6]. The mechanical properties of the sand surface not only determines the performance of an athlete, be they human or a tetrapod, but also as an important injury contributing factor [7,8]. There is still a significant gap in understanding the behaviour of sand surface under impact load [9,10]. Accordingly, understanding the mechanical properties of sand surface, variables that alter the sand surface dynamic behavior, and methods to measure these variables, are of paramount importance.

The characteristics of the sand are identified through the shape, size and percentage of the sand particles. The shape of the sand particles can vary from a 'very angular' to a 'well rounded' shape [11] and is a key influence on the dynamic behaviour of the sand [12]. There are two key variables used to classify sand particles, namely 'roundedness' and 'sphericity' [13]. Figure 1 provides a pictorial representation of the various sand particle shapes. As much as roundedness is desirable in terms of the impact attenuation properties, angularity is not. When the particles are very angular, they tend to pack tightly as the sharp corners interlock and will resist the movement of the particles when subjected to an impact. In contrast, well-rounded particles tend to smoothly transit, or flow, to different locations upon impact [14].

The amount of water retained on the sand (the sand moisture contents) and the compaction rate (sand density), also determine the sand surface dynamic behavior [15,16]. Accordingly, in a sport arena, where it is assumed the characteristics of the sand is controlled, the sand moisture content and density should be measured and compared against the safety benchmarks (The safety benchmark differs depending on the industry.) to avoid injuries. However, current safety benchmark, mainly those used in racing greyhound's arena, are not backed-up with science and research and are solely based on the experience of the track curators [2,17].

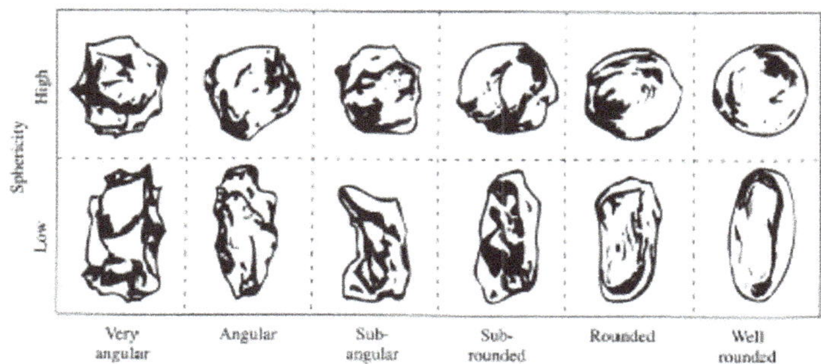

Figure 1. Sand particleshape in terms of sphericity, angularity and roundness [13].

An example of investigation on the effect on the sand moisture content and density on sports arena is a work conducted by Holt et al. [3], where the effect of sand moisture levels and rates of compaction of two different drainage systems (Limestone gravel and permavoid^{TM} drainage), on the dynamic performance of synthetic equestrian surfaces (93.84% sand, 5.15% fibre and 1.01% binding polymer) was studied. They used the Orono Biomechanical Surface Tester (OBST) [18], a 2.25 kg Clegg hammer [19] and a 30 kg traction device equipped with a horseshoe. The OBST, which simulates the collision of horse forelegs and the ground, was dropped four times on each surface for each treatment. The Clegg hammer was dropped four times based on the protocol recommended by ASTM Standard [20]. A 30 kg traction device, which was also used to measure the traction of the surface, was dropped once in four different locations of the test chamber for each treatment from a height of 200 mm.

Thiel et al. [1] mainly focused on the dry sand and designed a penetrometer to measure the stiffness of beach dry sand in-situ. To validate their method, their results were compared with that of a in-laboratory study, where the penetrometer test were conducted on a sand box [21]. It is claimed that their results are similar to the in-lab study and their method can be used to measure the stiffness coefficient of the sand, prior to a sporting event on the dry sand.

Force transducer, mainly the wearable sensor, are extensively used for gait analysis as they are cheap, easy to use, and user friendly [22]. Therefore, inertial inertial measurement units (IMU) have been used in different applications, mainly in clinical setting for gait analysis [22].

IMU technology can be also used to study the limb-surface interaction. In a recent studies conducted by the same author of this work, a single IMU was used to study the impact of different sports surfacing (grass vs wet sand) on the locomotion dynamics of galloping greyhounds [4–6]. Details of the most recent work [4] are discussed in the following sections. Worsey et al. [16], also used IMU technology (9 degrees of freedom (DOF) inertial-magnetic sensors, incorporating an 16 G-accelerometer, gyroscope, and magnetometer) to compare athletes running over three different surfacing (running track, hard sand, and soft sand). The purpose of this work was to provide more insight on a previously observed fact that athletes alter their gait mechanics to accommodate different running surfaces [23].

Mathematical modelling, mainly Spring-Loaded-Inverted-Pendulum (SLIP) models, firstly introduce by Blickhan et al. [24] are extensively used for gait analysis in different fields of science and engineering. SLIP models are simple and easy to interpret, yet provide substantial information about the under-studied subjects [24]. There are numerous off-the-shelf SLIP models, which one can modify based on their application. For instance, a SLIP model of a greyhound galloping on sport sand surfacing was modelled to study the effect of the sand surface with different moisture content levels and rate of compaction on the canine locomotion dynamics [25]. The results showed that small changes in sand surface mechanical properties can significantly affect the amount of force acting on the greyhound hind-leg which well correlates with the high rate of hind-leg severe injuries in this industry.

It was discussed previously that the ideal track surface should have enough impact attenuation properties to damp the initial impact shock, as well as providing enough traction for a stable gallop [26–28]. The surface with ideal mechanical properties has a low amount of energy loss and low impact acceleration (G_{max}) when the foot comes into contact with the surface. The low energy loss, would also increase the performance of the animal in the race [29].

The surface with high performance was associated with a higher risk of injuries. By contrast, the surface with impact attenuation properties tended to increase the muscular effort of the runner which affected their running performance [2].

The low density of the sand or the rates of compaction are also associated with the low rates of injury [30]. In practice, For sand sport surfacing, 'harrowing' is suggested, which can reduce the density or the rates of surface compaction [31]. However, a very low density surface may have a detrimental effect on locomotion efficiency as it affects the support needed for grip and propelling the body forward [2].

Surface traction is another variable identifying a safe surface composition. High traction will increase the bending moment applied to the bones, mainly the tarsal bones, and increase the risk of injuries [7]. However, not enough traction, usually seen in drier sands, will cause the surface not to sufficiently support the limb during the stance. Accordingly, as suggested by Holt et al. [3], increasing the moisture content of the sand while keeping its density low would result in a surface ideal in both race performance and injury reduction [3]. Overall, apart from acting as a supporting surface, the sand layer also acts as an energy absorbing layer to mitigate the impact shock. The optimal condition of the sand layer should have enough energy absorbing capacity (reflected as energy loss and contact time) while providing acceptable surface traction [32,33].

Contact time is another critical variable that affects the safety performance of the surface. The shorter the contact time, the higher the risk of injuries, because of an increased load rate to the musculoskeletal system [34,35]. Accordingly, this variable is considered as one of the primary safety thresholds in different applications such as playground surfacing tests [36,37].

The purpose of this work is to introduce methods to study the dynamic behaviour of athletic sand surfaces, with the aim of improving athletes performance while minimising the risk of injuries. The methods introduced here were originally designed for greyhound racing arenas but are adaptable to other sports such as horse racing [3], beach relay and sand volleyball [1].

2. Materials and Methods

2.1. A Drop Test to Study the Dynamic Behaviour of the Sport Sand Surfaces

It is discussed above that the sand characteristics contribute to the dynamic behaviour of the sand, mainly under impact loads. Below the sand particle sizes and percentages recommended for greyhound racing arena are given in Table 1.

The sample was taken from a typical greyhound racing arena. The sample was then oven dried for 24 h based on the AS 1289 Part 2.1.1 Standard [38]. As per the Standard, the sand sample should be heated up in an oven, between 105 to 110° for 16 to 24 h.

The sample was then loaded on the sieve shaker. The procedure adopted for this test was following the AS 1289 Part 3.6.1 2009 Standard [38]. To do so a sieve shaker, model EFL 2000, was used. As per the Standard, the size of the sieve tray was selected from 4.75 mm to 75 μm. The procedure was done as such the EFL 2000 were not overloaded. In the case of any sieve being overloaded, the overloaded sieve sample was further sieved into two or more portions. The sieve shaker was set to shake for a time duration in between 5–10 min so that the sand was completely separated according to their sizes. The same procedure was repeated for 6 samples of soil.

Table 1. Recommended sand particle sizes and percentage for greyhound racing surfaces.

Fraction	Size (mm)	Percentage (%)
Fine gravel	2.00	0
Very coarse sand	1.00	<5%
Coarse sand	0.50	10–20%
Medium sand	0.25	30–40%
Fine sand	0.15	40–50%
Very fine sand	0.05	40–50%
Silt/clay	<0.05	<5%

The calculations for generating the grading curve plots, which is based on AS 1289 Part 3.6.1 2009 Standard [38] is given below Equations (1) and (2):

$$Percent\ retained\ (\%) = \frac{Mass\ of\ the\ particles\ retained\ on\ selected\ sieve}{\Sigma Total\ retained\ mass} \times 100\ \% \quad (1)$$

$$Mass\ of\ soil\ retained = mass\ of\ the\ selected\ sieve\ with\ soil - mass\ of\ empty\ sieve \quad (2)$$

Once the percent retained is calculated. The cumulative percent retained for each sieve tray is calculated by adding the percent retained from the largest size sieve to the current size sieve, and then the percentage of the sand passing through the current sieve size can be obtained through Equation (3):

$$Percent\ passing\ (\%) = 100 - cumulative\ percent\ retained. \quad (3)$$

The sand grading curve is plotted and given in Figure 2. The cumulative percent passing is plotted against the sieve size (in logarithmic scale).

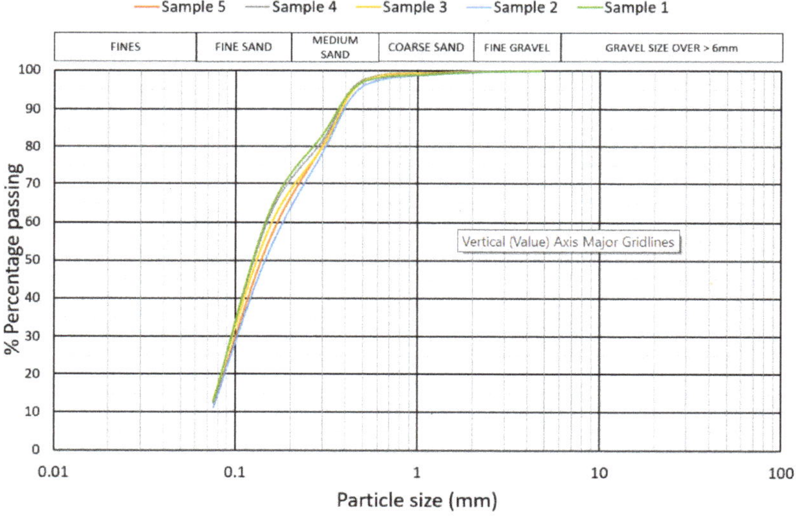

Figure 2. Typical soil grading curves for a greyhound racing surface.

The soil grading curve given above proved the fact that the soil used in the greyhound racing track is loamy sand, combination of sand with traces of clay [39]. The slight difference in each test can be attributed to the loss of soil during the test and therefore caution should be taken while conducting the test.

The sand moisture content and compaction rate are two important parameters that alter the mechanical behaviour of the sand. To study the effect of these two parameters on the dynamic behavior of the sand, collected from a typical sport arena (in our case, it was collected from a typical greyhound racing arena), an impact test which complies with AS 1289 Part 2.1.1 Standard [38], can be applied.

To perform the impact test, a conventional Clegg hammer was modified, by mounting two calibrated laboratory-grade Endevco high-G accelerometers. Adding the high-G precision accelerometers allowed a higher degree of experimental accuracy than that offered by the standard Clegg hammer. The reliability of the system was tested in previous studies on children's playgrounds for impact attenuation of surfacing [36,37].

The dynamic behaviour of the sand sample was studied by analysing the impact data, namely the maximum acceleration (G_{max}), the maximum rate of change of acceleration (Jerk) (J_{max}), the impact duration (contact time), and the energy loss. Before treating the sand sample, it should be again air or oven dried. The AS 1289 Part 2.1.1 Standard [38] is used to dry the sand sample. Based on the Standard, the sand sample was heated up in an oven, between 105 to 110° for 16 to 24 h.

The effect of three moisture levels—dry (12%), medium to ideal (17%), and ideal (20%); and three rates of compaction: low traffic (1.35 g/cm^3), medium traffic (1.45 g/cm^3), and high traffic (1.55 g/cm^3), on the dynamic behaviour of the sand sample, were studied. The density of the sand to replicate the traffic condition of the surface was previously used by Holt et al. [3].

For all three conditions, we used a cylindrical container with an inner diameter of 15.6 cm. The sand was filled at 3.0 cm increments until reaching the depth of 12.0 cm. The applied tampering here was manual. Preferably, the tamper should be equipped with an accelerometer which can provide a measure of the applied force. However, achieving a certain depth was the only possible control we could apply. The average of sand density (the mass of the sand sample divided by its volume) for the simulated traffic conditions was also calculated and given as follows:

- **Low traffic condition:** The top 3.0 cm layer was raked. The average of sand sample density for all moisture contents was 1.35 g/cm^3. This traffic condition is pictured in Figure 3A.
- **Medium traffic condition:** The top 3 cm top layer was struck with a tamper to achieve the depth of 14 cm. The average of sand sample density for all moisture contents was 1.45 g/cm^3. This traffic condition is pictured in Figure 3B.
- **High traffic condition:** The top 3 cm top layer was struck with a tamper to achieve the depth of 13 cm. The average of sand sample density for all moisture contents was 1.55 g/cm^3. This traffic condition is pictured in Figure 3C.

Figure 3. (**A**) Low traffic condition with the density of 1.35 g/cm^3. (**B**) Medium traffic condition with the density of 1.45 g/cm^3. (**C**) High traffic condition with the density of 1.55 g/cm^3.

After preparing the sand sample, an impact attenuation test, which complied with the ASTM-F3146 Standard [20], was conducted from three different heights, namely: 400 mm, 500 mm and 600 mm. Based on the Standard, the test was repeated four times from each height, and the maximum value was reported that is, the maximum value for G_{max}, J_{max} and contact time. After the fourth drop at each height in the same location, the sand sample was reconstructed to avoid the effect of over compacting of the lower layers on the results. The impact attenuation data were then post-processed using LabVIEW software and plotted in MATLAB R18. An ANOVA test (two-factor with replication) was conducted. Values of $p \leq 0.05$ were considered statistically significant. The experimental setup is illustrated below in Figure 4.

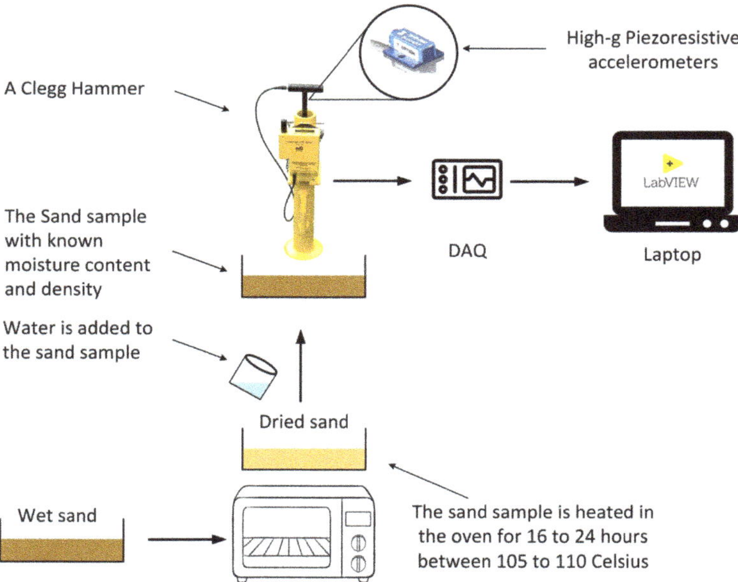

Figure 4. The process of drying the sand sample, altering the moisture content and impact testing using a modified Clegg hammer.

2.2. Pre-Surface Condition Data to Test Whether There Is a Correlation with Injuries

It is argued above that the sand moisture content and density affect the dynamics behavior of sports sand surfaces. It is also seen in a mathematical model of greyhounds that subtle changes in these two values significantly change the forces exposed to animals limb (mainly the hind-leg) [25]. Accordingly, measures should be in-place to correlate the sports sand surfacing moisture content and firmness with the probability of the injuries, which is also advised for other sport arenas, such as horse racing [40].

In Australian greyhound racing industry the sand moisture content and firmness are measured using a portable moisture meter (The instrument to measure track moisture content is typically the TDR 350 Soil Moisture Meter which measures the Volumetric Water Content of the sand as a percentage of retained moisture.) and penetrometer device (The instrument to measure track firmness is usually the FieldScout TruFirm turf firmness meter. The unit on measurement is depth of travel in either inches or mm.). Twenty-four readings are taken at 8 equidistant positions around the track and 3 locations across the track. The three locations across the track are 0.5 m from the inner rail, middle of the track and 0.5 m from the outside rail. These data mUst be collected pre-race, based on the compulsory minimum standards. The moisture content range should fall within the 26.0–33.6%. The sand firmness value should fall within the sand firmness range 15–40 mm. Both values are subjective to the type of the sand used in the sport arena.

In this section, surface condition data for a de-identified greyhound racing track are analysed for a duration of one year, July 2019 to July 2020, when an increase in the rate of catastrophic incidents was observed. The hypothesis was that the moisture and firmness range would not fall within the recommended range.

Moreover, any inconsistency on the track surface is dangerous and can cause an injury [8,26,28] as the greyhound is not capable of adjusting its gait based on changing surface conditions [27]. Apart from assessing whether the moisture and firmness data fall within the recommended range, the fluctuation between inside and middle track readings should be calculated at different vicinity of the track. It is hypothesised that high fluctuation in theses values suggests irregular surface properties, which might contribute to injuries.

The injury heat-map (the approximate locations on each track for each race distance where clusters of injuries occurred) are generated, based on the injury data provided from the industry, race video and the Stewart reports, and given in the later section. This would assist in finding a correlation between the surface condition data and locations of the track with high rate of injuries.

2.3. Use of Accelerometry to Study the Limb-Surface Interactions of Sprinters

Accelerometry or in other words use of accelerometer to record the locomotion dynamics of athletes are gaining attention as they are cheap, user friendly and provides fundamental information about the gaits. They are usually attached to subject joints and fused with each other for post-processing. In this section, the most recent accelerometry study on racing greyhounds is reviewed [4].

To study the effect of surface compliance on the galloping dynamics of racing greyhounds, an IMU, equipped with tri-axis accelerometer (sampling rate of 185 Hz , was used on two tracks), was used to measure the associated galloping accelerations in racing greyhounds. It was hypothesised the greyhound galloping dynamics are different on different surface types (sand surface vs grass surface).

The Anterior-Posterior (fore-aft) and Dorsal-ventral signals (vertical) acceleration signals, recorded via the IMU, were analysed, to see whether the surface type affect the locomotion dynamics of greyhounds. To do so, signals of galloping on straight sections of the sport arena, are compared with each other.

The recorded Dorsal-Ventral acceleration due to hind-leg strikes was more than triple that of the fore-leg strikes (15 G vs. 5 G). These results were in consistent with the role of hind-legs in powering the locomotion as well as their higher rates of injuries than fore-legs.

The mechanical properties of the sand and grass surface, mainly the impact deceleration (G_{max}) measured via a Clegg hammer of the sand surface were three times higher than that of the grass track [4]. Accordingly, it was expected to see higher acceleration on running on the sand surfacing than the grass one. However, the IMU data (the average of peaks of dorsal-ventral and anterior-posterior acceleration) for sand versus grass surfaces were not significantly different.

There might be different reason associated to the observed results that is, not significant difference in IMU signals despite the significant difference on surface type. Firstly, the IMU in this study is mounted on animal's neck (Figure 5a) and the signals would be damped while traveling through the body of the animal. Ideally, the IMU should be attached on animal's foot to sense the real impact load. Secondly, the applied signal processing method in this work are those usually used for linear time-series signals. It is hypothesised that applying nonlinear-time-series-analysis would identify different features in galloping over sand and grass. These methods are currently under the investigation by the same author of this work.

2.4. Results and Discussion

2.5. A Drop Test to Study the Dynamic Behaviour of the Sport Sand Surfaces

Figure 6A–I shows the impact acceleration versus time of the sand sample with three different moisture levels and rates of compaction. The peak of each impact acceleration is the maximum acceleration (G_{max}). The red, blue and black lines represent 400 mm, 500 mm and 600 mm drop heights, respectively.

The main observation from Figure 6 and impact data given in Table 2 is that, regardless of the moisture content, increasing the compaction rate of the sand sample, has resulted in an significant increase in the G_{max} [$p = 0.0003$ F = 12.9], J_{max} [$p = 0.00001$ F = 22.6] and Contact time [$p = 0.023$ F = 4.6].

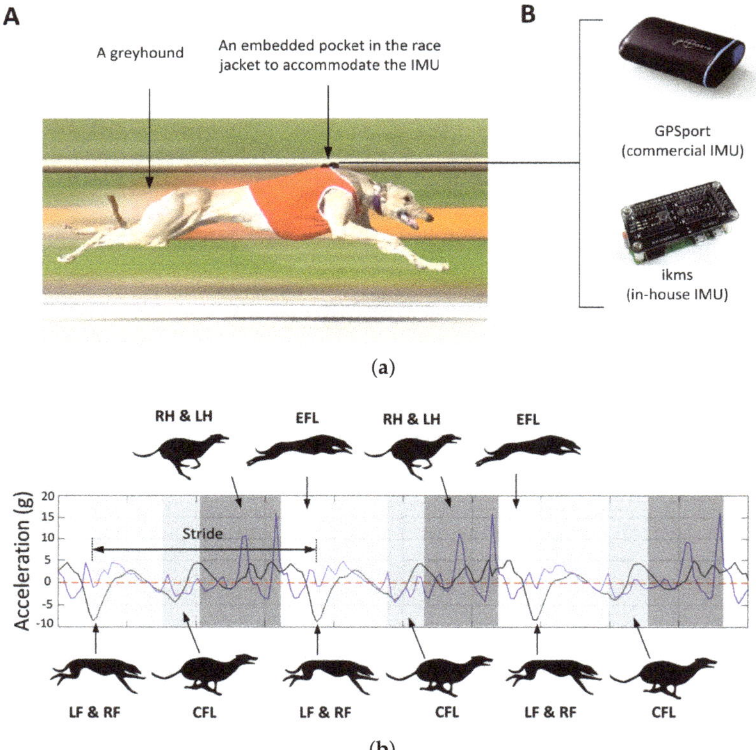

Figure 5. Accelerometry used on racing greyhounds. (**a**) (**A**) A greyhound galloping on the straight section of a track with sand surface and wearing the modified jacket with IMU pocket. (**B**) An integrated kinematic measurement system (ikms), developed in house, was used to record the acceleration signals. Data validation of the ikms was achieved using a commercial IMU (GPSport) device. (**b**) Forward (Black trace) and vertical (blue trace) acceleration vs time of three consecutive strides of a greyhound and the corresponding galloping gait events. Rotary gallop is a four-beat gait with two flight phases [41]. The limb impact in this gait moves from limb to limb in a circular pattern that is, left fore-leg (LF), right fore-leg (RF), compressed flight phase (CFL), right hind-leg (RH), left hind-leg (LH) and extended flight phase (EFL).

The same effect is also seen when the moisture content is increased (while the sand density is kept constant), mainly when the moisture level increased from 12% to 17%, which was statistically significant [$p = 0.054$ F = 4.21].

Increasing the drop height increased the velocity at the time of the impact and the higher the initial impact velocity, the higher the value of the G_{max}. This reveals the rate dependency of the sand [29].

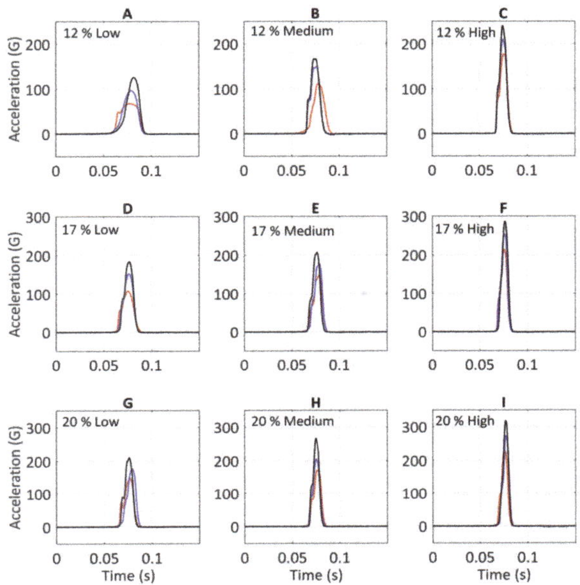

Figure 6. The G_{max} versus time of the sand samples with different moisture levels and rates of compaction. The red, blue and black lines represent the drop height of 400 mm, 500 mm and 600 mm, sequentially.

Figure 7A–I shows load-deformation plots of the sand sample with three different moisture levels and rates of compaction. The slopes of the fitted dashed lines to the loading phases of the superimposed plots is the stiffness coefficient of the sand sample (based on the method adopted by Aerts & Clercq [42] in analysing the performance of athletic shoes with hard and soft soles). The red, blue and black lines represent 400 mm, 500 mm and 600 mm drop heights, respectively.

To see whether the moisture content affects the stiffness coefficient of the sand, sand sample with the same density but different moisture content were compared with each other. It is observed that increasing the moisture contents (Here after moisture content) within a 12–20% range, increases the stiffness coefficient.

In the low traffic condition, this increase is 87% when the water content is changed from 12% to 17%, and only 24% when it is changed from 17% to 20%. In the medium traffic condition, this increase is 26% when the water content is changed from 12% to 17% and 55% when the moisture content increases from 17% to 20%. Similarly, in the high traffic condition altering the moisture content from 12% to 17% increases the stiffness by 47% and increasing the water content from 17% to 20%, increases the stiffness coefficient by 16%. This behaviour suggest there is a nonlinear positive relationship between the moisture content and the stiffness coefficient.

To see whether the sand density affect the stiffness coefficient of the sand sample, the samples stiffness coefficient are compared with each other while keeping the moisture content constant.

For a sand sample with 12% moisture content, the stiffness coefficient increases up to 95% and 84%, as the rate of compaction is altered from the low to medium traffic condition and from medium to high traffic condition, respectively. For a sand sample with 17% moisture, this increase is up to 41% and 92%, as the sand samples are compacted from the low to medium traffic condition and from the medium to high traffic condition, respectively. For a sand sample with 20% moisture content, this increase is up to 76% and 45% increase, as the sand samples are compacted from the low to medium traffic condition and from medium to high traffic condition, respectively.

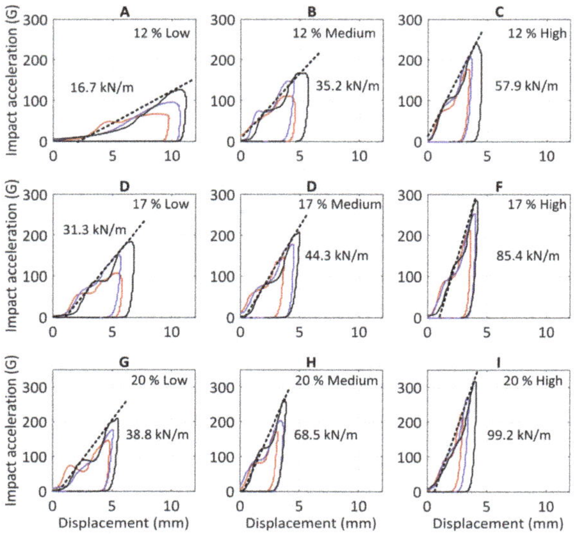

Figure 7. The load-deformation plots of the sand samples with different moisture levels and rates of compaction. The red, blue and black lines represent a drop height of 400 mm, 500 mm and 600 mm, respectively.

Increasing the sand density increases the stiffness coefficient of the samples. This is because when increasing the sand density, the interlock between sand particles will increase, hence increasing the stiffness [43].

The moisture content and traffic conditions of the sand samples, G_{max}, J_{max}, contact time (ms), energy loss (by calculating the area under the load-deformation plots), and the calculated stiffness coefficients (K), are tabulated in below Table 2.

Table 2. Impactdata from conducting a drop test on the sand sample.

Treatments	G_{max} (G)	J_{max} (kG/s)	Time (ms)	Energy loss (J)	K (kN/m)
12%-Low	96.2	43.4	6.7	46.1	16.7
12%-Medium	149.8	87.1	4.2	32.9	35.2
12%-High	176.9	11.07	3.5	34.4	57.9
17%-Low	152.7	75.9	4.7	36.4	31.3
17%-Medium	178.2	93.1	4.8	39.5	44.3
17%-High	253.6	174.0	3.1	40.9	84.5
20%-Low	176.7	91.7	5.0	31.9	38.8
20%-Medium	204.2	129.0	3.5	37.3	68.5
20%-High	272.2	205.0	3.7	32.2	99.2

In the provided results in Table 2, the contact time was not affected by the moisture level of the sand samples, but it significantly decreased with increases of the density of the sample. Thus, low to medium density of the sand sample was found to provide the favourable range of contact time with regards to injury prevention.

It is observed that altering the moisture content, significantly increased the G_{max} and J_{max} with no substantial change seen in the contact time. Moreover, the rates of compaction significantly increased all the impact data. It is also argued that the high G_{max} and J_{max} and short contact time were associated with high injury rate. Accordingly, comparing all the impact data it seems that the sand sample with 20% moisture content in a low traffic condition resulted in the most favourable behaviour with regards to both the injury prevention and race performance. The sand sample in this condition had the lowest energy loss compared to all other cases. The contact time was also in the favourable range as mentioned above. Finally, the G_{max} and J_{max} values were relatively low.

2.6. Use of Pre-Surface Condition Data to See If They Correlate with Injuries

The first step to analyse whether sand moisture content and firmness data correlate with the injury, for those race events with catastrophic incidents, the range for moisture content and the range for sand firmness were checked to see if they all fall within the recommended range. It is hypothesised that sand moisture and firmness range should not fall within the recommended range.

The second step to analyse whether sand moisture content and firmness data correlate with the injury, the fluctuation between the inside and middle track data (both moisture content and firmness values) should be calculated and compared with the overall fluctuation between the inside and middle track readings. It is hypothesised that any noticeable fluctuation between the inside and middle track readings contribute to injuries. To test this hypothesis, the injury location of the catastrophic incidents, determined using the injury heat-map given in Figure 8, were compared with high fluctuation arenas and the results are provided below.

A de-identified track heat map is presented in Figure 8. This heat map was generated using race videos, the Steward's reports and the injury data recorded by on-track veterinarians. The red circles on heat map represent the number of injuries at each specific injury location around the track. The larger the radius on the heat map, the higher the injury rate. It should be noted that only catastrophic incidents resulting in death were used to generate the injury heat-map.

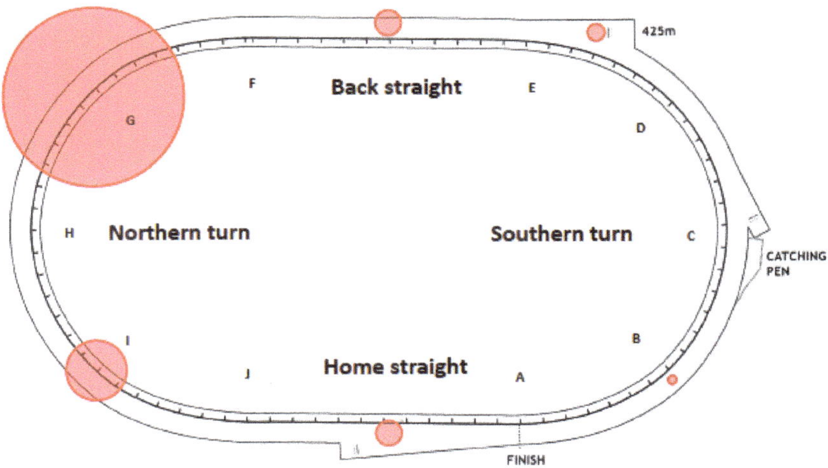

Figure 8. The de-identified greyhound track injury location *heat map* for the 425 m start. The magnitude of the normalised injury rate is depicted by the relative radius of the red circles.

The injury location, the sand surface moisture content range, the sand firmness value range, and the high fluctuation vicinity on the track are given in below Table 3.

Comparing the injury heat-map with Table 3 the injury heat-map correlates with the high fluctuation areas. The fluctuation between inside and middle track moisture of approximately 80% of the catastrophic injuries had the highest percentage at the injury vicinity among other locations of the track; The sand moisture content of approximately 80% of the races with catastrophic injuries fell within the recommended range. The sand firmness data of all (those that were available) of the races with catastrophic injuries fell within the recommended range.

Having high fluctuation between the inside and the middle track surface properties would expose trailing greyhounds to a running surface with different properties as they tend to jostle and change direction to avoid bumping and checking. Any sudden change in the surface condition will contribute to injuries.

Table 3. Injury location, surface condition range and the zones on the track with of high fluctuation.

Number	Moisture Content	Firmness	Injury Location	High Fluctuation Vicinity	Matching
1	27.2–31.1%	not available	F-G	Moisture (F-G), Firmness (N/A)	YES
2	24.2–31.6%	not available	F-G	Moisture (F-G), Firmness (N/A)	YES
3	26.7–31.1%	not available	B-C	Moisture (B-C, H), Firmness (N/A)	YES
4	23.2–34.6%	not available	C-D	Moisture (C-D), Firmness (N/A)	YES
5	27.2–32.2%	not available	F	Moisture (F), Firmness (N/A)	YES
6	26.8–31.2%	not available	C-D	Moisture (C-D), Firmness (N/A)	YES
7	24.7–35.1%	15–30	F-G	Moisture (F), Firmness (F-G)	YES
8	28.7–32.4%	15–25	F-G	Moisture (F), Firmness (G)	YES
9	29.2–32.6%	20–30	H-J	Moisture (E), Firmness (G)	NO
10	29.2–32.6%	20–30	H-J	Moisture (E), Firmness (G)	NO

The main maintenance practice which can assist in having a homogeneous surface condition is called harrowing. The depth and frequency of the harrowing practice is subjective to the sport arena, frequency of races and trials, season of the year, weather condition and rain fall, and more importantly, should be accompanied with an appropriate irrigation management. As discussed above, the sand moisture content and density are two important factors affecting the dynamic of the race track. It is also seen that, a relatively wet (20% for the under-studied sand in the laboratory condition) sand with low-traffic condition (3 cm raked top layer) was ideal in terms of both performance and safety. Accordingly, it is recommended that the harrowing should be conducted on a regular basis and with sufficient depth on a surface which is consistently irrigated through an appropriate irrigation system.

3. Conclusions

The sports sand surfacing is used in different sports and is proven to contribute to both increased athletic performance and a decrease in the risk of injuries. The first step to engineer an optimum sand surface is understanding the dynamic behaviour of sand surfacing. In this work, different methods to study the dynamic behaviour of sand surfacing used in the sporting industry are provided and where applicable, were backed-up with empirical data. Analysing the the impact data of laboratory-based experiments provided insights into how subtle alternation between the sand moisture content and density can significantly affect its dynamic behaviour upon impact force. Analysing the sand moisture content and firmness data, collected via portable moisture meter and penetrometer device, prior to a greyhound racing race, showed that high fluctuation between these values along the width of the track (mainly the inside and middle regions) can contribute to catastrophic incidents. The study provided in this work can contribute to the standardising of sport sand surfacing in sports other then greyhound racing such as volleyball.

Author Contributions: Conceptualization, H.H. and D.E.; methodology, H.H. and D.E.; software, H.H.; validation, H.H., D.E. and C.P.; formal analysis, H.H.; investigation, H.H.; resources, D.E.; data curation, H.H.; writing–original draft preparation, H.H.; writing–review and editing, X.H., D.E., Y.Q.; visualization, H.H.; supervision, D.E.; project administration, H.E.; funding acquisition, D.E. All authors have read and agreed to the published version of the manuscript.

Funding: The work is funded by Greyhound Racing New South Wales with UTS institution reference of PRO17-3051.

Conflicts of Interest: The authors declare no conflict of interest.

References

1. Thiel, D.V.; Worsey, M.T.; Klodzinski, F.; Emerson, N.; Espinosa, H.G. A penetrometer for quantifying the surface stiffness of sport sand surfaces. *Proceedings* **2020**, *49*, 64. [CrossRef]
2. Chateau, H.; Holden, L.; Robin, D.; Falala, S.; Pourcelot, P.; Estoup, P.; Denoix, J.; Crevier-Denoix, N. Biomechanical analysis of hoof landing and stride parameters in harness trotter horses running on different tracks of a sand beach (from wet to dry) and on an asphalt road. *Equine Vet. J.* **2010**, *42*, 488–495. [CrossRef] [PubMed]
3. Holt, D.; Northrop, A.; Owen, A.; Martin, J.; Hobbs, S.J. Use of surface testing devices to identify potential risk factors for synthetic equestrian surfaces. *Procedia Eng.* **2014**, *72*, 949–954. [CrossRef]
4. Hayati, H.; Mahdavi, F.; Eager, D. Analysis of agile canine gait characteristics using accelerometry. *Sensors* **2019**, *19*, 4379. [CrossRef]
5. Hayati, H.; Walker, P.; Mahdavi, F.; Stephenson, R.; Brown, T.; Eager, D. A comparative study of rapid quadrupedal sprinting and turning dynamics on different terrains and conditions: Racing greyhounds galloping dynamics. In Proceedings of the ASME 2018 International Mechanical Engineering Congress and Exposition, Pittsburgh, PA, USA, 9–15 November 2018.
6. Hayati, H.; Eager, D.; Jusufi, A.; Brown, T. A Study of Rapid Tetrapod Running and Turning Dynamics Utilizing Inertial Measurement Units in Greyhound Sprinting. In Proceedings of the ASME 2017 International Design Engineering Technical Conferences and Computers and Information in Engineering Conference, Cleveland, OH, USA, 6–9 August 2017.
7. Murray, R.C.; Walters, J.M.; Snart, H.; Dyson, S.J.; Parkin, T.D. Identification of risk factors for lameness in dressage horses. *Vet. J.* **2010**, *184*, 27–36. [CrossRef]
8. Swan, P.; Otago, L.; Finch, C.F.; Payne, W.R. The policies and practices of sports governing bodies in relation to assessing the safety of sports grounds. *J. Sci. Med. Sport* **2009**, *12*, 171–176. [CrossRef]
9. Beer, L.M. A Study of Injuries in Victorian Racing Greyhounds 2006–2011. Master' Thesis, University of Melbourne, Melbourne, Australia, 2014.
10. Cook, A. Literature survey of racing greyhound injuries, performance and track conditions. *J. Turfgrass Sci.* **1998**, *74*, 108–113.
11. Powers, M.C. A new roundness scale for sedimentary particles. *J. Sediment. Res.* **1953**, *23*, 117–119. [CrossRef]
12. Neville, A.M. *Properties of Concrete*; Pitman: London, UK, 1973.
13. Eager, D. Playground impact attenuation sands. *Australas. Parks Leis.* **2008**, 8–10.
14. Neville, A.M.; Brooks, J.J. *Concrete Technology*; Longman Scientific & Technical England: England, UK, 1987.
15. Hayati, H.; Eager, D.; Walker, P. An impact attenuation surfacing test to analyse the dynamic behaviour of greyhound racetrack sand surface. In Proceedings of the WEC2019: World Engineers Convention 2019, Melbourne, Australia, 20–22 November 2019; p. 391.
16. Worsey, M.T.; Espinosa, H.G.; Shepherd, J.B.; Lewerenz, J.; Klodzinski, F.S.; Thiel, D.V. Features Observed Using Multiple Inertial Sensors for Running Track and Hard-Soft Sand Running: A Comparison Study. *Proceedings* **2020**, *49*, 12. [CrossRef]
17. Peterson, M.; McIlwraith, C. Effect of track maintenance on mechanical properties of a dirt racetrack: A preliminary study. *Equine Vet. J.* **2008**, *40*, 602–605. [CrossRef] [PubMed]
18. Peterson, M.; McIlwraith, C.W.; Reiser II, R.F. Development of a system for the in-situ characterisation of thoroughbred horse racing track surfaces. *Biosyst. Eng.* **2008**, *101*, 260–269. [CrossRef]

19. Clegg, B. *An Impact Testing Device for In Situ Base Course Evaluation*; Australian Road Research Board Conference Proc: Melbourne, Australia, 1976; Volume 8.
20. ASTM. *Standard Test Method for Determination of the Impact Value (IV) of a Soil*; American Society for Testing and Materials: West Conshohocken, PA, USA, 2007.
21. Barrett, R.S.; Neal, R.J.; Roberts, L.J. The dynamic loading response of surfaces encountered in beach running. *J. Sci. Med. Sport* **1998**, *1*, 1–11. [CrossRef]
22. Muro-De-La-Herran, A.; Garcia-Zapirain, B.; Mendez-Zorrilla, A. Gait analysis methods: An overview of wearable and non-wearable systems, highlighting clinical applications. *Sensors* **2014**, *14*, 3362–3394. [CrossRef]
23. Ferris, D.P.; Louie, M.; Farley, C.T. Running in the real world: Adjusting leg stiffness for different surfaces. *Proc. R. Soc. Lond. Ser. B Biol. Sci.* **1998**, *265*, 989–994. [CrossRef]
24. Blickhan, R. The spring-mass model for running and hopping. *J. Biomech.* **1989**, *22*, 1217–1227. [CrossRef]
25. Hayati, H.; Eager, D.; Walker, P. The effects of surface compliance on greyhound galloping dynamics. *Proc. Inst. Mech. Eng. Part K J. Multi-Body Dyn.* **2019**, *223*, 1033–1043. [CrossRef]
26. Davis, P.E. Toe and muscle injuries of the racing greyhound. *N. Z. Vet. J.* **1973**, *21*, 133–146. [CrossRef]
27. Gillette, L. Track surface influences on the racing greyhound. *Greyhound Rev.* **1992**, *20*, 41–44.
28. Gillette, L. Optimizing Performance and Prevent Injuries of the Canine Sprint Athlete. In Proceedings of the North American Veterinary Conference, Orlando, FL, USA, 7–11 January 2007.
29. Ratzlaff, M.H.; Hyde, M.L.; Hutton, D.V.; Rathgeber, R.A.; Balch, O.K. Interrelationships between moisture content of the track, dynamic properties of the track and the locomotor forces exerted by galloping horses. *J. Equine Vet. Sci.* **1997**, *17*, 35–42.
30. Radin, E.L.; Ehrlich, M.G.; Chernack, R.; Abernethy, P.; Paul, I.L.; Rose, R.M. Effect of repetitive impulsive loading on the knee joints of rabbits. *Clin. Orthop. Relat. Res.* **1978**, *131*, 288–293.
31. Kai, M.; Takahashi, T.; Aoki, O.; Oki, H. Influence of rough track surfaces on components of vertical forces in cantering Thoroughbred horses. *Equine Vet. J.* **1999**, *31*, 214–217.
32. Qi, Y.; Indraratna, B. Energy-Based Approach to Assess the Performance of a Granular Matrix Consisting of Recycled Rubber, Steel-Furnace Slag, and Coal Wash. *J. Mater. Civ. Eng.* **2020**, *32*, 04020169.
33. Qi, Y.; Indraratna, B.; Heitor, A.; Vinod, J.S. The influence of rubber crumbs on the energy absorbing property of waste mixtures. In *Geotechnics for Transportation Infrastructure*; Springer: Berlin/Heidelberg, Germany, 2019; pp. 271–281.
34. Crevier-Denoix, N.; Audigié, F.; Emond, A.L.; Dupays, A.G.; Pourcelot, P.; Desquilbet, L.; Chateau, H.; Denoix, J.M. Effect of track surface firmness on the development of musculoskeletal injuries in French Trotters during four months of harness race training. *Am. J. Vet. Res.* **2017**, *78*, 1293–1304.
35. Eager, D. *Do Playground Surfacing Standards Reflect Reality*; International Conference Playground Fall Impacts; TüV Austria Academy: Brunn am Gebirge, Austria, 2013.
36. Eager, D.; Hayati, H.; Chapman, C. Impulse force as an additional safety criterion for improving the injury prevention performance of impact attenuation surfaces in children's playgrounds. In Proceedings of the ASME's International Mechanical Engineering Congress and Exposition (IMECE), Phoenix, AZ, USA, 13 November 2016.
37. Eager, D.; Hayati, H. Additional injury prevention criteria for impact attenuation surfacing within children's playgrounds. *ASCE-ASME J. Risk Uncertain. Eng. Syst. Part B Mech. Eng.* **2019**, *5*.
38. Standards Australia. *Methods of Testing Soils for Engineering Purposes*; Standards Australia: Sydney, Australia, 2014.
39. Wikipedia Contributors. Soil Texture—Wikipedia, The Free Encyclopedia. 2020. Available online: https://en.wikipedia.org/wiki/Soil_texture (accessed on 21 October 2020).
40. Dagg, L.A. The Effect of Two Maintenance Procedures on an Equine Arena Surface in Relation to Motion of the Hoof and Metacarpophalangeal Joint. Ph.D. Thesis, University of Central Lancashire, Preston, UK, 2012.
41. Hildebrand, M. The adaptive significance of tetrapod gait selection. *Am. Zool.* **1980**, *20*, 255–267. [CrossRef]

42. Aerts, P.; Clercq, D.D. Deformation characteristics of the heel region of the shod foot during a simulated heel strike: The effect of varying midsole hardness. *J. Sport. Sci.* **1993**, *11*, 449–461. [CrossRef]
43. Qi, Y.; Indraratna, B.; Coop, M.R. Predicted behavior of saturated granular waste blended with rubber crumbs. *Int. J. Geomech.* **2019**, *19*, 04019079. [CrossRef]

Review

Jerk within the Context of Science and Engineering—A Systematic Review

Hasti Hayati [1,*], David Eager [1], Ann-Marie Pendrill [2] and Hans Alberg [3]

[1] School of Mechanical and Mechatronic Engineering, University of Technology Sydney, P.O. Box 123, Broadway, Ultimo, NSW 2007, Australia; David.Eager@uts.edu.au
[2] National Resource Centre for Physics Education, Lund University, Box 118, SE 221 00 Lund, Sweden; Ann-Marie.Pendrill@fysik.lu.se
[3] Hägerstens Alle 20, lgh 1303, SE 129 37 Hägersten, Sweden; alberg.hans@gmail.com
* Correspondence: Hasti.Hayati@uts.edu.au

Received: 14 September 2020; Accepted: 10 October 2020; Published: 21 October 2020

Abstract: Rapid changes in forces and the resulting changes in acceleration, jerk and higher order derivatives can have undesired consequences beyond the effect of the forces themselves. Jerk can cause injuries in humans and racing animals and induce fatigue cracks in metals and other materials, which may ultimately lead to structure failures. This is a reason that it is used within standards for limits states. Examples of standards which use jerk include amusement rides and lifts. Despite its use in standards and many science and engineering applications, jerk is rarely discussed in university science and engineering textbooks and it remains a relatively unfamiliar concept even in engineering. This paper presents a literature review of the jerk and higher derivatives of displacement, from terminology and historical background to standards, measurements and current applications.

Keywords: jerk; acceleration onset; higher-order derivatives of acceleration; jounce; acceleration-dot

---

1. Introduction

Jerk—the time derivative of acceleration—is an important consideration for many applications in science and engineering. For example, jerk has long been used as a design factor to ensure ride comfort, e.g., in amusement rides [1–7], ships [8–10], lifts/elevators [11] and buses [12], and there are many reasons to believe that the relevance of jerk—and higher derivatives of displacement—will increase. A number of ISO standards also refer to jerk [13–19].

Displacement, velocity and acceleration are well known concepts for everyone who has studied physics at secondary level, whereas jerk—the time derivative of acceleration—and higher derivatives are rarely mentioned, let alone discussed, even in university physics or engineering textbooks. This omission was pointed out three decades ago by Sandin [20], who only found one reference to jerk [21] in a dozen reviewed text books. In addition, in an earlier article for university physics teachers [22], Schot presented the concept and also discussed the radial and tangential components of jerk. However, despite this early effort for jerk to be included in text books, jerk is not yet discussed in physics and engineering courses, except for a few textbooks [23–25]. The lack of detail in textbooks also contributes to some confusion concerning terminology.

Higher derivatives have been discussed by Thompson in a conference presentation [26], where he argued that since immediate acceleration onsets have a detrimental effect on equipment, acceleration should be ramped up by placing a limit on jerk. He then proposed an alternative strategy, claimed to be even better, which is ramping up jerk by placing a limit on its higher derivatives of snap, crackle and pop. However, the details of his studies are beyond the scope of this work.

Jerk and higher derivatives of acceleration are relevant for understanding the impact of motion and vibrations in a wide range of applications, as reviewed in this paper [1–12,27–195]

The considerable interest is also reflected in the large number of downloads (>100,000) of a 2016 paper on jerk and higher derivatives [1].

To support secondary school teachers and university lecturers who would like to introduce the concept of jerk, as well as higher derivatives, in their science or engineering courses, we have conducted a thorough systematic review, from terminology and historical background to standards, measurements and current applications. The articles are divided, based on the applications of jerk, into twenty-one categories. Each article is ranked against where it is published using Scientific Journal Rankings (SJR)—Scimago index, from the quartiles Q1 to Q4, where Q1 is occupied by the top 25% of journals, Q2 by the top 25% to 50% and so on. We used N/A for thesis, reports, and conferences and journals where we were unable find any information with regards to their quality. Other sources are also included in the reference list. The rationale for including these sources is the scarcity of research on jerk. It is then the responsibility of the reader to evaluate the articles against their own objective. A number of papers have been included which use the concept of jerk, even if they don't explicitly mention the term or only refer to it briefly.

Background

There is a limited number of good textbooks about the history of mechanics. The best are probably Szabo [196] (in German) and Dugas [197] (in French and translated into English [198]). Although these books contain a lot of interesting information, neither covers jerk or higher derivatives.

Newton's second law of motion is commonly written as $\mathbf{F} = m\mathbf{a}$, i.e., "force equals mass times acceleration", although Newton did not use the term acceleration in his equations but stated that the rate of change in momentum is equal to the applied force. The concept of acceleration was formalized by Pierre Varignon (1654–1722) [199]. Before the second world war German engineers, including Melchior [200] and Schlobach [201], pointed at applications of jerk for handheld machines.

One aspect that is frequently overlooked is terminology. By standardizing terminology, misunderstandings could be avoided, communication between scientists from different disciplines could be enabled, and searches on the global network would be facilitated. The term "jerk" for the first derivative of acceleration seems to be most widely used. However, alternative terms like "acceleration onset rate" are also used, e.g., Whinnery et al. [202], ISO 2041 [13] and a report by NASA [94].

The higher derivatives are less familiar, and different terminologies have been used by different groups. The term "jounce" is sometimes used to describe the fourth derivative of position, and in some Japanese articles it is referred to as "jerk-dot" [148,173,175].

In this work, we will use the more common terms "snap", "crackle" and "pop" for the 4th–6th derivatives (named after pictorial characters on Kellogs' Corn Flakes packages from the thirties). For the 7th–10th derivatives, the terms "lock", "drop", "shock" and "put" have been proposed informally, although we have been unable to find them in the literature.

2. Method

Articles in the period from 2015 to 2020 with the term "jerk" mentioned in the title were looked up in an electronic database (Scholar.google.com), excluding citations and patents. There were 550 results. The initial 550 articles was refined by excluding 129 based on title, non-English language and duplication. Of the remaining 421 articles, 147 articles were excluded in a more detailed subsequent assessment that included a review of the abstracts and conclusions, and their source, i.e., low-quality journals/conferences. Of the remaining 274 remaining articles, 139 articles that referred to jerk in the context of chaotic systems and nonlinear dynamics were excluded. Terms signaling this context include jerk system, hyper-jerk system, chaotic jerk-system, jerk attractors, jerk circuit, jerk dynamics, jerk map, jerk function, jerk oscillators and jerk equations, and traffic jerk model (n = 17) articles that used jerk in the context of weightlifting were removed, since this refers to a weightlifting technique rather than the jerk itself.

In addition to the articles emerging from this database search, the authors were aware of 84 articles before 2015, as well as articles where jerk was not mentioned in the title. These articles have been included in the review. The inclusion-exclusion criteria flow chart is given in Figure 1.

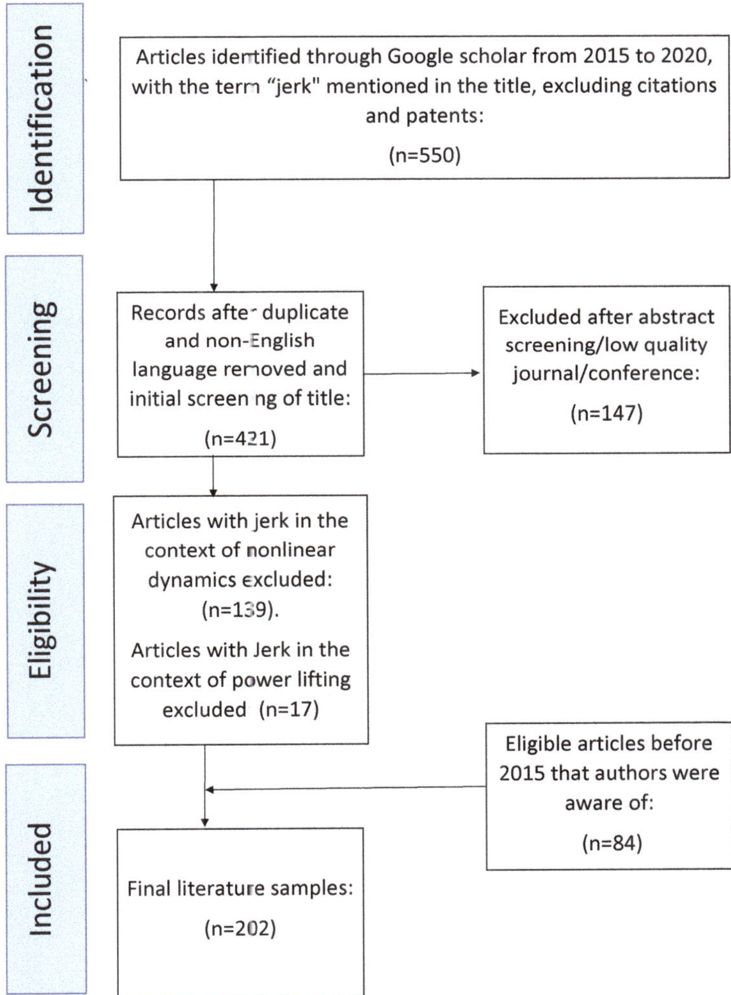

Figure 1. Systematic review 'inclusion-exclusion criteria' flow chart.

After reviewing the collected literature (n = 202) and those articles that the author were aware of, the applications of jerk were categorised into 20 categories.

Categorisation of Jerk Applications

The 21 jerk categories chosen with their respective references were:

1. Jerk in advanced manufacturing [27–90];

2. Jerk in amusement rides [1–7,14];
3. Jerk in cosmology and space technology [91–94];
4. Jerk in criteria for discomfort [1,4,5,9,10,12,19,25,95–131];
5. Jerk in equation of motion [132–136];
6. Jerk in global positioning systems (GPS) [137,138];
7. Jerk in human tolerance [202,203];
8. Jerk in kinesiology [139–147];
9. Jerk in measurement [148–151];
10. Jerk in motion analysis [152,153];
11. Jerk in ornithology [154];
12. Jerk in racing [155–160];
13. Jerk in sea-keeping [8–10,122,126,204];
14. Jerk in seismic analysis [134,161–170];
15. Jerk in shock response spectrum [8,205];
16. Jerk in sport science [1,171,172];
17. Jerk in standards [13–19]
18. Jerk in structural health monitoring [173–176];
19. Jerk in technical pain [177];
20. Jerk in unmanned aerial vehicle (UAV) [11,178–186]; and
21. Jerk in vehicles (ride comfort [12,15,17,19,25,45,95,98–106,108,109,111–113,115,118–121, 123–125,127–131], anti-jerk controller design [95,98–102,109,111,119,120,130], autonomous vehicles [12,16,114,123,187,188], and other [111,187,189–195].

The next section presents brief summaries of the papers in the different categories.

3. Results and Discussion

3.1. *Jerk in Advanced Manufacturing*

Jerk in advanced manufacturing can be categorised into jerk in industrial robots [27–63], machining [64–78], motors [79–83], and 3D printers [84–90].

3.1.1. Jerk in Industrial Robots

Jerk is mainly used to generate smooth trajectories in industrial robots (also referred to as manipulators) [27–63]. It is outside the scope and purpose of this paper to explain the jerk-controller. Details (author, source, SJR ranking and the title) of the Q1 articles are tabulated in Table 1.

As mentioned above, all of the cited reference used jerk as a limit for generating or controlling a smooth motion.

3.1.2. Jerk in 3D Printers

In recent years, jerk has been used in 3D printers [84–90]. One important consideration in 3D printing is that the term jerk, in the majority of 3D firmware, is defined in terms of the maximum "instantaneous" velocity change without consideration of the time required [84,90].

Inconsistent terminology causes confusion and needs to be addressed in the future. Notwithstanding, in an article on 3D printing by Hernandez [86], the actual jerk was considered (see Table 2).

3.1.3. Jerk in Machining

Jerk is recently used in generating smooth trajectories in machining [64–78]. Details of Q1 articles are tabulated in Table 3.

Table 1. Jerk in industrial robots.

Author	Source	SJR	Title
Lang (2015) [42]	IEEE Robotics Automation Letters	Q1	Path-accurate online trajectory generation for jerk-limited industrial robots real-time trajectory generation for industrial robot which complies with standards of smooth motion in robotic arms.
Chen (2015) [85]	IET Control Theory Applications	Q1	Minimum jerk norm scheme applied to obstacle avoidance of redundant robot arm with jerk bounded and feedback control.
Chen (2015) [29]	Automatica	Q1	Composite jerk feed-forward and disturbance observer for robust tracking of flexible systems.
Bianco (2017) [28]	Robotics Computer-Integrated Manufacturing	Q1	A scaling algorithm for the generation of jerk-limited trajectories in the operational space.
Besset (2017) [27]	Control Engineering Practice	Q1	Constraints and limits on velocity, acceleration, and jerk FIR filter-based online jerk-constrained trajectory generation.
Kaserer (2018) [41]	IEEE Robotics Automation Letters	Q1	Nearly optimal path following with jerk and torque rate limits using dynamic programming.
Kaserer (2018) [40]	IEEE Transactions on Robotics	Q1	Online robot–object synchronization with geometric constraints and limits on velocity, acceleration, and jerk.
Huang (2018) [38]	Mechanism Machine Theory	Q1	Optimal time-jerk trajectory planning for industrial robots.
Rojas (2019) [52]	IEEE Robotics Automation Letters	Q1	A variational approach to minimum-jerk trajectories for psychological safety in collaborative assembly stations.
Palleschi (2019) [49]	IEEE Robotics Automation Letters	Q1	Time-optimal path tracking for jerk controlled robots.
Dai (2020) [33]	IEEE Transactions on Automation Science and Engineering	Q1	Planning jerk-optimized trajectory with discrete time constraints for redundant robots.

Table 2. Jerk in 3D printers.

Author	Source	SJR	Purpose and Finding	Comments
Hernandez (2015) [86]	International Journal of Aviation, Aeronautics, Aerospace	Q4	To analyse factors which affect the dimensional precision of consumer 3D printing. The jerk was believed to play an insignificant role in printing quality but it was observed that a single instant of apparent error due to jerk could be accounted for, and should be considered by changing the the way infill was applied.	The maximum jerk is an input parameter that can be set by the user to control the dimensional precision of 3D print and have a cleaner print as a result.

Table 3. Jerk in machining.

Author	Source	SJR	Purpose and Findings	Comments
Zhang (2019) [78]	Computer-Aided Design	Q1	Generating a smooth curve, mainly at junctions, in computer numerical control (CNC) machining is a challenge. In this work, a new algorithm, which is claimed to have a jerk-smooth trajectory controller, is proposed.	Jerk as a measure to generate smooth and accurate trajectories.
Schroedter (2018) [74]	Mechatronics	Q1	A flatness-based feed forward control method using jerk-limited trajectories, which is based on a mechatronic micro mirror model, is proposed to reduce undesired oscillations in micro-scanners.	Jerk as a measure to generate smooth and accurate trajectories.
Dumanli (2018) [68]	Precision Engineering	Q1	Obtaining a smooth surface manufactured in a timely manner is a challenge in different industries, and, therefore, different smooth-trajectory controllers have been developed. In this study, a novel jerk value decision-making process is proposed for the parts-machining process.	Jerk as a measure to generate smooth and accurate trajectories.
Alzaydi (2019) [65]	Mechanical Systems Signal Processing	Q1	Current machine tool controllers cannot fully benefit from the speed of the drilling laser as they are not equipped with a proper trajectory function. In this work, a time-optimal and minimum jerk trajectory generator is designed and implemented on a gas turbine combustion chamber. The results showed 6% time-reduction as well as reduced oscillation due to controlled jerk.	Jerk as a measure to generate smooth and accurate trajectories.
Hashemian (2020) [69]	Computer-Aided Design	Q1	A novel jerk-minimised trajectory controller for multi-axis flank CNC machining is developed in this work.	Jerk as a measure to generate smooth and accurate trajectories.
Zhang et al. (2018) [64]	The International Journal of Advanced Manufacturing Technology	N/A	Corner smooth machining in high speed machining causes issues in advanced manufacturing. Current solution is based on the jerk-limited acceleration profile from the perspective of kinematics that generate continuous acceleration transition profiles. They offer the same approach but for a higher derivative of jerk (jounce).	They used Jerk onset (jounce) for smoothing algorithm for the corner motion in high speed machining.

3.1.4. Jerk in Motors

Jerk is also used in motors mostly a as a measure to control/generate smooth trajectories [79–83] (Table 4).

3.2. Jerk in Amusement Rides

Jerk is rarely mentioned in textbooks but is quite an important physical parameter, as we quite often experience it in daily life. One important aspect of jerk is that jerk is an element in the comfort and safety of amusement rides, as well as reducing the need for equipment maintenance (Table 5).

3.3. Jerk in Cosmology and Space Technology

Jerk is used in cosmology as well as space technology. Details of articles which were eligible based on the inclusion criteria of the current work are given in Table 6.

3.4. Jerk in Criteria for Discomfort

Ride comfort is an important parameter in amusement rides [1,4,5], sea-keeping [9,10,122,126] and traditional land-based vehicles [12,25,45,95,98–106,108,109,111–113,115,118–121,123–125,127–131]. There are still discussions ongoing regarding the significance of jerk regarding ride comfort for vehicles, and jerk is probably a better measure for driving comfort than acceleration, as pointed out by van Santen [128] and confirmed, e.g., in the study by Grant and Haycock [107]. Jerk monitoring also offers insurance companies a way to follow up the behaviour of drivers, as a basis for car insurance pricing [195].

The Dutch institutes TNO and MARIN have initiated a joint project to study the impact of jerk on the comfort of passengers in a master thesis conducted by Werkman [10]. Details of this work are given in Section 3.4.2 of this paper.

Förstberg [104] investigated ride comfort and motion sickness in trains. The results indicated that "it is motion dose from horizontal jerk or horizontal acceleration as well as roll acceleration or roll velocity that is the primary causes of provocation." Unfortunately, Förstberg passed away shortly after presenting his thesis.

Svensson and Ericsson [123] referred to concrete jerk values from American Association of State Highway and Transportation Official Standards (AASHTO; 2001) [206] in their master thesis. A value of lateral jerk ranging from 0.03 to 0.09 g/s has been used for highway design. Jia [25] claims that the jerk should be below 0.2 g/s in trains for passenger comfort. Martin and Litwhiler [118] investigated acceleration and jerk profiles in the metro-rail system in Washington DC and found jerk peaks of around 1.3 g/s.

Minimising the discomfort experienced during a journey between two points with the fixed travel time was studied by Anderson et al. [96]. They proposed jerk as a discomfort criteria. Their work was then revisited by Antonelli and Klotz [97] and Lemos [116] one and three years later, respectively. The jerk is proposed as a discomfort criteria. Antonelli and Klotz (2017) [97] and Lemos (2019) [116], commented on this work and mentioned that the integral of the square of acceleration and the integral of square of jerk should be considered as criteria for discomfort.

3.4.1. Jerk in Ride Comfort: Amusement Rides

Jerk is used as a ride comfort measure in amusement rides [1,4,5]. Details of studies mentioning jerk as a ride comfort parameter in amusement rides are given in Section 3.2, Table 5.

3.4.2. Jerk in Ride Comfort: Sea-Keeping

Jerk is used a ride comfort measure in sea-keeping [9,10,122,126]. Details of these studies are given in below Table 7.

Table 4. Jerk in motors.

Author	Source	SJR	Purpose and Findings	Comments
Masoudi (2016) [82]	IET Electric Power Applications	Q1	Although the linear switched reluctance motors (LSRMs) have many benefits such as being low cost, and having a comparatively high force-to-mass ratio and no need for mechanical rotary to linear motion, as converters, they have limited application because of their force ripple. To control this force ripple, a new control model based on a minimum jerk model is proposed, which can be used in elevator applications.	A controller based on a minimum jerk model for LSRM motors.
Jinhui (2017) [80]	Mechatronics	Q1	A minimum jerk trajectory controller to enhance the smoothness and stability of a rotor motor is developed in this work.	Jerk as a measure to generate smooth trajectories.

Table 5. Jerk in amusement rides.

Author	Source	SJR	Purpose and Findings	Comments
Pendrill (2005) [2]	Physics Education	Q3	Although textbook loops are often circular, real roller coaster loops are not. This paper looks into the mathematical description of various possible loop shapes, as well as their riding properties.	Jerk is mentioned indirectly, in the context of using clothoid curves for roller coaster loops.
Eager et al. (2016) [1]	European Journal of Physics	Q2	The concept of jerk is discussed using trampoline and amusement rides (roller coasters). The effect of jerk on human body is also discussed. The importance of jerk in amusement rides (roller coasters): safety (avoiding whiplash); ride comfort; when the safety of the passenger is not an issue, reducing the maintenance cost due to snap.	Jerk is important and is experienced in daily life, yet is not well-explained and understood.
Sicat et al. (2018) [4]	Proceedings for the Annual Occupational Ergonomics and Safety Conference	N/A	The design and validation protocol for wearable sensor technology used to collect acceleration and g-force exposure of a zip line rider is studied in this work.	They have briefly mentioned the importance of jerk (referred to it as rate of change of acceleration) in the ride comfort and safety of the passengers. Refers to Eager et al. [1].

Table 5. *Cont.*

Author	Source	SJR	Purpose and Findings	Comments
Väisänen (2018) [5]	Master thesis	N/A	A literature review to present general guidelines and principles of what is included in the design and engineering of roller coasters and other guest functions attached to them.	They have briefly mentioned jerk as a limit which may become standardized in future, since it affects the ride comfort and experience.
Pendrill et al. (2019) [6]	Physics Education	Q3	Assessed the first-year university understanding of the vertical motion on a roller coaster loop. It was found that students have partial conceptions about force and motion and contradictions in their responses and do not have a good understanding of abrupt changes in jerk	The importance of including jerk in physics textbooks.
Pendrill et al. (2020) [7]	Physics Education	Q3	The effect of acceleration, jerk, snap and vibration on the ride comfort experience and safety of roller coaster rides is analysed in this work, via authentic data from a dive coaster as an example.	Jerk and snap are considered as ride comfort criteria in amusement rides.

Table 6. Jerk in Cosmology and space technology.

Author	Source	SJR	Purpose and Finding	Comments
Visser (2004) [91]	Classical Quantum Gravity	N/A	To obtain a higher accuracy of jerk and snap parameters in cosmological Equations of States (EOS). It is found that although other parameters are known to high accuracy, the jerk and snap are still poorly estimated. This fact would cause direct observational constraints on cosmological EOS if not properly addressed.	The importance of the jerk and snap in cosmological EOS and how the lack of proper information has limited the researchers of the field.
Hur-Diaz et al. (2008) [92]	NASA technical report	N/A	In an article entitled "Three axis control of the Hubble Space Telescope using two reaction wheels and magnetic torquer bars for science observations", jerk has been used as a measure in maneuver planning.	They considered maximum jerk as they investigated possibilities for three-axis control if two of the four reaction wheels should fail .
STD, NASA . (2011) [94]	NASA technical report	N/A	Human-system standard. "The system shall limit crew exposure to acceleration rates of change larger than 500 g/s during any sustained (>0.5 s) acceleration event."	'Acceleration onset' or 'acceleration rate of change' is mentioned, but not jerk explicitly.
Ramon (2016) [93]	Master thesis	N/A	The touchdown dynamics of a lander was simulated. Jerk was a parameter in designing a honeycomb damper, which aimed to dampen the impact forces associated with touchdown dynamics of the lander.	Considered jerk in the analysis of touchdown dynamics of a lander in a masters' project.

Table 7. Jerk as a criteria for discomfort: sea-keeping.

Author	Source	SJR	Purpose and Findings	Comments
Tomi (1961) [126]	The Japan Society of Naval Architects and Ocean Engineering	N/A	Ride comfort of passengers on ships were studied. The influence of jerk on the rolling motion was discovered.	One of the first references considered jerk in ride comfort in sea-keeping. Jerk is referred to as the time derivative of acceleration.
Shigehiro et al. (2002) [122]	Fisheries science	Q3	The new evaluation method of passenger comfort is expressed by vertical and lateral accelerations and exposure duration represents the relationship between ship motions and seasickness. It is confirmed that the correlation coefficient between the results of the new method and the questionnaires given to the trainees show fairly similar results.	Jerk is not directly mentioned. Only the work conducted by Tomi [126] was briefly mentioned.
Sosa & Ooms (2016) [9]	Report	N/A	The role of the roll stabiliser in the ride comfort in the *yacht industry* is analysed. Jerk is mentioned as a ride comfort parameter but was not included in the ride comfort rating used in this study. Since it is seen in the literature that jerk is an important parameter in this respect, further research in including jerk in the yacht ride comfort rating is critical.	Points to the need for more studies of the effect of jerk in the yacht industry.
Werkman (2019) [10]	Master thesis	N/A	To study the impact of jerk on comfort of passengers in *sea-keeping*.	Points to the need for more studies of the effect of jerk. Importance of jerk in sea-keeping and ride comfort in general.

3.4.3. Jerk in Ride Comfort: Vehicles

Jerk has been used as a measure of ride comfort in vehicle engineering [9,10,122,126] and vehicle land-based traditional [12,25,45,95,98–106,108,109,111–113,115,118–121,123–125,127–131]. Details of studies in Q1–Q4 journals are given in Table 8.

3.5. Jerk in Equation of Motion (EOM)

Jerk is proposed to be used in the EOM, and details of studies explored this can be found in Table 9. The idea has been discussed earlier but, to date, no concrete examples where someone claims that adding higher-order derivatives might explain the discrepancies between theory and observations, have been seen.

3.6. Jerk in GPS Applications

The following articles discussed use of jerk in Global Positioning Systems (GPS), mainly those types that should perform in harsh dynamic conditions such as satellite launch vehicles (Table 10).

3.7. Jerk in Human Tolerance

Details of works studying human tolerance to jerk are given in Table 11.

3.8. Jerk In Kinesiology

Jerk has been used in different kinesiologies in different fields of clinical studies in sport science. Below, those works which studied jerk in kinesiology for different applications [139–147] have been captured (Table 12).

Recently, jerk has gained interest in non-clinical trials. For instance, in a recent study by Zhang [145–147], jerk was used as a measure to detect fatigue in workers.

3.9. Jerk in Measurement

Jerk is normally measured indirectly by calculating the derivative of acceleration, which is measured by accelerometers. Below, those studies that developed a device to measure jerk are tabulated (Table 13).

3.10. Jerk in Motion Analysis

In motion analysis application, jerk is used as s measure in machine learning classifiers [152,153] (Table 14).

3.11. Jerk in Ornithology

Ornithology is a branch of zoology which deals with birds. jerk has been used in one of the recent studies in this area, details of which are given in Table 15.

3.12. Jerk in Greyhound Racing

One of the interesting applications (refer to Table 16) of jerk is in high-speed sprint racing, such as greyhound racing. Recent articles on racing greyhounds, mainly the one in Nature's Scientific Report [160], clearly show its importance.

3.13. Jerk in Sea-Keeping

Jerk in sea-keeping is considered in both passengers' ride comfort and analysing shock spectrum in high-speed crafts [8–10,122,126,204] (Table 17).

Table 8. Jerk as a criteria for discomfort: vehicles.

Author	Source	SJR	Purpose and Findings	Comments
Deshmuck et al. (2016) [103]	International Journal of Passenger Cars-Mechanical Systems	Q3	The frequency characteristics of vehicle motion are studied to derive the inherent jerk. A new method is proposed, which, unlike the conventional method that uses the peak of jerk as a performance index, can differentiate the perceivable frequencies from the test data and quantify the actual jerk.	A new method to obtain inherent jerk was proposed.
Liu et al. (2016) [117]	Journal of Dynamic Systems, Measurement, Control	Q1	A new low-jerk suspension control system by mixing skyhook (SH), acceleration-driven-damper (ADD) is proposed in this work. The characteristics of the SH-ADD suspension system are compared both numerically and experimentally with other suspension control system.	Jerk is as a measure in designing suspension systems.
Sharma et al. (2016) [121]	Perspectives in Science	Q2	Since the ride comfort of the passengers on vehicles is of paramount importance, a mechanical model of a train subjected to external loads (rolling and longitudinal wheel resistance, and gravity) is considered. The jerk value are measured and analysed.	Jerk is considered as a criterion of discomfort in this work.
Huang et al. (2018) [109]	International Journal of Adaptive Control Signal Processing	Q1	An anti-jerk controller for electromechanical clutch engagement was developed in this work.	Jerk as a limiting factor in designing controllers in vehicles. Details of these controllers are beyond the scope of this work.
Zeng et al. (2018) [109]	Energies	Q2	An anti-jerk controller based on a data-driven vehicle dynamics model is proposed for a power-split electrical vehicle in this work.	Jerk as a limiting factor in designing controllers in vehicles. Details of these controllers are beyond the scope of this work.
Batra et al. (2018) [101]	Journal of Computational Nonlinear Dynamics	Q2	An anti-jerk controller based on road tests is proposed for an electric vehicle in this work.	Jerk as a limiting factor in designing controllers in vehicles. Details of these controllers are beyond the scope of this work.
Bae et al. (2019) [12]	journal of Electronics	Q3	To provide a comfortable driving experience, while not sabotaging the passengers' safety, on a self-driving shuttle bus. A time-optimal velocity planning method to guarantee a comfort criteria was developed. A better performance and comfortable passenger ride in a self-driving shuttle bus is experienced.	This is also aligned with the application of jerk in ride comfort and its potential to be considered in autonomous cars.

Table 8. *Cont.*

Author	Source	SJR	Purpose and Findings	Comments
Yamaguchi et al. (2019) [130]	IEEJ Journal of Industry Applications	Q2	A backlash based controller to reduce the jerk, while clutch is engaged in a hybrid car, is designed and numerically compared with other controllers. In this method, the magnitude of the shock is quantitatively evaluated against jerk to minimise the discomfort felt by the passengers.	Jerk as a measure to control the backlash caused by clutch engagement in hybrid cars.
Khorram et al. (2020) [112]	Theoretical Issues in Ergonomics Science	Q2	Longitudinal jerk and acceleration were considered as a measure of safety of the bus rapid transit buses. The results of this study can be used to predict dangerous driving style before severe accidents occur.	Longitudinal jerk can be used as a measure to identify dangerous driving styles.
Scamarcio et al. (2020) [120]	IEEE Transactions on Vehicular Technology	Q1	In this study, different types of jerk controller are compared with each other in terms of ride comfort and increased component life.	Jerk as a limiting factor in designing controllers in vehicles. Details of these controllers are beyond the scope of this work.
Tawadros et al. (2020) [124]	Proceedings of the Institution of Mechanical Engineers, Part D: Journal of Automobile Engineering	Q2	Since jerk is a criteria for discomfort, it was measured both experimentally and numerically (simulation) for a vehicle through a low-cost Bluetooth sensor among other variables such as acceleration.	Jerk as measure of discomfort in vehicles dynamics.

Table 9. Jerk in EOM.

Author	Source	SJR	Purpose and Findings	Comments
Muszynska & Bently (1990) [132,133]	Journal of Sound Vibration & NASA technical report	Q1	Developing fluid force models in rotor/bearing/seal systems and a fluid handling machine. Their results suggested that the fluid force contains terms of orders higher than two. Specifically, a third-order term (jerk) should be included.	Jerk, as a higher order term, is seen in the EOM of fluid model.
Inaudi et al. (1993) [134]	Earthquake engineering structural dynamics	Q1	An optimum hybrid isolation system to protect sensitive equipment from earthquake was designed, and the EOM of the ground motion obtained. Jerk was a parameter in the ground motion EOM.	Jerk was mentioned in the model of ground motion, which was numerically modeled.
Funakoshi et al. (2012) [135]	ASME 2012 5th Annual Dynamic Systems and Control Conference joint with the JSME 2012 11th Motion and Vibration Conference	N/A	A modeling method and a control system design procedure for a flexible rotor with many elastic modes using active magnetic bearings is presented. A local jerk feedback control system and stability analysis is developed by using root locus.	It has been proposed to include "jerk feedback" for rotors suspended in active magnetic bearings. In that case, jerk will occur in the EOM.
Eager (2018) [136]	Proceedings of ACOUSTICS	N/A	A novel way of measuring jerk, snap and crackle using accelerometers (while jumping on trampoline). The EOM of this dynamic is obtained where jerk appears in the EOM.	Jerk has appeared in the obtained EOM of jumping on a trampoline.

Table 10. Jerk in GPS application.

Author	Source	SJR	Purpose and Findings	Comments
Kwon et al. (2006) [137]	Proceedings of the Korean Institute of Navigation and Port Research Conference	N/A	GPS used in satellite launch vehicles should perform under a severe dynamic environment. Therefore, in this work, preliminary test results of a GPS receiver in such simulated high acceleration and jerk conditions are analysed.	Jerk as a test parameter in testing GPS used in satellite launch vehicles.
Kwon et al. (2008) [138]	Proceedings of the 17th World Congress The International Federation of Automatic Control	N/A	Comparative performance analyses of GPS receivers under high-dynamic conditions. They used jerk as one of the parameters of high-dynamic condition. The thresholds of GPS receiver performance under severe dynamic condition were found to be 12.54 m/s for maximum velocity, 16.07 g for maximum acceleration, and 202.14 g/s for maximum jerk. The jerk level is extremely high compared to the levels of velocity or acceleration.	Jerk as a test parameter in testing GPS used in satellite launch vehicles.

Table 11. Jerk in human tolerance.

Author	Source	SJR	Purpose and Findings	Comments
McKenney (1970) [203]	Report	N/A	A literature review to study the human tolerance to abrupt acceleration. Just as the slope of a velocity-time trace furnishes acceleration, so the slope, or tangent, of an acceleration-time pulse will yield rates of change in acceleration. They have studied higher rates of onset, meaning jerk, snap, etc., and mentioned in chapter 4 of the work that change of rate of acceleration is also called jerk.	A method to calculate jerk as the slope of the acceleration-time pulse plots was proposed.
Whinnery et al. (2013) [202]	Extreme physiology medicine	Q4	They referred to jerk as acceleration onset. This concussion curve provides a temporal prediction of when the concussion might occur. Jerk was referred to as acceleration onset rate.	The jerk loss of concussion curve was studied.

Table 12. Jerk in kinesiology.

Author	Source	SJR	Purpose and Findings	Comments
Fazio et al. (2013) [139]	Neurological Sciences	Q2	To evaluate the accelerometric parameters of human gait patterns in different neurological conditions with pathological gait impairment compared to healthy subjects. The acceleration and jerk data were used to compare different groups. Analyses of the basic accelerometric parameters associated with a jerk analysis could assist in differentiating between the population groups.	The use of P-value is currently heavily debated, see, e.g., The ASA Statement ([207]) on P-values. However, this work draws attention to using jerk in gait analysis.
Lapinski (2013) [140]	PhD thesis	N/A	In this PhD thesis titled *A platform for high-speed bio-mechanical analysis using wearable wireless sensors*, it is shown that jerk was a good monitor of a baseball pitcher's likelihood of tearing a tendon, while other parameters such as peak acceleration could not show this.	Jerk has the potential to monitor different injury scenarios or dynamics which other parameters such as peak acceleration are unable to detect.
Aguirre (2016) [141]	Master thesis	N/A	To evaluate the usage of measurement devices (accelerometer) in clinical setting, and to validate the recorded what kinematic parameters can be accurately measured by them. Measurement devices are a reliable tool in capturing the kinematic parameters such as movement time, maximum acceleration, mean acceleration, mean acceleration variability, and maximum jerk, and therefore should be considered in clinical setting as a low-cost technology compared with the more expensive motion-capturing technologies.	Expensive motion-capturing technologies can be replaced by measurement devices in clinical settings.
Zhang et al. (2019) [142]	Automation in Construction	Q1	In more recent times Zhang et al. [142] used jerk as an indicator of physical exertion and fatigue within the construction industry. They monitored the activity of a bricklayer using an inertial measurement device device that allowed jerk to be calculated.	Jerk as a measure to detect fatigue.
Washington et al. (2020) [143]	Journal of Sports Sciences	Q1	To determine the influence of pelvis and torso angular jerk on the hand velocity, which is a hitting performance indicator in female softball. Although the results did not show a significant relationship between pelvis and torso angular jerk and hand velocity, more research could be conducted on the "timing" of minimal jerk through the acceleration phase to predict the angular hand velocity.	Jerk, both its magnitude and duration, should be studied more in sport science.

Table 13. Jerk in measurement.

Author	Source	SJR	Purpose and Findings	Comments
Masuda et al. (2002) [148]	The Proceedings of the Symposium on Evaluation and Diagnosis	N/A	A jerk-dot sensor, which measures the second derivative of the acceleration, is developed. Its capability of detecting the local damage in the structural members is investigated through low-cycle fatigue tests, which prove its significant sensitivity to the abnormal responses due to the development of macroscopic damages. Further fracture tests are carried out to obtain the correlation between the measured jerk-dot and the crack length, which suggests that this sensor could provide an early alert before the crack grows to the fatal stage.	Comment by Iyama & Wakui on this paper [175]: "Although they found that the high-frequency components accompanying crack initiation by jerk or jerk-dot can be used for detection, only a qualitative relation was shown. The determination of threshold values for damage detection thus remains difficult."
Orsagh et al. (2002) [149]	Patent	N/A	A method and apparatus for measurement of the derivative of acceleration with respect to time (jerk) and the use of demodulation to analyze the jerk signal. The sensor used to measure jerk consists of a piezoelectric transducer coupled with an amplifier that produces a voltage or current signal that is proportional to jerk. In applications including rolling element bearing diagnostics, demodulation is used to measure changes in the jerk signal over time.	A patent containing information of a sensor to measure jerk.
Xueshan et al. (2008) [150]	Proceedings of the 14th World Conference on Earthquake Engineering	N/A	The principles and specifications of a new sensor for measuring jerk are given. Derivative of acceleration is the first-order differentiation of acceleration.	JW-3D tri-axial Jerk sensor. The specifications of the jerk sensor are provided. Xueshan Patents [150] exist for jerk-meter, but it is not known to what extent they are used in practice.
Manabe et al. (2018) [151]	2018 International Conference on Advanced Mechatronic Systems	N/A	A horizontal jerk sensor was produced by rebuilding a feedback circuit of a commercially available velocity sensor. It was then applied to a linear slider for mechanical impedance control.	A jerk sensor is rebuilt using a commercial velocity sensor.

Table 14. Jerk in motion analysis.

Author	Source	SJR	Purpose and Findings	Comments
Thompson (2012) [152]	Jyväskylä studies in humanities	N/A	Motion capture techniques were used to study music cognition, in which Jerk is used a measure in machine learning classifiers.	Jerk is considered in the embodied music cognition research.
Jongejan (2017) [153]	Proceedings of the 4th European and 7th Nordic Symposium on Multimodal Communication	N/A	Jerk is used a a measure for support vector machine classifier, used for classifying head movements, along with other measures of velocity and acceleration.	Jerk as a measure for machine learning classifiers in motion analysis application.

Table 15. Jerk in ornithology.

Author	Source	SJR	Purpose and Finding	Comments
Sharker et al. (2019) [154]	Bio-inspiration biometrics	Q1	To see why plunge diver birds can dive into water with high speed while surface bird divers cannot and to see if there is a correlation between this fact and physical geometry, mainly of the beak. Since real experiments could not be conducted, they used a 3D-printed model of three types of diving birds with an embedded accelerometer. Surface diving birds have high non-dimensional jerk $(J^* = (\Delta a \Delta t)[m/(\rho g v A)])$, which is higher than the recommended safe jerk limit (for humans/no data for birds).	The impact acceleration of the bird models were not distinguishable while the jerk value was quite different. This draws attention to the fact that some dynamical behaviour scenarios cannot be fully captured by acceleration and jerk should be considered in such scenarios.

Table 16. Jerk in greyhound racing.

Author	Source	SJR	Purpose and Finding	Comments
Hossain et al. (2016) [155]	Report	N/A	The jerk was calculated from race track survey plans to identify thresholds for a safe turn in racing greyhounds.	Jerk is studied as a safety threshold in greyhound racing track bend design.
Hayati et al. (2017) [156]	ASME 2017 International Design Engineering Technical Conferences	N/A	High-jerk turning, which is caused by an inappropriate bend and camber of race tracks, was given among the potential risk factors causing injuries in racing greyhounds.	Jerk is briefly mentioned in turn transitions.
Hayati et al. (2017) [157]	9th Australasian Congress on Applied Mechanics	N/A	High rates of injuries in racing greyhounds and the potential risk factors were discussed. Turns with inappropriate transitions with high jerks were mentioned as an important risk factor.	Jerk is briefly mentioned as a risk factor in the greyhound racing industry.
Hayati et al. (2018) [158]	ASME 2018 International Mechanical Engineering Congress and Exposition	N/A	High-jerk turning, which is caused by an inappropriate bend and camber of race tracks, was given among the potential risk factors causing injuries in racing greyhounds.	Jerk is briefly mentioned in turn transitions.
Mahdavi et al. (2018) [159]	ASME 2018 International Mechanical Engineering Congress and Exposition	N/A	High-jerk turning, which is caused by an inappropriate bend and camber of race tracks, was given among the potential risk factors causing injuries in racing greyhounds.	Jerk is briefly mentioned in turn transitions.
Hossain et al. (2020) [160]	Scientific Reports	Q1	Designing a racing greyhound ideal trajectory path to minimise the injuries. This ideal trajectory path is based on minimum jerk rate.	Confirms the significance of jerk in racing.

Table 17. Jerk in sea-keeping.

Author	Source	SJR	Purpose and Findings	Comments
Tomi (1961) [126]	The Japan Society of Naval Architects and Ocean Engineering	N/A	Ride comfort of passengers on ships was studied. The influence of jerk on the rolling motion was discovered.	One of the first references considered jerk in ride comfort in sea-keeping. Jerk is referred to as the time derivative of acceleration.
Shigehiro et al. (2002) [122]	Fisheries science	Q3	The new evaluation method of passenger comfort is expressed by vertical and lateral accelerations, and exposure duration represents the relationship between ship motions and seasickness. It is confirmed that the correlation coefficient between the results of the new method and the questionnaires given to the trainees show fairly similar results.	Jerk is not directly mentioned. Only the work conducted by Tomi [126] was briefly mentioned.
Railey et al. (2016) [8]	Report	N/A	The aim of this work was to provide a universal guidance for the measurement and analysis of recorded acceleration data in high-speed crafts for different organisation. Four visual observations to conclude that acceptable low-pass filtering of an acceleration record has been achieved were proposed (i.e., no over-filtering). One of the proposed observations was that the rate of acceleration application (i.e., jerk) of the filtered record is approximately the same as the unfiltered record.	The value of jerk was used to assess the accountability of filter acceleration data using a low pass filter.
Sosa & Ooms (2016) [9]	Report	N/A	The role of roll stabiliser in the ride comfort in the *yacht industry* is analysed. Jerk is mentioned as a ride comfort parameter but was not included in the ride comfort rating used in this study. Since it is seen in the literature that jerk is an important parameter in this respected, further research in including jerk in the yacht ride comfort rating is critical.	Points at the need for more studies of the effect of jerk in yacht industry.
Coats & Riley (2018) [204]	Naval Surface Warfare Center technical report	N/A	The preliminary guidance for laboratory testing of marine shock isolation seats is given in this report. Jerk was among those parameters considered in the shock severity analysis.	The importance of jerk in the shock severity analysis in high-speed craft.
Werkman (2019) [10]	Master thesis	N/A	To study the impact of jerk on comfort of passengers in *sea-keeping*.	Points at the need for more studies of the effect of jerk. Importance of jerk in sea keeping and ride comfort in general.

3.14. Jerk in Seismic Analysis

Seismic analysis is a subset of structural analysis and is the calculation of the response of a building structure to earthquakes. It is part of the process of structural design, earthquake engineering or structural assessment and retrofit in regions where earthquakes are prevalent [134,161–170] (Table 18).

3.15. Jerk in Shock Response Spectrum

Equipment that is delivered to naval ships needs to be shockproof. The basic philosophy is that the equipment should withstand the same explosions as the ship itself (it would be unfortunate if the ship survived the blast, while it were impossible to operate since all equipment has been destroyed). Preferably, the shock-proofness of the equipment should be verified both theoretically and experimentally.

There are essentially three methods for theoretical verification, including:

- Static calculation, where you assume inertia forces of the mass times g in various directions;
- Calculation with shock response spectrum;
- Integration in the time domain.

Static calculations are easiest to carry out. In these calculations, neither the shape of the time history of the shock (e.g., blast or earthquake) nor the dynamic properties of the equipment under consideration (e.g., resonance frequencies) are taken into account. This means that you should normally have a considerable safety margin.

The shock response spectrum was invented by the Flemish-American engineer Maurice Anthony Binot (1905–1985). The basic idea is to draw a diagram that shows the maximum acceleration and relative displacement for a number of single-mass oscillators with various eigen frequencies for a certain shock time history. Normally, a simplified curve is drawn that might be an envelope for several time histories. This curve can be used to calculate the maximum amplitude for each eigenmode of the real equipment, and finally the contributions of each eigenmode can be added. There are different ways to add the contributions from the eigenmodes depending on how cautious you are. The very conservative approach is that you assume that, at some point in time, all the contributions of the eigenmodes are pointing in the same direction. This curve can not be used for testing, so a new time history curve has to be synthesised as input to that activity. It is out of the scope of this article to discuss details of this method, but it is interesting to point out that in these calculations jerk, etc., are implicitly taken into account.

The most straightforward method to carry out shock calculations is by integration in the time domain. Since the calculations are based on the time history, the derivatives of the acceleration are implicitly taken into account. The time history can, of course, directly be used as input to the test activity and comparison between calculations and measurements should be straightforward. The main drawback of this method is that it consumes a lot of computer power.

Studies that studied jerk in shock responses are listed in Table 19.

3.16. Jerk in Sport Science

Sport science is a general term which can include the science behind designing running shoes and studying the impact attenuation properties of greyhound's surface. Those articles that passed the inclusion criteria of this work are tabulated in Table 20.

3.17. Jerk in Structural Health Monitoring

Vibration monitoring is another challenging area, e.g., it would be beneficial to be able to be able to identify the wear on gearboxes of wind power stations in due time. However, the application of higher-order derivatives of acceleration for damage detection requires further investigation. Using jerk as one parameter to be monitored has been suggested, e.g., by Zhang et al. (2012) [174] (Table 21).

Table 18. Jerk in seismic analysis.

Author	Source	SJR	Purpose and Findings	Comments
Malushte (1987) [161]	Master thesis	N/A	To predict the seismic design response spectra using ground characteristics. A single-degrees-of-freedom (SDOF) is used to study the ground motion characteristics, such as as peak displacement, velocity, acceleration and root mean square acceleration. The average value of the ratio of the maximum jerk to the maximum ground acceleration and maximum ground velocity to the maximum ground displacement are given. These values are required when the prediction of the relative velocity spectra from the pseudo velocity spectra is of interest.	Jerk is briefly mentioned in this work.
Inaudi et al. (1993) [134]	Earthquake engineering structural dynamics	Q1	An optimum hybrid isolation system to protect sensitive equipment from earthquake was designed and the EOM of the ground motion obtained. Jerk was a parameter in the ground motion EOM.	Jerk was mentioned in the model of of ground motion, which was numerically modeled.
Bertero et al. (2002) [162]	Earthquake engineering structural dynamics	Q1	Different objectives were given for this work but one of the main objective was to review the understanding of performance-based-seismic-design (PBSD), the requirements for a reliable PBSD, and finally to study why some designs fail to satisfy those requirement. Jerk was mentioned as a useful parameter which should be taken into consideration for such designs, mainly in the case of frequent minor earthquake ground motions.	Jerk is mentioned as a useful parameter for reliable PBSD design, mainly in the case of frequent minor earthquake ground motion.
Geoffrey et al. (2003) [163]	Earthquake engineering structural dynamics	Q1	To regulate the total structural jerk to manage the structural energy and enhancing the performance of civil structures undergoing large seismic events. This new method used in this work is preferred over conventional methods. Their proposed jerk regulation control method is shown to have better performance than typical structural control methods for near-field seismic events where the response is dominated by a large impulse.	Importance of jerk in seismic control of a civil structure which has benefits over the conventional method.
Tong et al. (2005) [164]	Earthquake Engineering and Engineering Vibration	Q2	The importance of jerk in seismic motion analysis is not extensively studied. Therefore, they studied the basic characteristics of time derivatives of acceleration (TdoA) on records from the 1999 Chi-Chi, earthquake (Mw 7.6) and one of its aftershocks (Mw 6.2).	Highlighted the importance of jerk (referred to it as TDoA) is seismic motion analysis.

Table 18. *Cont.*

Author	Source	SJR	Purpose and Findings	Comments
He et al. (2015) [165]	Shock and Vibration	Q2	Jerk and its response spectra can improve the recognition of non-stationary ground motion. Therefore, in this work, the jerk response spectra for elastic and in-elastic system are characterised.	It is mentioned that though jerk is an important characteristic of ground motion, it is not thoroughly studied in seismic motion analysis.
Chakraborty and Ray-Chaudhuri (2017) [166]	Journal of Engineering Mechanics	Q2	To study the energy transfer to a high-frequency mode of a building due to a sudden change in stiffness at its base during seismic excitation. Jerk was used to excite the structure.	Jerk is used for exciting the structure due to the fact that it is one of the characteristics of ground motion.
Sofronie et al. (2017) [170]	Journal of Geological Resource Engineering with Computers	N/A	The concept of jerk is extended in seismic engineering.	Jerk in the context of seismic analysis is discussed.
Taushanov (2018) [167]	Journal of Engineering Mechanics	N/A	Excessive "jerky motion" affects comfort in building and bridges, and therefore attempts to reduce this phenomenon should be taken into consideration in engineering design. A jerk response spectra is given, which should be considered in seismic analysis of a structure.	Jerk spectra features is studied for elastic and in-elastic systems.
Papandreou and Papagianno-poulos (2019) [168]	Soil Dynamics Earthquake Engineering	Q1	To extract/study the general feature of jerk spectera for different seismic motion scenarios. They used this method for in-elastic SDOF systems. An empirical formula that provides jerk estimate for in-elastic SDOF system is proposed for the first time in the literature (by the Bilinear, the Ramberg-Osgood and the Takeda hysteresis rules).	This shows the importance of jerk in earthquake engineering.
Yaseen et al. (2020 [169]	Soil Dynamics Earthquake Engineering	N/A	The performance of jerk and its higher derivatives (referred to as relevant derived parameters) to address the ground motion intensity in masonry building was studied.	Jerk was used as an intensity measure in Seismic analysis of masonry buildings.

Table 19. Jerk in shock response spectrum.

Author	Source	SJR	Purpose and Findings	Comments
Railey et al. (2016) [8]	Report	N/A	The aim of this work was to provide a universal guidance for measurement and analysis of recorded acceleration data in high-speed crafts for different organisation. Four visual observations to conclude that acceptable low-pass filtering of an acceleration record has been achieved were proposed (i.e., no over-filtering). One of the proposed observations was that the jerk of the filtered record is approximately the same as the unfiltered record.	The value of jerk was used to assess the accountability of filter acceleration data using a low pass filter.
Riley et al. (2018) [205]	Report	N/A	This report characterizes the shock severity in the response domain, so that the effects of shock pulse shape, peak amplitude, jerk, and pulse duration can be taken into account for systems across a broad range of natural frequencies.	It is mentioned that it might be useful to include jerk in shock investigations.

Table 20. Jerk in sport science.

Author	Source	SJR	Purpose and Findings	Comments
Savage [171]	Report	N/A	Although different types are running shoes have been designed, they seemed not to be based on the current science on injury prevention. Different types of shoe design and injury mechanisms were discussed in this report. It is mentioned that the importance of jerk should be considered in reducing the impact injuries while running.	It might be worthwhile to study the impact of jerk in designing running shoes.
Eager et al. (2016) [1]	European Journal of Physics	Q2	The concept of jerk is discussed using trampoline and amusement rides (roller coasters). The effect of jerk on the human body is also discussed. The importance of jerk in amusement rides (roller coasters) for safety (avoiding whiplash), ride comfort, and when the safety of the passenger is not an issue reducing the maintenance cost due to snap.	Jerk is important and is experienced in daily life, yet it is not well-explained and understood.
Hayati et al. (2019) [172]	World Engineers Convention 2019	N/A	A modified Clegg hammer was used to assess the safety of greyhound's race track sand and grass surfaces. The maximum acceleration, the impact duration, the energy loss and the maximum jerk, were considered as dynamic parameters that determine the safety of the race track surface.	Jerk is considered as a dynamic parameter to assess the safety of the greyhound race track surfaces.

Table 21. Jerk in structural health monitoring.

Author	Source	SJR	Purpose and Findings	Comments
Sone (2004) [173]	Journal of Transactions of Nippon Kikai Gakkai Ronbunshu C Hen	N/A	A statistical and data mining approach are applied on time-domain jerk data in wind turbine gearboxes to detect damage/failure. The appoach seemed promising in health monitoring and is suggested to be expanded to the other parts of the turbine as well.	Jerkdot should be another name for jounce/snap and is mentioned in this work. This work contains a description of a "jounce meter". It is not clear to what extent they are used in practice.
Zhang et al. (2012) [174]	IEEE transactions on energy conversion	Q1	Fault analysis and health monitoring of the wind turbine gearbox were the interests of this work. A statistical and data mining approach are applied on time-domain jerk data in wind turbine gearboxes to detect damage/failure.	The approach seemed promising in health monitoring and is suggested to be expanded to the other parts of the turbine as well.
Iyama et al. (2019) [175]	Japan Architectural Review	N/A	To extend the nonlinear behaviour detection by determining the mathematical model relation between the snap, stiffness change, and the velocity of the vibration system. It is discussed that there is lack of proper understanding of jerk and its higher derivative in the field.	Higher derivatives of jerk, mainly snap, can be used to detect nonlinearity in structural health monitoring methods.
Sumathy et al. (2019) [176]	Cogent Engineering	Q2	Fault diagnosis of wind turbine gear was investigated in this work. They proposed a novel method of analysing the stability of Interacting Multiple Model (IMM) algorithm for a linear system (the details of this methodare out of scope of this review article.) They mentioned that analysing jerk data coming from the vibration of acceleration data via the IMM Kalman filter is a novel endeavor to the best of their knowledge.	Jerk can be used in health monitoring of dynamical system.

3.18. Jerk in Technical Pain

Jerk was used to expand the understanding of researcher of sensed pain via a pain sensor where jerk was used as a biometric measure (Table 22).

3.19. Jerk in UAV

Jerk, and even higher derivatives of it (aka jounce/snap), have been recently used in designing, testing and controlling UAVs [11,178–186] (Table 23).

3.20. Jerk in Vehicles-Land Based Traditional

Jerk in vehicles is mainly used for measuring the ride comfort [12,25,45,95,98–106,108,109,111–113, 115,118–121,123–125,127–131]. It is also used as a controller design, referred to as an anti-jerk controller design [95,98–102,109,111,119,120,130]. Jerk is also used in autonomous vehicles [12,114,123,187,188], which is explained in the below sections.

3.20.1. Jerk in Vehicles: Ride Comfort

One of the main applications of jerk in vehicles is as criteria for discomfort. Details of these studies were given in Table 7.

3.20.2. Jerk in Vehicles: Autonomous Vehicles

Jerk in autonomous driving is used to evaluate the comfortable ride and safety of the passengers [12,114,123,187,188] (Table 24).

Table 22. Jerk in technical pain.

Author	Source	SJR	Purpose and Findings	Comments
Ostermeyer et al. (2008) [177]	Applied Mechanics and Materials	N/A	The concept of technical pain, which is based on the analysis of mesoscopic systems, was introduced in this work. The system was modeled as damped oscillators and the heat transferred by an impact. Pain was defined as an integral over the square of jerk. They developed a pain sensor and used it to analyse the pain caused by different types of signals.	Jerk used to obtain a measure of pain and used in the development of a pain sensor.

Table 23. Jerk in UAV.

Author	Source	SJR	Purpose and Finding	Comments
Luukkonen (2011) [181]	Independent research project in applied mathematics	N/A	To model and control the quadcopter. It is found that high snap values will contribute to high control input values, and therefore the jounces have to be considered closely when generating the accelerations.	Jounce/snap is considered in modelling and controlling the quadcopter. Not only jerk, but also snap (jounce), may be relevant in the design and control of the drone.
Rakgowa et al. (2015) [183]	EEE International Symposium on Robotics and Intelligent Sensors	N/A	A minimum-jerk trajectory controllers was developed for a quadrotor during high-acceleration dynamics, e.g., lift-off.	Jerk as a measure to generate trajectories in UAVs.
Phang et al. (2015) [182]	Mechatronics	Q1	In an UAV calligraphy project, having a smooth trajectory control was desired. Accordingly, jerk was used as a limiting parameter to generate smooth trajectories.	Jerk as a measure to generate trajectories in UAVs.
Fiori et al. (2016) [178]	The 10th International Conference on Circuits, Systems, Signals and Telecommunications	N/A	Use of jerk (referred to as lurch index) to assess the drone's attitude fluency maneuverability. It is found that the geometric lurch index is fairly sensitive to the fluency of attitude maneuvering.	Uses the term "lurch index".

Table 23. *Cont.*

Author	Source	SJR	Purpose and Finding	Comments
Nemes & Mester (2016) [11]	The 4th International Scientific Conference on Advances in Mechanical, Engineering	N/A	Jerk is used as a parameter in controlling a drone. It is discussed that an abrupt jerk induces vibrations and, in the case of vehicles, it would negatively affect the ride comfort.	Use of jerk in controlling a drone.
Silva et al. (2018) [185]	Unmanned Systems	Q2	Controlling drones in a high-wind scenario and during thrust was a challenge in this study. Accordingly, a jerk-minimum trajectory controller was developed to optimise the motion.	Jerk as a limiting parameter in trajectory controllers.
Guye (2018) [179]	Masters thesis	N/A	An Indoor multi-rotor test-bed for experimentation on autonomous guidance strategies is developed. It is found that the jerk would minimize the product of the control inputs, which are the major force that consume the power energy of the drone. Consequently, minimizing these control inputs will also reduce the power consumption. As a result, minimising the jerk results in minimising the power consumption.	Use of jerk in designing a test bed for a drone. Since jerk minimisation minimises the energy consumption, it should be considered in drone designs and controlling.
Rousseau et al. (2018) [184]	European Control Conference	N/A	Minimum jerk trajectories and piece-wise polynomial trajectories are used for cinematographic flight plans of a quadcopter.	Jerk as a measure to generate trajectories.
Tal & Karaman (2018) [186]	2018 IEEE Conference on Decision and Control	N/A	Tracking of aggressive (high-speed and high-acceleration) quadrotor trajectories. The main contribution of this work is a trajectory tracking control design that achieves accurate tracking aggressive maneuvers without depending on modeling or estimation of aerodynamic drag parameters. The design exploits the differential flatness of the quadcopter dynamics to generate feed-forward control terms based on the reference trajectory and its derivatives up to the fourth order: velocity, acceleration, jerk, and snap.	Use of jerk in controlling a drone. Not only jerk, but also snap (jounce), may be relevant in the design and controlling of the drone.
Lai et al. (2019) [180]	Frontiers of Information Technology & Electronic Engineering	Q2	To have a safe flying corridor, jerk limit ted trajectories were used in real-time scenarios.	Jerk as a limiting parameter in trajectory controllers.

Table 24. Jerk in autonomous vehicles.

Author	Source	SJR	Purpose and Findings	Comments
Bae et al. (2019) [12]	journal of Electronics	Q3	To provide a comfortable driving experience, while not sabotaging the passengers' safety on a self-driving shuttle bus, a time-optimal velocity planning method to guarantee a comfort criteria was developed. A better performance and comfortable passenger ride in the self-driving shuttle bus is experienced.	This is also aligned with the application of jerk in ride comfort and its potential to be considered in autonomous cars.
Krüger (2019) [114]	Master thesis	N/A	To implement and evaluate a motion planner to find a rough speed profile in autonomous cars. The value of jerk was considered as passengers' ride comfort and safety measures.	Points at the importance of low-jerk trajectories to guarantee passenger's safety and ride comfort.

4. Conclusions

The purpose of this work was to show the importance of jerk in the context of science and engineering, by conducting a thorough systematic review of recent academic articles (2015–2020) where the term 'jerk' was mentioned in the title. The quality of papers was assessed based on Scientific Journal Rankings (SJR)—Scimago index, from the quartiles Q1 to Q4, where Q1 is occupied by the top 25% of journals, Q2 by the top 25% to 50%, and so on. The articles were then categorised based on the application of jerk in twenty categories. The result of this systematic review showed that, although jerk is overlooked in secondary and higher education, jerk is ubiquitous. Road, rail and sea all have examples of a jerk, from crack initiation to ride comfort. Traditional printing and 3D-printing control systems all contain examples of a jerk. This review has provided a solid foundation for future research on the importance of jerk in different fields. It has identified research gaps which will assist researchers in creating a concise road map toward a more comprehensive study on jerk. The authors also emphasize that jerk is still essentially overlooked in secondary and higher education. This review provides support for teachers and textbook authors who may wish to include examples of jerk in their lessons and textbooks.

Jerk is all around as if we care to listen, feel, open our eyes and observe—from greyhound tracks to roller coasters.

Author Contributions: H.A. performed an initial literature search, including historical sources, and also followed up on some references in more detail through contacts with authors. A.-M.P. and D.E., following a collaboration of jerk in trampolines and roller coasters, initiated the research into jerk with a STEM educational context. D.E. suggested a more systematic literature review. H.H. did most of the searching, categorization and writing Conceptualization, H.H., D.E., A.-M.P. and H.A.; methodology, H.H. and D.E.; validation, H.H., D.E., A.-M.P. and H.A.; formal analysis, D.E., A.-M.P. and H.A.; investigation, H.H. and H.A.; resources, D.E.; data curation, H.H.; writing—original draft preparation, H.H. and H.A.; writing—review and editing, H.H., D.E., A.-M.P. and H.A.; visualization, H.H., D.E., A.-M.P. and H.A.; supervision, D.E.; project administration, H.H.; funding acquisition, D.E. All authors have read and agreed to the published version of the manuscript.

Funding: The work is funded by Greyhound Racing New South Wales with UTS institution reference of PRO17-3051.

Conflicts of Interest: The authors declare no conflict of interest.

References

1. Eager, D.; Pendrill, A.M.; Reistad, N. Beyond velocity and acceleration: Jerk, snap and higher derivatives. *Eur. J. Phys.* **2016**, *37*, 65–68. [CrossRef]
2. Pendrill, A.M. Rollercoaster loop shapes. *Phys. Educ.* **2005**, *40*, 517. [CrossRef]
3. Gierlak, P.; Szybicki, D.; Kurc, K.; Burghardt, A.; Wydrzyński, D.; Sitek, R.; Goczał, M. Design and dynamic testing of a roller coaster running wheel with a passive vibration damping system. *J. Vibroeng.* **2018**, *20*, 1129–1143. [CrossRef]
4. Sicat, S.; Woodcock, K.; Ferworn, A. Wearable Technology for Design and Safety Evaluation of Rider Acceleration Exposure on Aerial Adventure Attractions. In Proceedings of the Annual Occupational Ergonomics and Safety Conference, Pittsburgh, PA, USA, 7–8 June 2018.
5. Vaisanen, A. Design of Roller Coasters. Master's Thesis, Aalto University, Espoo, Finland, 2018.
6. Pendrill, A.M.; Eriksson, M.; Eriksson, U.; Svensson, K.; Ouattara, L. Students making sense of motion in a vertical roller coaster loop. *Phys. Educ.* **2019**, *54*, 065017. [CrossRef]
7. Pendrill, A.M.; Eager, D. Velocity, acceleration, jerk, snap and vibration: Forces in our bodies during a roller coaster ride. *Phys. Educ.* **2020**, *55*, 065012. [CrossRef]
8. Coats, T.W.; Haupt, K.D.; Murphy, H.P.; Ganey, N.C.; Riley, M.R. *A Guide for Measuring, Analyzing, and Evaluating Accelerations Recorded During Seakeeping Trials of High-Speed Craft*; Report; Naval Surface Warfare Center Carderock Division: Norfolk, VA, USA, 2016.
9. Sosa, L.; Ooms, J. *A Comfort Analysis of an 86 m Yacht Fitted with Fin Stabilizers Vs. Magnus-Effect Rotors*; Report; The Society of Naval Architects and Marine Engineers: Alexandria, VA, USA, 2016.
10. Werkman, J. Determining and Predicting the Seakeeping Performance of Ships Based on Jerk in the Ship Motions. Master's Thesis, Delft University of Technology, Delft, The Netherlands, 2019.

11. Nemes, A.; Mester, G. Energy Efficient Feasible Autonomous Multi-Rotor Unmanned Aerial Vehicles Trajectories. In Proceedings of the 4th International Scientific Conference on Advances in Mechanical, Engineering, Debrecen, Hungary, 13–15 October 2016; pp. 369–376.
12. Bae, I.; Moon, J.; Seo, J. Toward a Comfortable Driving Experience for a Self-Driving Shuttle Bus. *Electronics* **2019**, *8*, 943. [CrossRef]
13. *ISO-2041 Mechanical Vibration, Shock and Condition Monitoring—Vocabulary*; Standard; International Organization for Standardization: Geneva, Switzerland, 2009.
14. *ISO/TC 17929 Biomechanical Effects on Amusement Ride Passengers*; Standard; International Organization for Standardization: Geneva, Switzerland, 2014.
15. *ISO 11026 Heavy Commercial Vehicles and Buses—Test Method for Roll Stability—Closing-Curve Test*; Standard; International Organization for Standardization: Geneva, Switzerland, 2010.
16. *ISO 15623 Intelligent Transport Systems—Forward Vehicle Collision Warning Systems—Performance Requirements and Test Procedures*; Standard; International Organization for Standardization: Geneva, Switzerland, 2013.
17. *ISO 18737-1: Measurement of Ride Quality—Part 1: Lifts (Elevators)*; Standard; International Organization for Standardization: Geneva, Switzerland, 2012.
18. *ISO/TS 14649-201: Industrial Automation and Integration—Physical Device Control—Data Model for Computerized Numerical Controllers—Part 201: Machine Tool Data for Cutting Processes*; Standard; International Organization for Standardization: Geneva, Switzerland, 2012.
19. *ISO 25745-2: Energy Performance of Lifts, Escalators and Moving Walks—Part 2: Energy Calculation and Classification for Lifts (Elevators)*; Standard; International Organization for Standardization: Geneva, Switzerland, 2015.
20. Sandin, T. The jerk. *Phys. Teach.* **1990**, *28*, 36–40. [CrossRef]
21. Sears, F.W.; Zemansky, M.W.; Young, H.D. *University Physics*; Addison-Wesley: Boston, MA, USA, 1987.
22. Schot, S.H. Jerk: the time rate of change of acceleration. *Am. J. Phys.* **1978**, *46*, 1090–1094. [CrossRef]
23. Lalanne, C. *Mechanical Vibration and Shock Analysis, Fatigue Damage*; John Wiley & Sons: Hoboken, NJ, USA, 2010.
24. Smith, J.D. *Vibration Measurement and Analysis*; Butterworth-Heinemann: Oxford, UK, 2013.
25. Jia, J. *Essentials of Applied Dynamic Analysis*; Springer: Berlin/Heidelberg, Germany, 2014.
26. Thompson, P.M. Snap crackle and pop. In Proceedings of the AIAA Southern California Aerospace Systems and Technology Conference, Hawthorne, CA, USA, 2011. Available online: http://www.justuslearning.com/wp-content/uploads/2013/12/AIAAOC_SnapCracklePop_docx.pdf (accessed on 21 October 2020).
27. Besset, P.; Béarée, R. FIR filter-based online jerk-constrained trajectory generation. *Control Eng. Pract.* **2017**, *66*, 169–180. [CrossRef]
28. Bianco, C.G.L.; Ghilardelli, F. A scaling algorithm for the generation of jerk-limited trajectories in the operational space. *Robot. Comput.-Integr. Manuf.* **2017**, *44*, 284–295. [CrossRef]
29. Chen, S.L.; Li, X.; Teo, C.S.; Tan, K.K. Composite jerk feedforward and disturbance observer for robust tracking of flexible systems. *Automatica* **2017**, *80*, 253–260. [CrossRef]
30. Chen, D.; Li, S.; Li, W.; Wu, Q. A multi-level simultaneous minimization scheme applied to jerk-bounded redundant robot manipulators. *IEEE Trans. Autom. Sci. Eng. Comput.* **2019**, *17*, 463–474. [CrossRef]
31. Chen, D.; Zhang, Y. Minimum jerk norm scheme applied to obstacle avoidance of redundant robot arm with jerk bounded and feedback control. *IET Control Theory Appl.* **2016**, *10*, 1896–1903. [CrossRef]
32. Chen, D.; Zhang, Y. Jerk-level synchronous repetitive motion scheme with gradient-type and zeroing-type dynamics algorithms applied to dual-arm redundant robot system control. *Int. J. Syst. Sci.* **2017**, *48*, 2713–2727. [CrossRef]
33. Dai, C.; Lefebvre, S.; Yu, K.M.; Geraedts, J.M.; Wang, C.C. Planning Jerk-Optimized Trajectory With Discrete Time Constraints for Redundant Robots. *IEEE Trans. Autom. Sci. Eng.* **2020**, *17*, 1711–1724. [CrossRef]
34. Dong, H.; Cong, M.; Liu, D.; Wang, G. An effective technique to find a robot joint trajectory of minimum global jerk and distance. In Proceedings of the 2015 IEEE International Conference on Information and Automation, Lijiang, China, 8–10 August 2015; pp. 1327–1330.
35. Duan, H.; Zhang, R.; Yu, F.; Gao, J.; Chen, Y. Optimal trajectory planning for glass-handing robot based on execution time acceleration and jerk. *J. Robot.* **2016**, *2016*, 9329131. [CrossRef]
36. Feifei, L.; Fei, L. Time-jerk optimal planning of industrial robot trajectories. *Int. J. Robot. Autom.* **2016**, *31*, 1–7. [CrossRef]

37. Glorieux, E.; Svensson, B.; Danielsson, F.; Lennartson, B. Simulation-based time and jerk optimisation for robotic press tending. In Proceedings of the 29th European Simulation and Modelling Conference, EUROSIS, Leicester, UK, 7–8 October 2015; pp. 377–384.
38. Huang, J.; Hu, P.; Wu, K.; Zeng, M. Optimal time-jerk trajectory planning for industrial robots. *Mech. Mach. Theory* **2018**, *121*, 530–544. [CrossRef]
39. Jiang, L.; Lu, S.; Gu, Y.; Zhao, J. Time-Jerk Optimal Trajectory Planning for a 7-DOF Redundant Robot Using the Sequential Quadratic Programming Method. In *International Conference on Intelligent Robotics and Applications*; Springer: Berlin/Heidelberg, Germany, 2017; pp. 343–353.
40. Kaserer, D.; Gattringer, H.; Müller, A. Online Robot-Object Synchronization With Geometric Constraints and Limits on Velocity, Acceleration, and Jerk. *IEEE Robot. Autom. Lett.* **2018**, *3*, 3169–3176. [CrossRef]
41. Kaserer, D.; Gattringer, H.; Müller, A. Nearly optimal path following with jerk and torque rate limits using dynamic programming. *IEEE Trans. Robot.* **2018**, *35*, 521–528. [CrossRef]
42. Lange, F.; Albu-Schäffer, A. Path-accurate online trajectory generation for jerk-limited industrial robots. *IEEE Robot. Autom. Lett.* **2015**, *1*, 82–89. [CrossRef]
43. Lange, F.; Suppa, M. Trajectory generation for immediate path-accurate jerk-limited stopping of industrial robots. In Proceedings of the 2015 IEEE International Conference on Robotics and Automation (ICRA), Seattle, WA, USA, 26–30 May 2015; pp. 2021–2026.
44. Lange, F.; Suppa, M. Trajectory Generation for Path-Accurate Jerk-Limited Sensor-Based Path Corrections of Robot Arms. In Proceedings of the IEEE International Conference on Robotics and Automation (ICRA), Seattle, WA, USA, 26–30 May 2015; pp. 4087–4087.
45. Liu, L.; Chen, C.; Zhao, X.; Li, Y. Smooth trajectory planning for a parallel manipulator with joint friction and jerk constraints. *Int. J. Control Autom. Syst.* **2016**, *14*, 1022–1036. [CrossRef]
46. Lu, S.; Li, Y. Minimum-Jerk Trajectory Planning of a 3-DOF Translational Parallel Manipulator. In Proceedings of the ASME 2015 International Design Engineering Technical Conferences and Computers and Information in Engineering Conference, Boston, MA, USA, 2–5 August 2015.
47. Lu, S.; Ding, B.; Li, Y. Minimum-jerk trajectory planning pertaining to a translational 3-degree-of-freedom parallel manipulator through piecewise quintic polynomials interpolation. *Adv. Mech. Eng.* **2020**, *12*, 1687814020913667. [CrossRef]
48. Lu, S.; Zhao, J.; Jiang, L.; Liu, H. Solving the time-jerk optimal trajectory planning problem of a robot using augmented lagrange constrained particle swarm optimization. *Math. Probl. Eng.* **2017**, *2017*. [CrossRef]
49. Palleschi, A.; Garabini, M.; Caporale, D.; Pallottino, L. Time-Optimal Path Tracking for Jerk Controlled Robots. *IEEE Robot. Autom. Lett.* **2019**, *4*, 3932–3939. [CrossRef]
50. Park, B.J.; Lee, H.J.; Oh, K.K.; Moon, C.J. Jerk-Limited Time-Optimal Reference Trajectory Generation for Robot Actuators. *Int. J. Fuzzy Log. Intell. Syst.* **2017**, *17*, 264–271. [CrossRef]
51. Rezaeifar, H.; Najafi, F. Path planning using via-points and Jerk-minimum method with static obstacles for a 7 DOF manipulator. *Modares Mech. Eng.* **2015**, *15*, 153–163.
52. Rojas, R.A.; Garcia, M.A.R.; Wehrle, E.; Vidoni, R. A Variational Approach to Minimum-Jerk Trajectories for Psychological Safety in Collaborative Assembly Stations. *IEEE Robot. Autom. Lett.* **2019**, *4*, 823–829. [CrossRef]
53. Rout, A.; Dileep, M.; Mohanta, G.B.; Deepak, B.; Biswal, B. Optimal time-jerk trajectory planning of 6 axis welding robot using TLBO method. *Procedia Comput. Sci.* **2018**, *133*, 537–544. [CrossRef]
54. Rout, A.; Mohanta, G.B.; Gunji, B.M.; Deepak, B.; Biswal, B.B. Optimal time-jerk-torque trajectory planning of industrial robot under kinematic and dynamic constraints. In Proceedings of the 2019 9th Annual Information Technology, Electromechanical Engineering and Microelectronics Conference (IEMECON), Jaipur, India, 13–15 March 2015; pp. 36–42.
55. Shi, X.; Fang, H.; Guo, W. Time-Energy-Jerk Optimal Trajectory Planning of Manipulators Based on Quintic NURBS. *Mach. Des. Res.* **2017**, *2017*, 45. [CrossRef]
56. Shi, X.L.; Fang, H.G. Time-Energy-Jerk Optimal Planning of Industrial Robot Trajectories. *Mach. Des. Manuf.* **2018**, 66. [CrossRef]
57. Shimada, N.; Yoshioka, T.; Ohishi, K.; Miyazaki, T.; Yokokura, Y. Variable dynamic threshold of jerk signal for contact detection in industrial robots without force sensor. *Electr. Eng. Jpn.* **2015**, *193*, 43–54. [CrossRef]
58. Wang, P.; Yang, H.; Xue, K. Jerk-optimal trajectory planning for stewart platform in joint space. In Proceedings of the 2015 IEEE International Conference on Mechatronics and Automation (ICMA), Beijing, China, 2–5 August 2015; pp. 1932–1937.

59. Yue, S.; Yinya, L.; Guoqing, Q.; Ardong, S. Time-jerk Optimal Trajectory Planning for Industrial Robots Based on PSO Algorithm. *Comput. Meas. Control* **2017**, *2017*, 45.
60. Zeeshan, M.; Xu, H. Jerk-Bounded Trajectory Planning of Industrial Manipulators. In Proceedings of the 2019 IEEE International Conference on Robotics and Biomimetics (ROBIO), Dali, China, 6–8 December 2019; pp. 1089–1096.
61. Zhang, D.; Wu, F.; Li, R. Time-optimal and minimum-jerk trajectory planning of 3-DOF PM spherical motor. In Proceedings of the 2017 IEEE International Conference on Mechatronics and Automation (ICMA), Takamatsu, Japan, 6–9 August 2017; pp. 1843–1847.
62. Zhang, Y.; Yang, M.; Qiu, B.; Luo, J.; Tan, H. Jerk-level solutions to manipulator inverse kinematics with mathematical equivalence of operations discovered. In Proceedings of the 2016 12th International Conference on Natural Computation, Fuzzy Systems and Knowledge Discovery (ICNC-FSKD), Changsha, China, 13–15 August 2016; pp. 2121–2126.
63. Zhao, R.; Sidobre, D. Trajectory smoothing using jerk bounded shortcuts for service manipulator robots. In Proceedings of the 2015 IEEE/RSJ International Conference on Intelligent Robots and Systems (IROS), Hamburg, Germany, 28 September–2 October 2015; pp. 4929–4934.
64. Zhang, L.; Du, J. Acceleration smoothing algorithm based on jounce limited for corner motion in high-speed machining. *Int. J. Adv. Manuf. Technol.* **2018**, *95*, 1487–1504. [CrossRef]
65. Alzaydi, A. Time-optimal, minimum-jerk, and acceleration continuous looping and stitching trajectory generation for 5-axis on-the-fly laser drilling. *Mech. Syst. Signal Process.* **2019**, *121*, 532–550. [CrossRef]
66. Besset, P.; Béarée, R.; Gibaru, O. FIR filter-based online jerk-controlled trajectory generation. In Proceedings of the 2016 IEEE International Conference on Industrial Technology (ICIT), Taipei, Taiwan, 14–17 March 2016; pp. 84–89.
67. Bosetti, P.; Ragni, M. Milling Part Program Preprocessing for Jerk-limited, Minimum-time Tool Paths Based on Optimal Control Theory. *IEEJ J. Ind. Appl.* **2016**, *5*, 53–60. [CrossRef]
68. Dumanli, A.; Sencer, B. Optimal high-bandwidth control of ball-screw drives with acceleration and jerk feedback. *Precis. Eng.* **2018**, *54*, 254–268. [CrossRef]
69. Hashemian, A.; Bo, P.; Bartoň, M. Reparameterization of ruled surfaces: Toward generating smooth jerk-minimized toolpaths for multi-axis flank CNC milling. *Comput.-Aided Des.* **2020**, *127*, 102868. [CrossRef]
70. Huang, J.; Zhu, L.M. Feedrate scheduling for interpolation of parametric tool path using the sine series representation of jerk profile. *Proc. Inst. Mech. Eng. Part B J. Eng. Manuf.* **2017**, *231*, 2359–2371. [CrossRef]
71. Li, H.; Wu, W.J.; Rastegar, J.; Guo, A. A real-time and look-ahead interpolation algorithm with axial jerk-smooth transition scheme for computer numerical control machining of micro-line segments. *Proc. Inst. Mech. Eng. Part B J. Eng. Manuf.* **2019**, *233*, 2007–2019. [CrossRef]
72. Lin, M.T.; Yu, N.T.; Chiu, W.T.; Lee, C.Y.; Lu, Y.M. A master-axis-based feedrate scheduling with jerk constraints for five-axis tool center point trajectory. In Proceedings of the 2015 IEEE International Conference on Automation Science and Engineering (CASE), Gothenburg, Sweden, 24–28 August 2015; pp. 111–116.
73. Mutlu, M.K.; Keysan, O.; Ulutas, B. Limited-Jerk Sinusoidal Trajectory Design for FOC of PMSM with H-Infinity Optimal Controller. In Proceedings of the 2018 IEEE 18th International Power Electronics and Motion Control Conference (PEMC), Budapest, Hungary, 26–30 August 2018; pp. 704–710.
74. Schroedter, R.; Roth, M.; Janschek, K.; Sandner, T. Flatness-based open-loop and closed-loop control for electrostatic quasi-static microscanners using jerk-limited trajectory design. *Mechatronics* **2018**, *56*, 318–331. [CrossRef]
75. Weng, W.H.; Kuo, C.F.J. Jerk decision for free-form surface effects in multi-axis synchronization manufacturing. *Int. J. Adv. Manuf. Technol.* **2019**, *105*, 799–812. [CrossRef]
76. Zhang, Y.; Ye, P.; Wu, J.; Zhang, H. An optimal curvature-smooth transition algorithm with axis jerk limitations along linear segments. *Int. J. Adv. Manuf. Technol.* **2018**, *95*, 875–888. [CrossRef]
77. Zhang, Y.; Ye, P.; Zhang, H.; Zhao, M. A local and analytical curvature-smooth method with jerk-continuous feedrate scheduling along linear toolpath. *Int. J. Precis. Eng. Manuf.* **2018**, *19*, 1529–1538. [CrossRef]
78. Zhang, Y.; Zhao, M.; Ye, P.; Zhang, H. A G4 continuous B-spline transition algorithm for CNC machining with jerk-smooth feedrate scheduling along linear segments. *Comput.-Aided Des.* **2019**, *115*, 231–243. [CrossRef]
79. Ansoategui, I.; Campa, F.J. Mechatronic Model Based Jerk Optimization in Servodrives with Compliant Load. In *European Conference on Mechanism Science*; Springer: Berlin/Heidelberg, Germany, 2018; pp. 45–52.

80. Fang, J.H.; Guo, F.; Chen, Z.; Wei, J.H. Improved sliding-mode control for servo-solenoid valve with novel switching surface under acceleration and jerk constraints. *Mechatronics* **2017**, *43*, 66–75. [CrossRef]
81. Liu, J.; Cai, S.; Chen, W.; Chen, I.M. Minimum-jerk trajectory generation and global optimal control for permanent magnet spherical actuator. In Proceedings of the 2017 IEEE International Conference on Robotics and Biomimetics (ROBIO), Macau, China, 5–8 December 2017; pp. 2249–2254.
82. Masoudi, S.; Feyzi, M.R.; Sharifian, M.B.B. Force ripple and jerk minimisation in double sided linear switched reluctance motor used in elevator application. *IET Electr. Power Appl.* **2016**, *10*, 508–516. [CrossRef]
83. Wu, F.; Zhai, X.; Zhang, D.; Li, R. Minimum-Jerk Trajectory Planning of 3-DOF PM Spherical Motor. *Small Spec. Electr. Mach.* **2016**, *10*, 82–85.
84. Bui, H.T. Toolpath Planning Methodology for Multi-Gantry Fused Filament Fabrication 3D Printing. Master's Thesis, University of Arkansas, Fayetteville, AR, USA, 2019.
85. Chen, S.L.; Ma, J.; Teo, C.S.; Kong, C.J.; Lin, W.; Tay, A.; Al Mamun, A. A constrained linear quadratic optimization approach to jerk decoupling cartridge design for vibration suppression. In Proceedings of the 2015 IEEE International Conference on Advanced Intelligent Mechatronics (AIM), Busan, Korea, 7–11 July 2015; pp. 1472–1477.
86. Hernandez, D.D. Factors affecting dimensional precision of consumer 3D printing. *Int. J. Aviat. Aeronaut. Aerosp.* **2015**, *2*, 2. [CrossRef]
87. Kamaldin, N.; Chen, S.L.; Teo, C.S.; Lin, W.; Tan, K.K. A novel adaptive jerk control with application to large workspace tracking on a flexure-linked dual-drive gantry. *IEEE Trans. Ind. Electron. Autom.* **2018**, *66*, 5353–5363. [CrossRef]
88. Ma, J.; Chen, S.L.; Teo, C.S.; Kong, C.J.; Tay, A.; Lin, W.; Al Mamun, A. A constrained linear quadratic optimization algorithm toward jerk-decoupling cartridge design. *J. Frankl. Inst.* **2017**, *354*, 479–500. [CrossRef]
89. Mab, J.; Chena, S.L.; SingTeoc, C.; Kongc, C.J.; Tayb, A.; Linc, W.; Al Mamunb, A. Constrained linear quadratic optimization for jerk-decoupling cartridge design. *Precis. Motion Syst. Model. Control Appl.* **2019**, *2019*, 13.
90. Whyman, S.; Arif, K.M.; Potgieter, J. Design and development of an extrusion system for 3D printing biopolymer pellets. *Int. J. Adv. Manuf. Technol.* **2018**, *96*, 3417–3428. [CrossRef]
91. Visser, M. Jerk, snap and the cosmological equation of state. *Class. Quantum Gravity* **2004**, *21*, 2603. [CrossRef]
92. Hur-Diaz, S.; Wirzburger, J.; Smith, D. *Three Axis Control of the Hubble Space Telescope Using Two Reaction Wheels and Magnetic Torquer Bars for Science Observations*; Report; The National Aeronautics and Space Administration: Washington, DC, USA, 2008 .
93. Ramon, R.G. Model for Touchdown Dynamics of a Lander on the Solar Power Sail Mission. Master's Thesis, Luleå University of Technology, Kingston, Jamaica, 2016.
94. STD, N. *3001. NASA Space Flight Human-System Standard, Volume 2: Human Factors, Habitability, and Environmental Health*; The National Aeronautics and Space Administration: Washington, DC, USA, 2011.
95. Abuasaker, S. Anti-Jerk Controller with Optimisation-Based Self-Tuning. Ph.D. Thesis, University of Surrey, Guildford, UK, 2016.
96. Anderson, D.; Desaix, M.; Nyqvist, R. The least uncomfortable journey from A to B. *Am. J. Phys.* **2016**, *84*, 690–695. [CrossRef]
97. Antonelli, R.; Klotz, A.R. A smooth trip to Alpha Centauri: The least uncomfortable journey from A to B. *Am. J. Phys.* **2017**, *85*, 469–472. [CrossRef]
98. Batra, M. Dynamics and Model-Predictive Anti-Jerk Control of Connected Electric Vehicles. Ph.D. Thesis, University of Waterloo, Waterloo, ON, Canada, 2018.
99. Batra, M.; Maitland, A.; McPhee, J.; Azad, N.L. Non-linear model predictive anti-jerk cruise control for electric vehicles with slip-based constraints. In Proceedings of the 2018 Annual American Control Conference (ACC), Milwaukee, WI, USA, 27–29 June 2018; pp. 3915–3920.
100. Batra, M.; McPhee, J.; Azad, N.L. Anti-jerk model predictive cruise control for connected electric vehicles with changing road conditions. In Proceedings of the 2017 11th Asian Control Conference (ASCC), Gold Coast, QLD, Australia, 17–20 December 2017; pp. 49–54.
101. Batra, M.; McPhee, J.; Azad, N.L. Anti-jerk dynamic modeling and parameter identification of an electric vehicle based on road tests. *J. Comput. Nonlinear Dyn.* **2018**, *13*, 101005. [CrossRef]
102. Darokar, K.K. Automotive Driveline Backlash State and Size Estimator Design for Anti-Jerk Control. Master's Thesis, Michigan Technology University, Houghton, MI, USA, 2019.

103. Deshmukh, A.; Mulani, B.; Jadhav, N.; Parihar, A.S. Study of frequency characteristics of vehicle motions for the derivation of inherent jerk. *Int. J. Passeng. Cars-Mech. Syst.* **2016**, *9*, 419–423. [CrossRef]
104. Förstberg, J. Ride Comfort and Motion Sickness in Tilting Trains. Ph.D. Thesis, Institutionen för Farkostteknik, Lund, Sweden, 2000.
105. Gangadharan, K.; Sujatha, C.; Ramamurti, V. Experimental and analytical ride comfort evaluation of a railway coach. In Proceedings of the SEM ORG IMAC XXII, 2004, Conf; pp. 1–15. Available online: https://www.semanticscholar.org/paper/Experimental-and-Analytical-Ride-Comfort-Evaluation-Gangadharan/671e4c731e0ec2b4da32b0482373d61531730a9d?p2df (accessed on 21 October 2020).
106. George, T.K.; Gadhia, H.M.; Sukumar, R.; Cabibihan, J.J. Sensing discomfort of standing passengers in public rail transportation systems using a smart phone. In Proceedings of the 2013 10th IEEE International Conference on Control and Automation (ICCA), Hangzhou, China, 12–14 June 2013; pp. 1509–1513.
107. Grant, P.R.; Haycock, B. Effect of jerk and acceleration on the perception of motion strength. *J. Aircr.* **2008**, *45*, 1190–1197. [CrossRef]
108. Hoberock, L.L. *A Survey of Longitudinal Acceleration Comfort Studies in Ground Transportation Vehicles*; Report; Council for Advanced Transportation Studies: Austin, TX, USA, 1976.
109. Huang, W.; Wong, P.K.; Zhao, J.; Ma, X. Output-feedback model-reference adaptive calibration for map-based anti-jerk control of electromechanical automotive clutches. *Int. J. Adapt. Control Signal Process.* **2018**, *32*, 265–285. [CrossRef]
110. Ikuhisa, K.; Lu, X.; Ota, T.; Hamada, H.; Kida, N.; Goto, A. Evaluation of Comfortable Using Jerk Method During Transfer Caring. In *Congress of the International Ergonomics Association*; Springer: Berlin/Heidelberg, Germany, 2018; pp. 178–188.
111. Ismail, K.; Susanto, B.; Sholahuddin, U.; Sabar, M. Design of Motor Control Electric Push-Scooter using Accelerometer as Jerk Sensor. In Proceedings of the 2019 International Conference on Sustainable Energy Engineering and Application (ICSEEA), Tangerang, Indonesia, 23–24 October 2019; pp. 69–73.
112. Khorram, B.; Af Wåhlberg, A.; Tavakoli Kashani, A. Longitudinal jerk and celeration as measures of safety in bus rapid transit drivers in Tehran. *Theor. Issues Ergon. Sci.* **2020**, *21*, 577–594. [CrossRef]
113. Kilinc, A.S.; Baybura, T. Determination of minimum horizontal curve radius used in the design of transportation structures, depending on the limit value of comfort criterion lateral jerk. In *TS06G-Engineering Surveying, Machine Control and Guidance*; Rome, Italy, 2012. Available online: https://www.fig.net/resources/proceedings/fig_proceedings/fig2012/papers/ts06g/TS06G_kilinc_baybura_5563.pdf (accessed on 21 October 2020).
114. Kruger, T.J. Graph-Based Speed Planning for Autonomous Driving. Master's Thesis, Free University of Berlin, Berlin, Germany, 2019.
115. Kushiro, I.; Suzuki, K. Mathematical model of skilled driver's steering pattern based on minimum jerk model. *Trans. Soc. Automot. Eng. Jpn.* **2016**, *47*, 1103–1110.
116. Lemos, N.A. On the least uncomfortable journey from A to B. *Eur. J. Phys.* **2019**, *40*, 055802. [CrossRef]
117. Liu, Y.; Zuo, L. Mixed skyhook and power-driven-damper: A new low-jerk semi-active suspension control based on power flow analysis. *J. Dyn. Syst. Meas. Control* **2016**, *138*, 081009. [CrossRef]
118. Martin, D.; Litwhiler, D.H. An Investigation of acceleration and jerk profiles of public transportation vehicles. In Proceedings of the ASEE Annual Conference and Exposition, Pittsburgh, PA, USA, 22–25 June 2008.
119. Reddy, G.V.P. Control Oriented Modeling of an Automotive Drivetrain for Anti-Jerk Control. Master's Thesis, Michigan Technological University, Houghton, MI, USA, 2018.
120. Scamarcio, A.; Metzler, M.; Gruber, P.; De Pinto, S.; Sorniotti, A. Comparison of anti-jerk controllers for electric vehicles with on-board motors. *IEEE Trans. Veh. Technol.* **2020**. [CrossRef]
121. Sharma, S.K.; Chaturvedi, S. Jerk analysis in rail vehicle dynamics. *Perspect. Sci.* **2016**, *8*, 648–650. [CrossRef]
122. Shigehiro, R.; Aguilar, G.D.; Kuroda, T. Evaluation method of seakeeping performance for training ships from the viewpoint of passenger comfort. *Fish. Sci.* **2002**, *68*, 1827–1830. [CrossRef]
123. Svensson, L.; Eriksson, J. Tuning for Ride Quality in Autonomous Vehicle: Application to Linear Quadratic Path Planning Algorithm. Master's Thesis, Uppsala University, Uppsala, Sweden, 2015.
124. Tawadros, P.; Awadallah, M.; Walker, P.; Zhang, N. Using a low-cost bluetooth torque sensor for vehicle jerk and transient torque measurement. *Proc. Inst. Mech. Eng. Part D J. Automob. Eng.* **2020**, *234*, 423–437. [CrossRef]

125. Tawadros, P.S. Powertrain Electrification for Jerk Reduction and Continuous Torque Delivery. Ph.D. Thesis, Universoty of Technology Sydney, Sydney, Australia, 2019.
126. Tomi, T. A study on ship vibration and oscillation limits from the viewpoint of unpleasant feelings of passengers (1st report-The susceptibility of human beings to motions). *Jpn. Soc. Nav. Archit. Ocean Eng.* **1961**, *104*, 18–30.
127. Vallee, P.; Robert, T. A numerical model to assess the risk of fall in public transportation—application to the influence of the Jerk in emergency braking. In Proceedings of the IRCOBI Conference, Lyon, France, 9–11 September 2015.
128. Van Santen, G.W. *Introduction to a Study of Mechanical Vibration: UDC No. 534.1: 621–752*; Philips Technical Library: London, UK, 1953. Available online: https://science.sciencemag.org/content/120/3109/179.2 (accessed on 21 October 2020).
129. Veerapaneni, N.V.M. Real-Time Minimum Jerk Optimal Trajectory Synthesis and Tracking for Ground Vehicle Applications. Master's Thesis, University of Texas, Austin, TX, USA, 2018.
130. Yamaguchi, A.; Ohishi, K.; Yokokura, Y.; Miyazaki, T.; Sasazaki, K. Backlash-based Shock Isolation Control for Jerk Reduction in Clutch Engagement. *IEEJ J. Ind. Appl.* **2019**, *8*, 160–169. [CrossRef]
131. Zeng, X.; Cui, H.; Song, D.; Yang, N.; Liu, T.; Chen, H.; Wang, Y.; Lei, Y. Jerk analysis of a power-split hybrid electric vehicle based on a data-driven vehicle dynamics model. *Energies* **2018**, *11*, 1537. [CrossRef]
132. Muszynska, A.; Bently, D. Frequency-swept rotating input perturbation techniques and identification of the fluid force models in rotor/bearing/seal systems and fluid handling machines. *J. Sound Vib.* **1990**, *143*, 103–124. [CrossRef]
133. Muszynska, A.; Bently, D.E. *Comments on Frequency Swept Rotating Input Perturbation Techniques and Identification of the Fluid Force Models in Rotor/bearing/seal Systems and Fluid Handling Machines*; Report; 1991. Available online: https://ntrs.nasa.gov/citations/19920005152 (accessed on 21 October 2020).
134. Inaudi, J.A.; Kelly, J.M. Hybrid isolation systems for equipment protection. *Earthq. Eng. Struct. Dyn.* **1993**, *22*, 297–313. [CrossRef]
135. Funakoshi, D.; Okada, S.; Watanabe, T.; Seto, K. Levitation and Vibration Supression of an Elastic Rotor by Using Active Magnetic Bearings. In Proceedings of the ASME 2012 5th Annual Dynamic Systems and Control Conference Joint with the JSME 2012 11th Motion and Vibration Conference, Fort Lauderdale, FL, USA, 17–19 October 2012; pp. 61–65.
136. Eager, D. Accelerometers used in the measurement of jerk, snap, and crackle. In Proceedings of the Australian Acoustical Society 2018 Annual Conference, Adelaide, Australia, 6–9 November 2018.
137. Kwon, B.M.; Moon, J.H.; Choi, H.D. Performance analysis of the GPS receiver under high acceleration and jerk environments. In Proceedings of the Korean Institute of Navigation and Port Research Conference, 2006; Volume 2, pp. 279–283. Available online: https://www.koreascience.or.kr/article/CFKO200636035497527.page (accessed on 21 October 2020)
138. Kwon, B.M.; Moon, J.H.; Choi, H.D.; Cho, G.R. Comparative Performance Analyses of GPS Receivers under High-Dynamic Conditions. In Proceedings of the 17th World Congress The International Federation of Automatic Control, Seoul, Korea, 6–11 July 2008; Volume 41, pp. 4725–4730.
139. Fazio, P.; Granieri, G.; Casetta, I.; Cesnik, E.; Mazzacane, S.; Caliandro, P.; Pedrielli, F.; Granieri, E. Gait measures with a triaxial accelerometer among patients with neurological impairment. *Neurol. Sci.* **2013**, *34*, 435–440. [CrossRef]
140. Lapinski, M.T. A Platform for High-Speed Bio-Mechanical Analysis Using Wearable Wireless Sensors. Ph.D. Thesis, Massachusetts Institute of Technology, Cambridge, MA, USA, 2013.
141. Aguirre, A. Evaluation of a Technological Device for Upper-Limb Motor Assessment. Ph.D. Thesis, San Francisco State University, San Francisco, CA, USA, 2016.
142. Zhang, L.; Diraneyya, M.M.; Ryu, J.; Haas, C.T. Jerk as an indicator of physical exertion and fatigue. *Autom. Constr.* **2019**, *104*, 120–128. [CrossRef]
143. Washington, J.K.; Oliver, G.D. Relationship of pelvis and torso angular jerk to hand velocity in female softball hitting. *J. Sport. Sci.* **2020**, *38*, 46–52. [CrossRef]
144. de Lucena, D.S.; Stoller, O.; Rowe, J.B.; Chan, V.; Reinkensmeyer, D.J. Wearable sensing for rehabilitation after stroke: Bimanual jerk asymmetry encodes unique information about the variability of upper extremity recovery. In Proceedings of the 2017 International Conference on Rehabilitation Robotics (ICORR), London, UK, 17–20 July 2017; pp. 1603–1608.

145. Zhang, L.; Diraneyya, M.M.; Ryu, J.; Haas, C.T.; Abdel-Rahman, E. Assessment of Jerk as a Method of Physical Fatigue Detection. In Proceedings of the ASME 2018 International Design Engineering Technical Conferences and Computers and Information in Engineering Conference, Quebec City, QC, Canada, 26–29 August 2018.
146. Zhang, L.; Diraneyya, M.; Ryu, J.; Haas, C.; Abdel-Rahman, E. Automated Monitoring of Physical Fatigue Using Jerk. In Proceedings of the International Symposium on Automation and Robotics in Construction, Banff, AB, Canada, 21–24 May 2019; Volume 36, pp. 989–997.
147. Zhang, L. Jerk as a Method of Identifying Physical Fatigue and Skill Level in Construction Work. Master's Thesis, University of Waterloo, Waterloo, ON, Canada, 2019.
148. Masuda, A.; Sone, A.; Matsuura, T. Development of Jerk Dot Sensor and Its Application to Condition Monitoring. In Proceedings of the Symposium on Evaluation and Diagnosis; pp. 118–121. Available online: https://www.jstage.jst.go.jp/article/jsmesed/2002.1/0/2002.1_118/_article/-char/ja/ (accessed on 21 October 2020).
149. Orsagh, R.; Brown, D. Sensor for Measuring Jerk and a Method for Use Thereof. U.S. Patent Application 11/191, 2 February 2006.
150. Xueshan, Y.; Xiaozhai, Q.; Lee, G.C.; Tong, M.; Jinming, C. Jerk and jerk sensor. In Proceedings of the 14th World Conference on Earthquake Engineering, Beijing, China, 12–17 October 2008.
151. Manabe, T.; Wakui, S. Production and Application of Horizontal Jerk Sensor. In Proceedings of the 2018 International Conference on Advanced Mechatronic Systems (ICAMechS), Zhengzhou, China, 30 August–2 September 2018; pp. 298–303.
152. Thompson, M. The application of motion capture to embodied music cognition research. *Jyväskylä Stud. Humanit.* **2012**. *176*, 146493785.
153. Jongejan, B.; Paggio, P.; Navarretta, C. Classifying head movements in video-recorded conversations based on movement velocity, acceleration and jerk. In Proceedings of the 4th European and 7th Nordic Symposium on Multimodal Communication (MMSYM 2016), Copenhagen, Denmark, 29–30 September 2016; pp. 10–17.
154. Sharker, S.I.; Holekamp, S.; Mansoor, M.M.; Fish, F.E.; Truscott, T.T. Water entry impact dynamics of diving birds. *Bioinspiration Biomimetics* **2019**, *14*, 056013. [CrossRef]
155. Hossain, M.; Hayati, H.; Eager, D. *A Comparison of the Track Shape of Wentworth Park and Proposed Murray Bridge*; Report; University of Technology Sydney: Sydney, Australia, 2016.
156. Hayati, H.; Eager, D.; Jusufi, A.; Brown, T. A Study of Rapid Tetrapod Running and Turning Dynamics Utilizing Inertial Measurement Units in Greyhound Sprinting. In Proceedings of the ASME 2017 International Design Engineering Technical Conferences and Computers and Information in Engineering Conference, Cleveland, OH, USA, 6–9 August 2017.
157. Hayati, H.; Eager, D.; Stephenson, R.; Brown, T.; Arnott, E. The impact of track related parameters on catastrophic injury rate of racing greyhounds. In Proceedings of the 9th Australasian Congress on Applied Mechanics, Sydney, Australia, 27–29 November 2017; p. 311.
158. Hayati, H.; Walker, P.; Mahdavi, F.; Stephenson, R.; Brown, T.; Eager, D. A comparative study of rapid quadrupedal sprinting and turning dynamics on different terrains and conditions: Racing greyhounds galloping dynamics. In Proceedings of the ASME 2018 International Mechanical Engineering Congress and Exposition, Pittsburgh, PA, USA, 11–14 November 2018.
159. Mahdavi, F.; Hossain, M.I.; Hayati, H.; Eager, D.; Kennedy, P. Track Shape, Resulting Dynamics and Injury Rates of Greyhounds. In Proceedings of the ASME 2018 International Mechanical Engineering Congress and Exposition, Pittsburgh, PA, USA, 11–14 November 2018; p. V013T05A018.
160. Hossain, I.; Eager, D.; Walker, P. Greyhound racing ideal trajectory path generation for straight to bend based on jerk rate minimization. *Sci. Rep.* **2020**, *10*, 7088. [CrossRef]
161. Malushte, S.R. Prediction of Seismic Design Response Spectra Using Ground Characteristics. Ph.D. Thesis, Virginia Tech, Blacksburg, VA, USA, 1987.
162. Bertero, R.D.; Bertero, V.V. Performance-based seismic engineering: The need for a reliable conceptual comprehensive approach. *Earthq. Eng. Struct. Dyn.* **2002**, *31*, 627–652. [CrossRef]
163. Geoffrey Chase, J.; Barroso, L.R.; Hunt, S. Quadratic jerk regulation and the seismic control of civil structures. *Earthq. Eng. Struct. Dyn.* **2003**, *32*, 2047–2062. [CrossRef]
164. Tong, M.; Wang, G.Q.; Lee, G.C. Time derivative of earthquake acceleration. *Earthq. Eng. Eng. Vib.* **2005**, *4*, 1–16. [CrossRef]

165. He, H.; Li, R.; Chen, K. Characteristics of jerk response spectra for elastic and inelastic systems. *Shock Vib.* **2015**, *2015*. Available online: https://www.hindawi.com/journals/sv/2015/782748/ (accessed on 21 October 2020). [CrossRef]
166. Chakraborty, S.; Ray-Chaudhuri, S. Energy Transfer to High-Frequency Modes of a Building due to Sudden Change in Stiffness at Its Base. *J. Eng. Mech.* **2017**, *143*, 04017050. [CrossRef]
167. Taushanov, A. *Jerk Response Spectrum*; Report; 2018. Available online: https://www.researchgate.net/profile/Alexander_Taushanov/publication/323150255_Jerk_Response_Spectrum/links/5a82e5040f7e9bda86a00d81/Jerk-Response-Spectrum.pdf (accessed on 21 October 2020).
168. Papandreou, I.; Papagiannopoulos, G. On the jerk spectra of some inelastic systems subjected to seismic motions. *Soil Dyn. Earthq. Eng.* **2019**, *126*, 105807. [CrossRef]
169. Yaseen, A.A.; Ahmed, M.S.; Al-Kamaki, Y.S.S. Jerk Performance as Seismic Intensity Measure. In Proceedings of the 3rd International Conference on Recent Innovations in Engineering, Duhok, Iraq, 9–10 September 2020.
170. Sofronie, R. On the Seismic Jerk. *J. Geol. Resour. Eng. Comput.* **2017**, *4*, 147–152. [CrossRef]
171. Savage, J. The Science of Running Shoes. Technical Report. Available online: http://fellrnr.com/wiki/The_Science_of_Running_Shoes (accessed on 20 October 2020).
172. Hayati, H.; Eager, D.; Walker, P. An impact attenuation surfacing test to analyse the dynamic behaviour of greyhound racetrack sand surface. In Proceedings of the World Engineers Convention, Melbourne, Australia, 20–22 November 2019; p. 391.
173. Sone, A.; Masuda, A.; Matsuura, T.; Yamamura, T.; Yamada, M.; Yamamoto, S. Detection of structural damages by jerk-dot sensors. *Trans. Nippon Kikai Gakkai Ronbunshu C Hen* **2004**, *16*, 1318–1323.
174. Zhang, Z.; Verma, A.; Kusiak, A. Fault analysis and condition monitoring of the wind turbine gearbox. *IEEE Trans. Energy Convers.* **2012**, *27*, 526–535. [CrossRef]
175. Iyama, J.; Wakui, M. Threshold value and applicable range of nonlinear behavior detection method using second derivative of acceleration. *Jpn. Archit. Rev.* **2019**, *2*, 153–165. [CrossRef]
176. Sumathy, M.; Kilicman, A.; Manuel, M.M.S.; Mary, J. Qualitative study of Riccati difference equation on maneuvering target tracking and fault diagnosis of wind turbine gearbox. *Cogent Eng.* **2019**, *6*, 1621423. [CrossRef]
177. Ostermeyer, G.P.; Schiefer, F. On Pain Detection in Multibody Systems. In *Applied Mechanics and Materials*; Trans Tech Publications: Stafa-Zurich, Switzerland, 2008; Volume 9, pp. 115–126. Available online: https://www.scientific.net/AMM.9.115 (accessed on 21 October 2020)
178. Fiori, S.; Sabino, N.; Bonci, A. In-Lab Drone's Attitude Maneuvering Fluency Evaluation by a Gyroscopic Lurch Index. In Proceedings of the 10th International Conference on Circuits, Systems, Signals and Telecommunications, Barcelona, Spain, 13–15 February 2016 pp. 37–46.
179. Guye, K. Development of an Indoor Multirotor Testbed for Experimentation on Autonomous Guidance Strategies. Master's Thesis, South Dakota State University, Brookings, SD, USA, 2018.
180. Lai, S.P.; Lan, M.l.; Li, Y.x.; Chen, B.M. Safe navigation of quadrotors with jerk limited trajectory. *Front. Inf. Technol. Electron. Eng.* **2019**, *20*, 107–119. [CrossRef]
181. Luukkonen, T. Modelling and control of quadcopter. *Indep. Res. Proj. Appl. Math.* **2011**, *22*. Available online: https://sal.aalto.fi/publications/pdf-files/eluu11_public.pdf (accessed on 21 October 2020).
182. Phang, S.K.; Lai, S.; Wang, F.; Lan, M.; Chen, B.M. Systems design and implementation with jerk-optimized trajectory generation for UAV calligraphy. *Mechatronics* **2015**, *30*, 65–75. [CrossRef]
183. Rakgowa, T.; Wong, E.K.; Sim, K.S.; Nia, M. Minimal jerk trajectory for quadrotor VTOL procedure. In Proceedings of the 2015 IEEE International Symposium on Robotics and Intelligent Sensors (IRIS), Langkawi, Malaysia, 18–20 October 2015; pp. 284–287.
184. Rousseau, G.; Maniu, C.S.; Tebbani, S.; Babel, M.; Martin, N. Quadcopter-performed cinematographic flight plans using minimum jerk trajectories and predictive camera control. In Proceedings of the 2018 European Control Conference (ECC), Limassol, Cyprus, 12–15 June 2018; pp. 2897–2903.
185. Silva, J.P.; De Wagter, C.; de Croon, G. Quadrotor Thrust Vectoring Control with Time and Jerk Optimal Trajectory Planning in Constant Wind Fields. *Unmanned Syst.* **2018**, *6*, 15–37. [CrossRef]
186. Tal, E.; Karaman, S. Accurate tracking of aggressive quadrotor trajectories using incremental nonlinear dynamic inversion and differential flatness. In Proceedings of the 2018 IEEE Conference on Decision and Control (CDC), Miami Beach, FL, USA, 17–19 December 2018; pp. 4282–4288.

187. Perri, S.; Bianco, C.G.L.; Locatelli, M. Jerk bounded velocity planner for the online management of autonomous vehicles. In Proceedings of the 2015 IEEE International Conference on Automation Science and Engineering (CASE), Gothenburg, Sweden, 24–28 August 2015; pp. 618–625.
188. Raineri, M.; Bianco, C.G.L. Jerk limited planner for real-time applications requiring variable velocity bounds. In Proceedings of the 2019 IEEE 15th International Conference on Automation Science and Engineering (CASE), Vancouver, BC, Canada, 22–26 August 2019; pp. 1611–1617.
189. Bisoffi, A.; Biral, F.; Da Lio, M.; Zaccarian, L. Longitudinal jerk estimation for identification of driver intention. In Proceedings of the 2015 IEEE 18th International Conference on Intelligent Transportation Systems, Las Palmas, Spain, 15–18 September 2015; pp. 1855–1861.
190. Bisoffi, A.; Biral, F.; Da Lio, M.; Zaccarian, L. Longitudinal jerk estimation of driver intentions for advanced driver assistance systems. *IEEE/ASME Trans. Mechatron.* **2017**, *22*, 1531–1541. [CrossRef]
191. Feng, F.; Bao, S.; Sayer, J.R.; Flannagan, C.; Manser, M.; Wunderlich, R. Can vehicle longitudinal jerk be used to identify aggressive drivers? An examination using naturalistic driving data. *Accid. Anal. Prev.* **2017**, *104*, 125–136. [CrossRef]
192. Hasan, N. Relationship between Twist, Jerk, and Speed: Twist-Tolerance Values and Measuring Chords. *J. Transp. Eng. Part A Syst.* **2020**, *146*. 04020005. [CrossRef]
193. Itkonen, T.H.; Pekkanen, J.; Lappi, O.; Kosonen, I.; Luttinen, T.; Summala, H. Trade-off between jerk and time headway as an indicator of driving style. *PLoS ONE* **2017**, *12*, e0185856. [CrossRef]
194. Mousavi, S.M. Identifying High Crash Risk Roadways through Jerk-Cluster Analysis. Master's Thesis, Louisiana State University, Baton Rouge, LA, USA, 2015.
195. Weidner, W.; Transchel, F.W.; Weidner, R. Telematic driving profile classification in car insurance pricing. *Ann. Actuar. Sci.* **2017**, *11*, 213–236. [CrossRef]
196. Szabó, I. *Geschichte der Mechanischen Prinzipien und Ihrer Wichtigsten Prinzipien*; Birkhäuser: Basel, Switzerland, 1977.
197. Dugas, R. *Histoire de la Mechanique*; Éditions de Griffon: Neuchâtel, Switzerland, 1955.
198. Dugas, R. *History of Mechanics, translation by John Maddox of Histoire de la Mechanique*; Dover: London, UK, 1988.
199. Wikipedia Contributors. Pierre Varignon—Wikipedia, The Free Encyclopedia. 2018. Available online: https://en.wikipedia.org/wiki/Pierre_Varignon (accessed on 28 May 2020).
200. Melchior, P. *Zeitschfrift Ver. Deutsche Ing.* **1928**, *72*, 1842.
201. Schlobach. Gluckauf. *Berg- und Huttenmännischen Zeitschrift* **1928**, *73*, 16.
202. Whinnery, T.; Forster, E.M. The+ Gz-induced loss of consciousness curve. *Extrem. Physiol. Med.* **2013**, *2*, 19. [CrossRef] [PubMed]
203. McKenney, W.R. *Human Tolerance to Abrupt Accelerations: A Summary of the Literature*; Report; Dynamic Science Inc. Avser Facility: Phoenix, AZ, USA, 1970.
204. Coats, T.W.; Riley, M.R. *A Comparison of Shock Isolated Seat and Rigid Seat Acceleration Responses to Wave Impacts in a High-Speed Craft*; Report; Naval Surface Warfare Center: Carderock, VA, USA, 2018.
205. Riley, M.R.; Haupt, K.D.; Ganey, H.C.; Coats, T.W. *Laboratory Test Requirements for Marine Shock Isolation Seats*; Report; Naval Surface Warfare Center Carderock Division: Norfolk, VA, USA, 2018.
206. AASHTO-2001. *American Association of State Highway and Transportation Officials*; Washington, DC, USA, 2001. Available online: https://www.transportation.org/ (accessed on 21 October 2020).
207. Wasserstein, R.L.; Lazar, N.A. The ASA statement on p-values: Context, process, and purpose. *Am. Stat.* **2016**, *70*, 129–133. [CrossRef]

Publisher's Note: MDPI stays neutral with regard to jurisdictional claims in published maps and institutional affiliations.

Article

A Comparison of Time-Frequency Methods for Real-Time Application to High-Rate Dynamic Systems

Jin Yan [1,*], Simon Laflamme [1], Premjeet Singh [2], Ayan Sadhu [2] and Jacob Dodson [3]

1 Department of Civil, Construction, and Environmental Engineering, Iowa State University, Ames, IA 50010, USA; laflamme@iastate.edu;

2 Department of Civil and Environmental Engineering, Western University, London, ON N6A 3K7, Canada; psing225@uwo.ca (P.S.); asadhu@uwo.ca (A.S.)

3 Air Force Research Laboratory, Eglin AFB, FL 32542, USA; jacob.dodson.2@us.af.mil

* Correspondence: yanjin@iastate.edu

Received: 24 July 2020; Accepted: 21 August 2020; Published: 24 August 2020

Abstract: High-rate dynamic systems are defined as engineering systems experiencing dynamic events of typical amplitudes higher than 100 g_n for a duration of less than 100 ms. The implementation of feedback decision mechanisms in high-rate systems could improve their operations and safety, and even be critical to their deployment. However, these systems are characterized by large uncertainties, high non-stationarities, and unmodeled dynamics, and it follows that the design of real-time state-estimators for such purpose is difficult. In this paper, we compare the promise of five time-frequency representation (TFR) methods at conducting real-time state estimation for high-rate systems, with the objective of providing a path to designing implementable algorithms. In particular, we examine the performance of the short-time Fourier transform (STFT), wavelet transformation (WT), Wigner–Ville distribution (WVD), synchrosqueezed transform (SST), and multi-synchrosqueezed transform (MSST) methods. This study is conducted using experimental data from the DROPBEAR (Dynamic Reproduction of Projectiles in Ballistic Environments for Advanced Research) testbed, consisting of a rapidly moving cart on a cantilever beam that acts as a moving boundary condition. The capability of each method at extracting the beam's fundamental frequency is evaluated in terms of precision, spectral energy concentration, computation speed, and convergence speed. It is found that both the STFT and WT methods are promising methods due to their fast computation speed, with the WT showing particular promise due to its faster convergence, but at the cost of lower precision on the estimation depending on circumstances.

Keywords: high-rate dynamics; structural health monitoring; time-frequency analysis; synchrosqueezing transform (SST)

1. Introduction

High-rate dynamic systems are defined as engineering systems experiencing high-amplitude disruptions (acceleration >100 g_n) within a very short duration (<100 ms) [1]. Examples of high-rate systems include blast mitigation mechanisms, hypersonic aircraft, and advanced weaponry. Generally, these systems experience rapid changes in their dynamics that can cause malfunctions and sudden failures. The capability to conduct real-time identification of such changes combined with real-time adaptation is critical in ensuring their continuous operation and safety [2]. In terms of system identification, the high-rate problem consists of being able to identify and quantify changes in dynamics over a very short period of time for dynamics containing (1) large uncertainties in the external loads; (2) high levels of non-stationarities and heavy disturbance, and (3) unmodeled dynamics from changes in system configuration.

Work directly addressing the problem of high-rate state estimation, or system identification is limited. In previous work, the authors have proposed an adaptive sequential neural network with a self-adapting input space enabling fast learning of nonstationary signals from high-rate systems [3]. Although the data-based technique showed great promise at high rate state estimation, it did not provide insight into the system's physical characteristics, as it is generally the case with data-based techniques. Physics-driven methods, such as those borrowing on model reference adaptive system (MRAS) theory, showed great promise in handling nonlinearities, uncertainties, and perturbations [4,5]. MRAS was applied to the problem of high-rate state estimation in [6], where the position of a moving cart was accurately identified under 172 ms through a time-based adaptive algorithm used in reaching the reference model with an average computing time of 93 μs per step, obtained through numerical simulations conducted in MATLAB. A frequency-based approach was proposed in [7] to identify the position of that same cart. Their algorithm consisted of extracting the dominating frequency through a Fourier transform over a finite set of data and matching that frequency to a set of pre-analyzed finite element models. The authors applied their algorithm experimentally using a field-programmable gate array (FPGA), and were able to accurately identify the position of the cart within 202 ms with a 4.04 ms processing time per step.

A net advantage of frequency-based methods over time-based methods is that they do not typically rely on the tuning of parameters such as adaptive gains. However, they are harder to apply in real time because they are inherently batch processing techniques. It follows that, to enable applications to high-rate systems, one must integrate a temporal approach to the frequency technique in order to extract the required real-time information, a method known as time-frequency representation (TFR). For example, this was done in [7] through the use of a non-overlapping sliding window of 198 ms length. The objective of this paper is to explore the applicability of TFRs to the high-rate problem.

TFRs are widely used for the detection and quantification of faults through vibration-based data [8]. Frequency domain characteristics, such as frequencies, damping ratios, energy in different frequency ranges, and time-frequency domain characteristics, such as time-frequency spread [9], can be used as key features to conduct structural health monitoring [9]. A number of TFRs for instant frequency recognition have been proposed. Popular approaches include linear non-parametric methods, such as short-time Fourier transform (STFT), wavelet transform (WT), and Wigner–Ville distribution (WVD) [9,10]. The application of these methods results in a trade-off between time and frequency resolutions [11]. An adaptive non-parametric method have also been proposed, including the Hilbert–Huang transform (HHT) [12–14], the Cauchy continuous wavelet transform (CCWT) [15], the instantaneous frequencies (IF) re-assignment methods, synchrosqueezing transform (SST) [16], and the multi-SST (MSST) [17].

In this paper, five of these methods are selected, namely the STFT, WT, WVD, SST, and MSST, and their real-time applicability are compared with a specific focus on weakly time-varying systems, here defined as systems with continuously time-varying frequencies, in opposition to abrupt changes (steps, jumps, shifts) in frequencies [18,19]. The objective is to provide a path in designing the next generation of real-time high-rate algorithms. The quantification of performance is conducted on experimental data from the DROPBEAR (Dynamic Reproduction of Projectiles in Ballistic Environments for Advanced Research) testbed [20], which includes the moving cart dynamics used in [6,7]. The remainder of the paper is organized as follows. First, the background on the five TFR methods is presented—after DROPBEAR is introduced and the performance of the methods is analyzed numerically on two different sets of experimental data. Lastly, the performance of each method is compared, and key conclusions are drawn.

2. Time-Frequency Response Methods

This section gives an introductory background on the TFR methods under comparison. These include the following traditional TFR methods: STFT, WT, WVD, and TF re-assignment methods: SST and MSST.

2.1. Traditional TFR Methods

2.1.1. Short-Time Fourier Transform

The Fourier transform of a sequence of an N discretely sampled signal $x[n]$ to the discrete frequency domain $X[k]$ is taken as:

$$X[k] = \sum_{n=0}^{N-1} x[n]e^{-j2\pi nk/N} \quad n = 0, 1, 2, ..., N-1 \tag{1}$$

where j is the imaginary unit, and k is the corresponding frequency. To integrate a temporal notion, the Fourier transform can be applied over short time segments through a moving window, where the signal can be assumed stationary between two consecutive segments, a method known as STFT. The local Fourier spectrum of each segment can be generated around the position of the window, and the frequency's temporal variation can be observed locally. This is done by modifying Equation (1) as follows [21,22]:

$$\text{STFT}_x[m, k] = \sum_{n=0}^{L-1} x[n]g[n-m]e^{-j2\pi nk/N} \tag{2}$$

where $g[n]$ is the windowing function of length L, and m denotes a time shift. With the STFT, a narrower window will improve the time domain resolution but will result in a lower frequency domain resolution. Conversely, a wider window will improve the frequency domain resolution but will result in a lower time domain resolution.

2.1.2. Wavelet Transform

The WT method is known for its superior spectral resolution by overcoming the STFT's requirement of predefining a window length. It provides a linear time-frequency representation based on a preselected mother wavelet ψ using simultaneous dilation and translation operations. The discretized version of the continuous WT of a signal $x[n]$ is written [23,24]:

$$\text{WT}_\psi[m, k] = \frac{1}{\sqrt{c}} \sum_{n=0}^{L-1} x[n]\psi\left[\left(\frac{n}{c} - m\right) T\right] \tag{3}$$

where c is scaling factor. Generally, the WT method has better temporal resolution and lower frequency resolution for higher frequency contents, and better frequency resolution and lower temporal resolution for lower frequency contents. The resulting time-frequency signal transform may be blurred and cannot achieve high resolution simultaneously in time and frequency.

2.1.3. Wigner–Ville Distribution

The WVD method is an approach based on the quadratic energy density obtained through an instantaneous autocorrelation function [25]:

$$\text{WVD}_x[m, k = \sum_{n} x\left[n + m/2\right] x^*\left[n - m/2\right] e^{-j2\pi mk/N} \tag{4}$$

where the asterisk denotes the complex conjugate. Challenges in applying the WVD include interferences and negative values [26]. There are times when cross-terms produce oscillatory interference with multiple frequency components, and the magnitude of the interference may range from extremely negative to extremely positive values.

2.2. Time-Frequency Reassignment

2.2.1. Synchrosqueezing Transform

The SST method is time-frequency reallocation method that yields finer time-frequency representations for a non-stationary multi-component signal. With SST, the energies of the time-frequency coefficients are reassigned to achieve higher energy around the trajectories of IF, resulting in more accurate tracking of these IFs [27]. The SST representation is conducted using:

$$\mathrm{SST}_{\phi}[m,k] = \frac{1}{\Delta k}\sum_{c_m} \mathrm{WT}_{\psi}[m,k]c^{-3/2}\Delta c_m \tag{5}$$

where c_m is the discrete scale for which the wavelet decomposition $\mathrm{WT}_{\psi}[m,k]$ is computed, and $\Delta c_m = c_{m-1} - c_m$ is a scaling step. The SST results in a more concentrated profile and unique IF curves but is inherently more computationally intensive comparing to the WT method. A limitation of the SST method is that it assumes a weakly time-varying system.

2.2.2. Multi-Synchrosqueezing Transform

The MSST method is an improvement of the SST that results in a sharper energy concentration of the TFR [17]. It does not require any other redundant parameter or a priori information to demodulate the frequency-modulated signals, and thus it can be applied beyond weakly time-varying systems. MSST is formulated to post-processes STFT:

$$\mathrm{MSST}^{1}[m,k] = \mathrm{STFT} - \mathrm{SST}_{x}[m,k] = \sum_{n=0}^{N-1} \mathrm{STFT}_{x}[m,k]\delta[k - \hat{k}[m,k]] \tag{6}$$

Multiple iterations of the process can be conducted with

$$\mathrm{MSST}^{N}[m,k] = \sum_{n=0}^{N-1} \mathrm{MSST}^{[N-1]}[m,k]\delta[k - \hat{k}[m,k]] \tag{7}$$

where $\mathrm{MSST}^{N}[m,k]$ is the SST at the Nth iteration for $N > 2$. Increasing N will yield better IF estimates, but at the cost of higher computational time.

3. Methodology

This section introduces the research methodology. It includes a description of the experimental setup, data collection process, and TFR performance investigation methodology.

3.1. Experiment Setup

DROPBEAR is an experimental testbed designed to conduct reproducible high-rate dynamic responses [20]. Shown in Figure 1, the testbed features a cantilever steel beam with a rolling cart moving along the beam. The movable cart functions as a variable pin, used to mimic sudden or gradual changes in stiffness due to a change in boundary conditions. The beam was equipped with one PCB 353B17 accelerometer connected to the beam located 300 mm away from the clamp to measure the beam's response to the moving cart. A modal hammer PCB 086C01 was used to excite the beam at the tip. Figure 1 also shows an electromagnet that can be used to simulate a sudden change in mass through a controlled drop. That feature was not utilized in generating test data used in this paper.

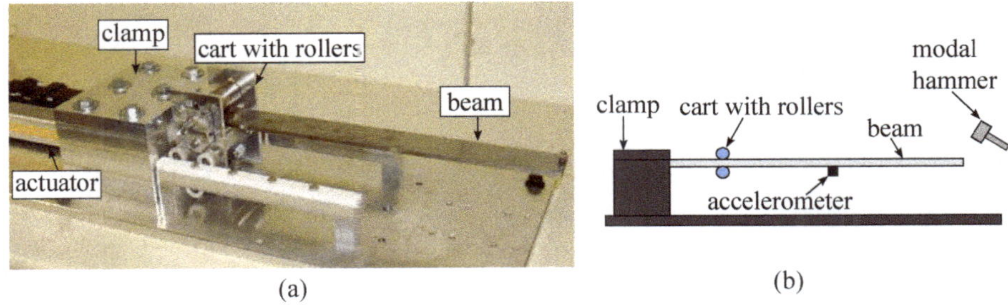

Figure 1. DROPBEAR testbed: (**a**) picture and (**b**) schematic of the setup.

Two types of experiments were conducted. First, the beam was excited with the cart fixed at various locations (i.e., "fixed cart configuration"): positions A, B, C, D (respectively 50 mm, 100 mm, 150 mm, and 200 mm away from the clamp). For each location, the cart was maintained in place for 2 s and the beam impacted at 0.5, 2.5, 4.5, and 6.5 s (Figure 2a). Data from the accelerometer were sampled at 1 kHz and used to compute frequency response functions (FRFs) using the H_1 estimation method [28] plotted in Figure 2c and extract the fundamental natural frequencies at 26.5, 31, 38, and 47.5 Hz (Positions A, B, C, and D, respectively). The fixed cart configuration was used to examine the performance of each TFR method over the entire dataset, without the use of sliding windows (except for the STFT). Second, the cart was moved (i.e., "moving cart configuration") starting at 50 mm from the clamp at 0.5 s to 200 mm from the clamp over 1 s, stayed for 1.39 s, and then returned to the initial position at 4.26 s. The acceleration was sampled at 25 kHz. The time series response of the beam is plotted in Figure 2b. The moving cart configuration was used to examine the real-time applicability of methods through the use of sliding or non-overlapping sliding windows. The type and size of windows, along with the type of wavelets, were selected heuristically to obtain the best overall performance.

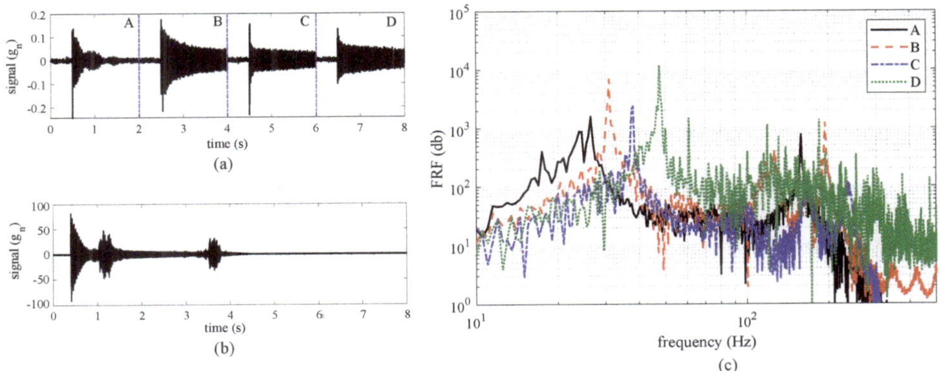

Figure 2. (**a**) signal from the fixed cart configuration at 50, 100, 150, and 200 mm; (**b**) signal from moving cart configuration between 50 mm to 200 mm; and (**c**) frequency response functions (FRFs) for DROPBEAR with fixed cart positions at 50, 100, 150, and 200 mm.

3.2. Performance Analysis

Analyses were performed in MATLAB 2019b, with an Intel(R) Core(TM) i7-7700 CPU 3.6 GHz. The performance of the TFR methods in the fixed cart configuration is assessed as a function of four performance metrics (J_1 to J_4). Metric J_1 is the mean absolute error between the estimated $\hat{\omega}$ and real ω

frequency over the length n of the signal. Metric J_2 is the energy concentration of the TFR using Renyi entropy [29]. Metric J_3 is the computation time per window. Metric J_4 is the mean convergence time when the estimation error falls and remains within an error threshold, here taken as 5%. Metrics J_1 and J_2 can be expressed mathematically as follows:

$$\begin{aligned} J_1 &= \sum_{i=1}^{n} \frac{|\hat{\omega}_i - \omega_i|}{n} \\ J_2 &= \iint \log|\text{TFR}(t,\omega)^3| dt d\omega \end{aligned} \tag{8}$$

In terms of performance assessment, small values for J_1, J_3, and J_4 are desired, while a high value for J_2 is desired. The performance of the TFR methods in the moving cart configuration is only assessed using performance metrics J_1 and J_3, given that the problem of interest is frequency tracking.

4. Results and Discussion

This section presents results from the fixed and moving cart configurations and discusses the performance of the different TFR methods for applicability to high-rate system identification.

4.1. Fixed Cart Configuration

A parametric study was first conducted to study the influence of TFR parameters and select the optimal parameters in comparing performance across TFRs. The study starts with the STFT, where the window length and type are investigated. In the investigation, the window size ranged from 128 to 768 at an interval size of 64, window length overlaps were a half and a quarter of the window length, and windowing functions were Hanning, Gaussian, and Blackman, as shown in Figure 3. Figure 4a plots the results. To facilitate the comparison, the four metrics were normalized to the highest value at 1. It can be observed that both J_1 and J_2 converge after a window length of 512. In addition, under the J_2 and J_3 metrics, the half overlapped window performs better than the quarter overlapped windows. Under J_4, the half overlapped Gaussian window performs similarly to the quarter overlapped windows. However, the other half overlapped window functions generally perform worse, and the window length of 512 appears to yield optimal results. From these results, a window length of 512 with a half-overlapping Gaussian windowing function was selected as the optimal set of parameters.

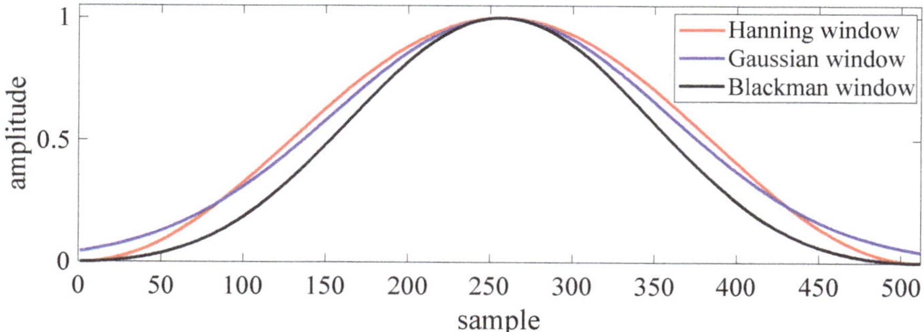

Figure 3. Window functions used in STFT.

The parametric study for the WT consisted of evaluating the performance under different wavelets, including the Morse, Morlet, and Bump wavelets [30,31]. Figure 4a plots the results. Results show a similar performance across all wavelet types, with slightly better performance for the Morlet wavelet observable under J_3. The Morlet wavelet was selected as the optimal parameter For the MSST, the number of 2 to 5 iterations were studied. Results were also similar across all iteration numbers, and thus a total of two iterations were selected as the optimal parameter.

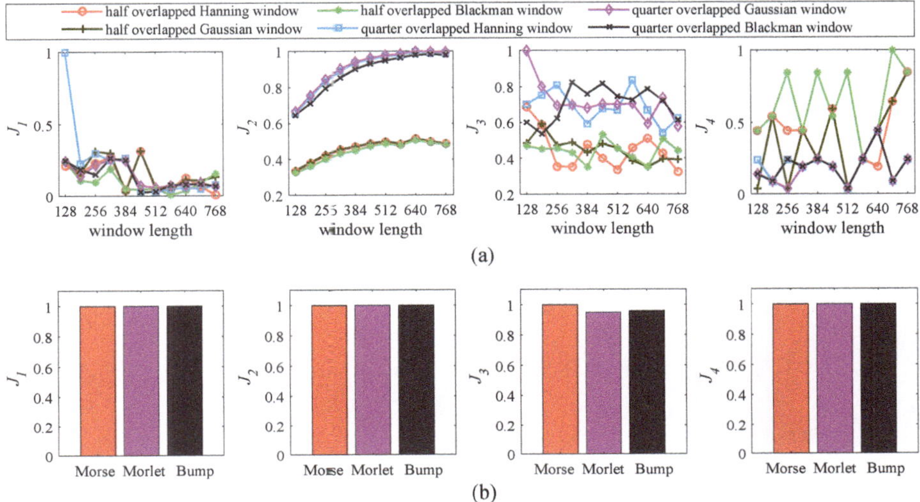

Figure 4. Parametric investigation for (**a**) STFT; and (**b**) WT.

Results from the fixed cart configuration experiment using the selected optimal parameters are shown in Figure 5. Figure 5a is the time series response over 2 s under each location. Figure 5b–f show the time-frequency content extracted by the STFT, WT, WVD, SST, and MSST methods, respectively. Ridge detection is used to identify the first natural frequency (shown as a solid red line) by extracting the maximum-energy time-frequency ridge of the spectrograms [32]. The measured frequencies from the FRF (Figure 2c) are also shown in blue dashed lines. Table 1 lists performance under metrics J_1–J_4.

Visual observation of the time-frequency plane (Figure 5b–f) shows that the WVD provides the best frequency identification at position A, followed by the SST, while the WT yielded high variance in results. This may be attributed to the weaker acceleration response from the beam compared to other positions (Figure 5a). Over other positions, all methods appear to have identified a stable frequency, with the SST and MSST converging faster than the other methods, followed by the STFT. An examination of the performance metrics listed in Table 1 confirms these observations. The SST provided the most precise estimate of frequency, while the WT was the least precise (J_1). The SST and MSST methods showed the highest energy concentrations (J_2), as expected from time-frequency reassignment methods. However, their computation time (J_3) was significantly higher compared to the other methods, with the STFT and WT being the fastest. In terms of convergence (J_4), the SST and MSST were the fastest, while the STFT was the slowest. From the static cart experiment results, it appears that the WT is one of the most applicable methods to the high-rate problem given its fast convergence speed and lower computation time, although at the cost of precision on the estimation of frequency over weak signals.

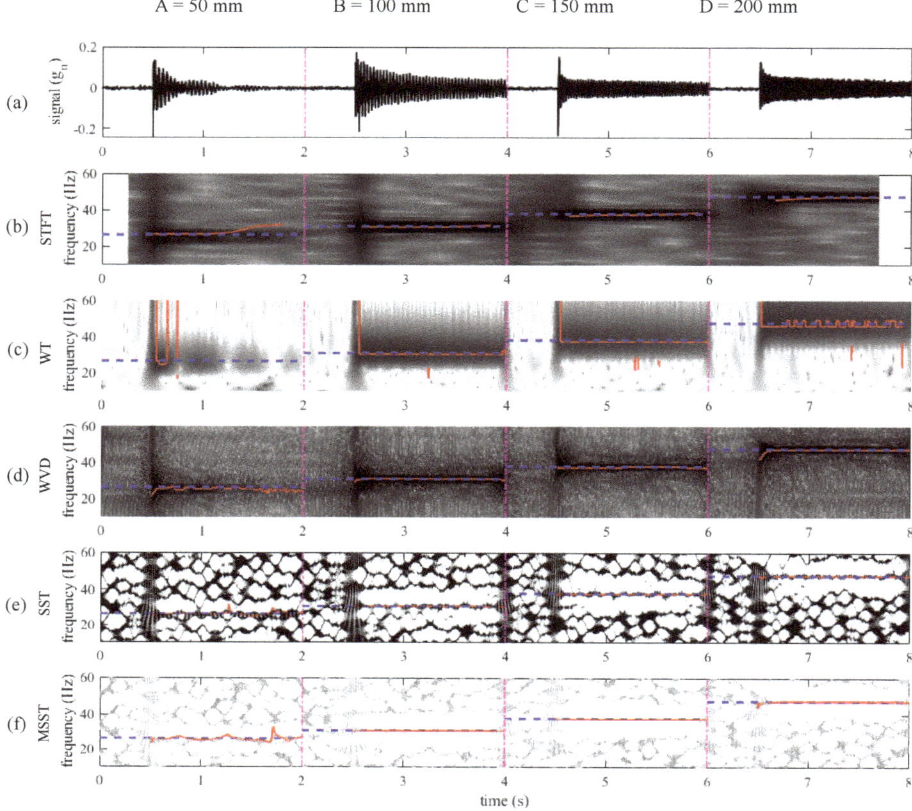

Figure 5. Time-frequency planes from the fixed cart configuration obtained using the (**a**) signal from the fixed cart configuration at 50, 100, 150, and 200 mm; (**b**) STFT; (**c**) WT; (**d**) WVD; (**e**) SST; and (**f**) MSST method with extracted frequencies (red solid lines) and estimated true frequencies (dashed blue lines).

Table 1. Fixed cart configuration time-frequency analysis comparison.

TFR Method	J_1 **(Hz)**	J_2 **(**$/J_{2\max}$**)**	J_3 **(ms)**	J_4 **(ms)**
STFT	0.384	0.012	6.9	286
WT	0.729	0.0220	242	44
WVD	0.316	0.0573	1093	51
SST	0.095	1	10,505	1
MSST	0.155	0.6667	12,236	1

4.2. Moving Cart Configuration

Time-frequency planes obtained from the moving cart experiment are shown in Figure 6. Results also show the estimated true temporal variation of the beam's fundamental frequency (in dashed blue), obtained by assuming linearity of the system and interpolating between the measured frequencies from the FRF (Figure 2c) at the 50 mm and 200 mm positions. The STFT and WT were conducted with a sliding window of length 4096, corresponding to 164 ms, and the overlap size is half of the window length. Data were down-sampled from 25 kHz to 1.25 kHz to conduct the WVD in order to reduce the computational burden by maintaining a low-size window of 512 data points and improve the frequency resolution to bin sizes of 2.4 Hz, instead of 12.2 Hz under 25 kHz. The SST and MSST were processed with a non-overlapping sliding window size of 1024 data points at a reduced sampling rate of 1.25 kHz, corresponding to a duration of 819 ms. The WT was conducted using Morse wavelets, the SST used a Gaussian window, and the MSST was iterated five times, as for the fixed cart configuration. Table 2 lists results for performance metrics J_1 and J_3, along with a summary of the processing window lengths used in the study.

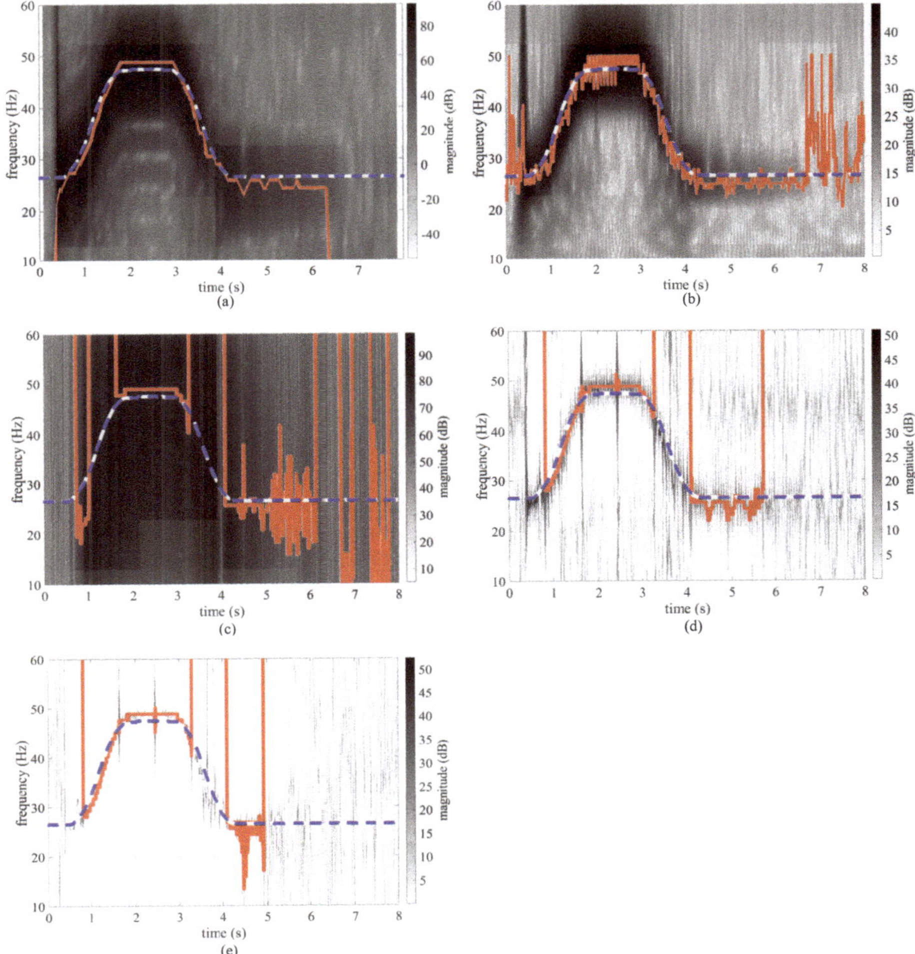

Figure 6. Time-frequency planes from the fixed cart configuration obtained using the (**a**) STFT; (**b**) WT; (**c**) WVD; (**d**) SST; and (**e**) MSST method with extracted frequencies (red solid lines) and estimated true frequencies (dashed blue lines).

A visual comparison of the fundamental frequency extracted by the TFR methods (solid red line) with the estimated true frequency (dashed blue line) in Figure 6 shows that the SFTF provided the most precise estimation of the beam's fundamental frequency that can be linked to the cart's location, followed by the WT. The WT showed more chattering in the results, but with a better adaptation to the varying frequency. This is confirmed by performance metric J_1 (Table 2), which also shows that the SST underperformed with respect to the other TFR methods. The computation time per iteration (J_3) was significantly faster for the STFT and WT methods, smaller than the window hopping time (82 ms). For the WVD, the down-sampling strategy enabled a computation time of 262 ms per window, instead of approximately 10 s using a window size of 2048 data points. However, despite such improvement in the frequency resolution and computational time, the WVD failed at identifying the beam frequency during the movement of the cart, as observable in Figure 6. The MSST's computation time is significantly longer than for the other methods, attributable to the longer window lengths that were necessary in implementing the methods.

Table 2. Moving cart configuration time-frequency analysis comparison.

TFR Method	J_1 (Hz)	J_3 (ms)	Window (ms)
STFT	0.54	3.8	82
WT	0.56	9.3	82
WVD	1.88	262	819
SST	0.91	288	819
MSST	0.66	543	819

Results from the moving cart experiment showed that both the STFT and WT were adequate methods through their fast computational speed and acceptable precision on the frequency estimation. The three other methods did not provide adequate performance in terms of computation time. Moreover, the WVD did not succeed at extracting the fundamental frequency with acceptable precision. Overall, compared with results obtained under the fixed cart configuration experiment, it can be concluded that both the STFT and WT methods have good promise for real-time application to high-rate state estimation due to their fast computation time and level of precision. It is worth remarking that the WT's precision relative to the STFT is approximately the same, unlike results seen under the fixed cart configuration where the WT's estimation error was close to three times that of the STFT. This can be attributed to the faster convergence speed of the WT, whereas the WT was capable of adapting more quickly to a change in the system's frequency under the moving cart configuration. Thus, it appears that the WT is more applicable to the high-rate problem, given its faster convergence speed, but this may come at the cost of lower precision on the estimation depending on circumstances.

5. Conclusions

This paper examined the promise of various time-frequency representation methods at conducting real-time high-rate state estimation. In particular, five methods were compared: the short-term Fourier transform (STFT), wavelet transform (WT), Wigner–Ville distribution (WVD), synchrosqueezing transform (SST), and multi-SST (MSST). The performance of the methods was assessed using high-rate experimental data produced on the DROPBEAR (Dynamic Reproduction of Projectiles in Ballistic Environments for Advanced Research) testbed. Such data included acceleration measurements of a beam with a cart located at fixed positions ("fixed cart configuration") sampled at 1 kHz, and with a cart moving between two locations ("moving cart configuration"), sampled at 25 kHz.

Results from the fixed cart configuration show that both the STFT and WT methods could be performed significantly faster than the three other methods, with the WT outperforming other methods in terms of convergence speed. Under the moving cart experiment, both the STFT performed similarly in terms of frequency estimation precision, but with the STFT being computationally faster to implement. The WVD failed at identifying the fundamental frequency, while the SST and MSST had unacceptable computation times, attributable to the longer window lengths that were necessary for implementing the methods. The SST and MSST can achieve good energy concentration and estimation in fixed cart configuration, but not for the moving-cart configuration.

Overall, it appears that the WT would be a better candidate for real-time applicability to high-rate state-estimation given its relatively faster convergence, but this may come at the cost of lower precision on the estimation depending on circumstances. The performance of the WT is yet to be assessed for strongly time-varying systems to characterize high-rate mechanisms undergoing sudden and high amplitude changes in their dynamics. It should also be noted that this paper limited the investigation to only five methods over a very specific experimental dataset and that, while results point towards the WT method as a possible path to high-rate applications, different conclusions could be drawn in a different environment, in particular for systems dominated by higher frequencies. In general, it is envisioned that applications to the high-rate problem would come in the form of advanced yet computationally fast algorithms inspired by the STFT or WT methods and that their implementations in field-programmable gate arrays (FPGAs) would significantly improve their performance in terms of computation time.

Author Contributions: The authors made the following contributions: Conceptualization: J.Y., S.L., and A.S.; methodology: J.Y. and S.L.; software: J.Y.; resources: J.D.; writing—original draft preparation: J.Y. and P.S.; writing—review and editing: S.L. and A.S.; visualization: J.Y.; supervision: A.S. and S.L. All authors have read and agreed to the published version of the manuscript.

Funding: The work presented in this paper is partially funded by the National Science Foundation under award number CCSS-1937460. Their support is gratefully acknowledged. Any opinions, findings, and conclusions or recommendations expressed in this material are those of the authors and do not necessarily reflect the views of the sponsors.

Conflicts of Interest: The author(s) declared no potential conflicts of interest with respect to the research, authorship, and/or publication of this article.

Abbreviations

The following abbreviations are used in this manuscript:

TFR	Time-frequency representation
STFT	Short-time Fourier Transform
WT	Wavelet transform
WVD	Wigner–Ville distribution
SST	Synchrosqueezed transform
MSST	Multi-synchrosqueezed transform

References

1. Hong, J.; Laflamme, S.; Dodson, J.; Joyce, B. Introduction to State Estimation of High-Rate System Dynamics. *Sensors* **2018**, *18*, 217. [CrossRef]
2. Dodson, J.; Joyce, B.; Hong, J.; Laflamme, S.; Wolfson, J. Microsecond State Monitoring of Nonlinear Time-Varying Dynamic Systems. Volume 2: Modeling, Simulation and Control of Adaptive Systems Integrated System Design and Implementation Structural Health Monitoring. In Proceedings of the ASME 2017 Conference on Smart Materials, Adaptive Structures and Intelligent Systems, Snowbird, UT, USA, 18–20 September 2017; American Society of Mechanical Engineers: New York, NY, USA, 2017. [CrossRef]
3. Hong, J.; Laflamme, S.; Cao, L.; Dodson, J.; Joyce, B. Variable input observer for nonstationary high-rate dynamic systems. *Neural Comput. Appl.* **2018**, 32, 5015–5026. [CrossRef]
4. Kumar, R.; Syam, P.; Das, S.; Chattopadhyay, A.K. Review on model reference adaptive system for sensorless vector control of induction motor drives. *IET Electr. Power Appl.* **2015**, *9*, 496–511. [CrossRef]
5. Yan, J.; Laflamme, S.; Leifsson, L. Computational Framework for Dense Sensor Network Evaluation Based on Model-Assisted Probability of Detection. *Mater. Eval.* **2020**, *78*. [CrossRef]

6. Yan, J.; Laflamme, S.; Hong, J.; Dodson, J. Online Parameter Estimation under Non-Persistent Excitations for High-Rate Dynamic Systems. In Review.
7. Downey, A.; Hong, J.; Dodson, J.; Carroll, M.; Scheppegrell, J. Millisecond model updating for structures experiencing unmodeled high-rate dynamic events. *Mech. Syst. Signal Process.* **2020**, *138*, 106551. [CrossRef]
8. Boashash, B. Chapter 15 - Time-Frequency Diagnosis, Condition Monitoring, and Fault Detection. In *Time-Frequency Signal Analysis and Processing*, 2nd ed.; Boashash, B., Ed.; Academic Press: Oxford, UK, 2016; pp. 857–913. [CrossRef]
9. Goyal, D.; Pabla, B.S. The Vibration Monitoring Methods and Signal Processing Techniques for Structural Health Monitoring: A Review. *Arch. Comput. Methods Eng.* **2015**, *23*, 585–594. [CrossRef]
10. Dörfler, M.; Matusiak, E. Nonstationary Gabor frames-approximately dual frames and reconstruction errors. *Adv. Comput. Math.* **2015**, *41*, 293–316. [CrossRef]
11. Paul Busch, T.H.; Lahti, P. Heisenberg's uncertainty principle. *Phys. Rep.* **2007**, *452*, 155–176. [CrossRef]
12. Huang, N.E.; Shen, Z.; Long, S.R.; Wu, M.C.; Shih, H.H.; Zheng, Q.; Yen, N.C.; Tung, C.C.; Liu, H.H. The empirical mode decomposition and the Hilbert spectrum for nonlinear and non-stationary time series analysis. *Proc. R. Soc. Lond. Ser. Math. Phys. Eng. Sci.* **1998**. *454*, 903–995. [CrossRef]
13. Chen, Y.; Liu, T.; Chen, X.; Li, J.; Wang, E. Time-frequency analysis of seismic data using synchrosqueezing wavelet transform. *IEEE Geosci. Remote. Sens. Lett.* **2014**, *11*, 2042–2044. [CrossRef]
14. Yuan, M.; Sadhu, A.; Liu, K. Condition assessment of structure with tuned mass damper using empirical wavelet transform. *J. Vib. Control* **2017**, *24*, 4850–4867. [CrossRef]
15. Sadhu, A.; Sony, S.; Friesen, P. Evaluation of progressive damage in structures using tensor decomposition-based wavelet analysis. *J. Vib. Control* **2019**, *25*, 2595–2610. [CrossRef]
16. Mahato, S.; Chakraborty, A. Sequential clustering of synchrosqueezed wavelet transform coefficients for efficient modal identification. *J. Civ. Struct. Health Monit.* **2019**, *9*, 271–291. [CrossRef]
17. Yu, G.; Wang, Z.; Zhao, P. Multisynchrosqueezing Transform. *IEEE Trans. Ind. Electron.* **2019**, *66*, 5441–5455. [CrossRef]
18. Soudan, M.; Vogel, C. Correction Structures for Linear Weakly Time-Varying Systems. *IEEE Trans. Circuits Syst. Regul. Pap.* **2012**, *59*, 2075–2084. [CrossRef]
19. Pham, D.H.; Meignen, S. High-Order Synchrosqueezing Transform for Multicomponent Signals Analysis—With an Application to Gravitational-Wave Signal. *IEEE Trans. Signal Process.* **2017**, *65*, 3168–3178. [CrossRef]
20. Joyce, B.; Dodson, J.; Laflamme, S.; Hong, J. An Experimental Test Bed for Developing High-Rate Structural Health Monitoring Methods. *Shock Vib.* **2018**, *2018*, 1–10. [CrossRef]
21. Gabor, D. Theory of communication. Part 1: The analysis of information. *J. Inst. Electr.-Eng.-Part III Radio Commun. Eng.* **1946**, *93*, 429–441. [CrossRef]
22. Feng, Z.; Liang, M.; Chu, F. Recent advances in time–frequency analysis methods for machinery fault diagnosis: A review with application examples. *Mech. Syst. Signal Process.* **2013**, *38*, 165–205. [CrossRef]
23. Mallat, S. A wavelet tour of signal processing. In *The Sparse Way*; Elsevier: Amsterdam, The Netherlands, 1999.
24. Flandrin, P. *Time-Frequency/Time-Scale Analysis*; Academic press: Cambridge, MA, USA, 1998.
25. Boashash, B.; Black, P. An efficient real-time implementation of the Wigner–Ville distribution. *IEEE Trans. Acoust. Speech Signal Process.* **1987**, *35*, 1611–1618. [CrossRef]
26. Bradford, S.C. *Time-Frequency Analysis of Systems with Changing Dynamic Properties*; Earthquake Engineering Research Laboratory: Reno, NV, USA, 2006.
27. Thakur, G. The Synchrosqueezing transform for instantaneous spectral analysis. In *Excursions in Harmonic Analysis*; Springer: Berlin/Heidelberg, Germany, 2015; Volume 4, pp. 397–406. [CrossRef]
28. Ewins, D.J. *Modal Testing : Theory, Practice, and Application*; Research Studies Press: Letchworth, UK; Philadelphia, PA, USA, 2000.
29. Chen, G.; Chen, J.; Dong, G. Chirplet Wigner–Ville distribution for time–frequency representation and its application. *Mech. Syst. Signal Process.* **2013**, *41*, 1–13. [CrossRef]
30. Olhede, S.; Walden, A. Generalized Morse wavelets. *IEEE Trans. Signal Process.* **2002**, *50*, 2661–2670. [CrossRef]

31. Li, L.; Cai, H.; Jiang, Q. Adaptive synchrosqueezing transform with a time-varying parameter for non-stationary signal separation. *Appl. Comput. Harmon. Anal.* **2020**, *49* . [CrossRef]
32. Iatsenko, D.; McClintock, P.; Stefanovska, A. Extraction of instantaneous frequencies from ridges in time–frequency representations of signals. *Signal Process.* **2016**, *125*, 290–303. [CrossRef]

Article

Continuous Evaluation of Track Modulus from a Moving Railcar Using ANN-Based Techniques

Ngoan T. Do, Mustafa Gül * and Saeideh Fallah Nafari

Department of Civil & Environmental Engineering, University of Alberta, Edmonton, AB T6G 1H9, Canada; tndo@ualberta.ca (N.T.D.); fallahna@ualberta.ca (S.F.N.)

* Correspondence: mustafa.gul@ualberta.ca; Tel.: +1-780-492-3002

Received: 12 May 2020; Accepted: 18 June 2020; Published: 22 June 2020

Abstract: Track foundation stiffness (also referred as the track modulus) is one of the main parameters that affect the track performance, and thus, quantifying its magnitudes and variations along the track is widely accepted as a method for evaluating the track condition. In recent decades, the train-mounted vertical track deflection measurement system developed at the University of Nebraska–Lincoln (known as the MRail system) appears as a promising tool to assess track structures over long distances. Numerical methods with different levels of complexity have been proposed to simulate the MRail deflection measurements. These simulations facilitated the investigation and quantification of the relationship between the vertical deflections and the track modulus. In our previous study, finite element models (FEMs) with a stochastically varying track modulus were used for the simulation of the deflection measurements, and the relationships between the statistical properties of the track modulus and deflections were quantified over different track section lengths using curve-fitting approaches. The shortcoming is that decreasing the track section length resulted in a lower accuracy of estimations. In this study, the datasets from the same FEMs are used for the investigations, and the relationship between the measured deflection and track modulus averages and standard deviations are quantified using artificial neural networks (ANNs). Different approaches available for training the ANNs using FEM datasets are discussed. It is shown that the estimation accuracy can be significantly increased by using ANNs, especially when the estimations of track modulus and its variations are required over short track section lengths, ANNs result in more accurate estimations compared to the use of equations from curve-fitting approaches. Results also show that ANNs are effective for the estimations of track modulus even when the noisy datasets are used for training the ANNs.

Keywords: railroad tracks; track modulus; computer simulation; artificial neural networks

1. Introduction

It is widely accepted that a track modulus, and its variations, are indicators of subgrade conditions [1–5]. A track modulus is a measure of the vertical stiffness of the rail foundation and is defined as the ratio of the vertical supporting force per unit length of rail to the vertical deflection [1]. A practical way to assess the track modulus is to measure the rail deflection under specified loads [6–8]. Measured deflections can be correlated to the track modulus using mathematical equations. Two methods are available to measure rail deflections: trackside measurement techniques and on-train approaches. Trackside measurement techniques are used to measure the rail deflection at specific locations under specified static loads or moving loads [9]. Although these techniques provide accurate estimations of track stiffness, they are laborious and time-consuming, especially when multi-point measurements are required. On the other hand, on-train measurement systems allow the measurement of rail deflections over long distances and thus provide a good overall evaluation of the entire railway network [10–15]. Comprehensive analysis is typically needed to investigate the relationship between deflection measurements from on-train systems and track modulus [16,17].

The real-time vertical track deflection measurement system (known as MRail System) developed at the University of Nebraska–Lincoln, under the sponsorship of the Federal Railroad Administration (FRA), has become more popular in recent decades [10–12]. The system computes relative vertical deflection (Y_{rel}) between the rail/wheel contact plane, and the rail surface at a distance of 1.22 m from the nearest wheel to the sensor system. The MRail system has been tested over different railway lines in the USA and Canada for evaluating track conditions [18–21]. Results from the MRail field tests show that the system not only has the potential to identify the local track problems, i.e., muddy ballast, degraded joints, crushed rail head, broken ties, but also provides an opportunity to map the subgrade condition and assess the track performance along the railway line [22–25].

In addition to the experimental studies, different numerical models have been used to investigate the relationship between track modulus and Y_{rel} data, and numerical approaches have been proposed to estimate the track modulus from Y_{rel} [21,26]. The current study aims to propose a new and advanced approach for estimating track modulus statistical properties from Y_{rel} data more accurately compared to previous studies. First, the details of the MRail system are briefly presented, and the numerical models developed by others and their shortcomings are discussed. Then, artificial neural networks (ANNs) are explained as the main tool to investigate the relationship between track modulus and Y_{rel} data in this paper. Different methods for training the ANNs are used, and the effectiveness of the trained ANNs are investigated using error measurement parameters such as the coefficient of determination (R^2), the root mean square error (RMSE), and mean absolute percentage error (MAPE). Suitable signatures of Y_{rel} data are identified by conducting both statistical and frequency analysis. Feedforward neural networks are proposed as a function approximation technique to estimate the track modulus average (U_{Ave}) and standard deviation (U_{SD}) from Y_{rel} data. To further investigate the effectiveness of the ANNs for estimating the track modulus, noisy finite element models (FEM) datasets are employed for training the ANNs. The accuracy of the track modulus estimations using these ANNs is also investigated using R^2, RMSE and MAPE.

2. The Stiffness Measurement System and Numerical Simulations

2.1. MRail Measurement System

The MRail system was originally developed at the University of Nebraska–Lincoln under the sponsorship of the Federal Railroad Administration (FRA) [10–12]. The system measures the relative vertical deflection (Y_{rel}) between the rail surface and the rail/wheel contact plane, at a distance of 1.22 m from the nearest wheel to the acquisition system (Figure 1a). The sensors consist of two laser lines and a digital camera mounted on the side frame of the rail car (Figure 1b). The laser system projects two curves on the rail surface, whose minimum distance (d) is captured by the camera (Figure 1c). Subsequently, the distance between the camera and the rail surface (h) is computed by converting d. Finally, the relative deflection Y_{rel} is calculated by subtracting h from (Y_{rel} + h), the fixed distance between the rail/wheel contact plane and the camera.

The MRail system can measure the deflection at different sampling rates with the speed up to 96 km/h (60 mph). The Winkler model and the finite element models have been used to estimate the track modulus from Y_{rel} [24].

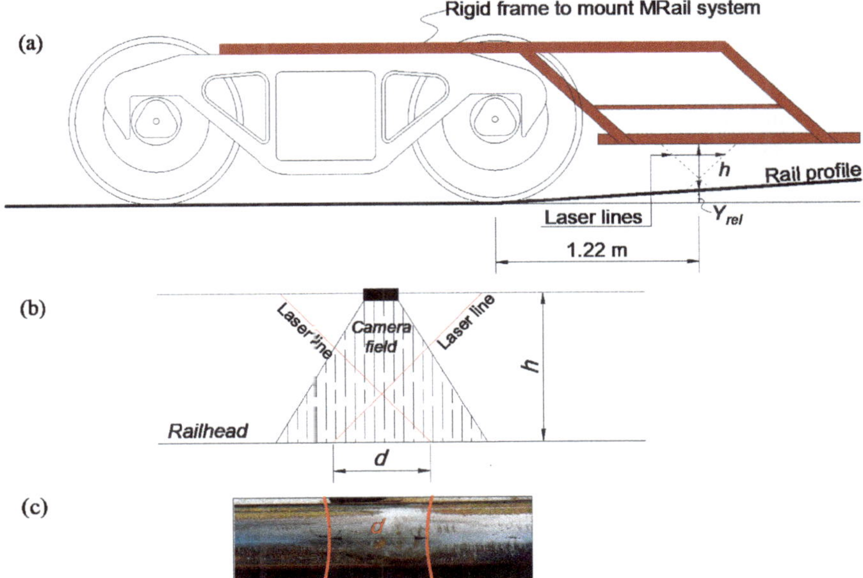

Figure 1. Demonstration of the MRail system (real-time vertical track deflection measurement system) for Y_{rel} measurements: (**a**) the measurement system on a rigid frame; (**b**) the sensor system; and (**c**) the projections of the laser lines on the railhead.

2.2. Winkler Model

Rail deformation and bending stress under specific loads are typically estimated using THE Winkler model, which considers the track as an infinite beam on a continuous elastic foundation [27–29]. Using the Winkler model (Equation (1)), the vertical rail deflection (y) at a distance x from the applied load (P) is computed as follows:

$$y(x) = \frac{P\beta e^{-\beta x}(\cos\beta x + \sin\beta x)}{2U} \quad (1)$$

where β is the stiffness ratio, which is equal to $(U/(4EI))^{0.25}$, U is the track modulus, E is the modulus of the elasticity of the rail, and I is the second moment of area of the rail.

From the Winkler model, the vertical deflection profile of a rail is only dependent on the track modulus value when the rail size and vertical loads are known. Once a value is assumed for track modulus, the rail vertical deflection profile can be estimated using Equation (1), and from the rail vertical deflection profile, Y_{rel} can be calculated as the relative vertical deflection between the rail surface and the rail/wheel contact plane at a distance of 1.22 m from the nearest wheel (Figure 1a) [11,30]. The main shortcoming in this method is that the Winkler model assumes a track modulus is constant along the track while the field data shows that a track modulus stochastically varies along the track [31,32]. Therefore, the estimation of the track modulus from the Y_{rel} measurements needs more advanced numerical models.

2.3. Finite Element Model

FEMs allow the simulation of a stochastically varying track modulus, and therefore, a more accurate simulation of Y_{rel} measurements. Fallah Nafari et al., developed 90 FEMs with a stochastically varying track modulus to facilitate a more detailed investigation of the relationship between the Y_{rel} and the track modulus [21]. Datasets from the 90 FEMs were used for the study in this paper.

Hence, the details of the models are discussed briefly. The models are developed using CSiBridge software, where each model includes a 180.8 m track structure with two rails, crossties, and spring supports [33]. To develop each model, a normal track modulus distribution is assumed and randomly selected numbers from this distribution are assigned to the spring supports along the track. Statistical properties of the assumed normal distributions are summarized in Table 1, and the applied loads are depicted in Figure 2. RE136 rail size and 0.508 m tie spacings are used in the models.

Table 1. Statistical properties of the track modulus in the FEMs.

Track Modulus Average (MPa)	Coefficient of Variation (COV)	No. of Simulations
41.4	0.25; 0.5; 0.75	30 (10 simulations for each COV)
27.6	0.25; 0.5; 0.75	30 (10 simulations for each COV)
12.8	0.25; 0.5; 0.75	30 (10 simulations for each COV)

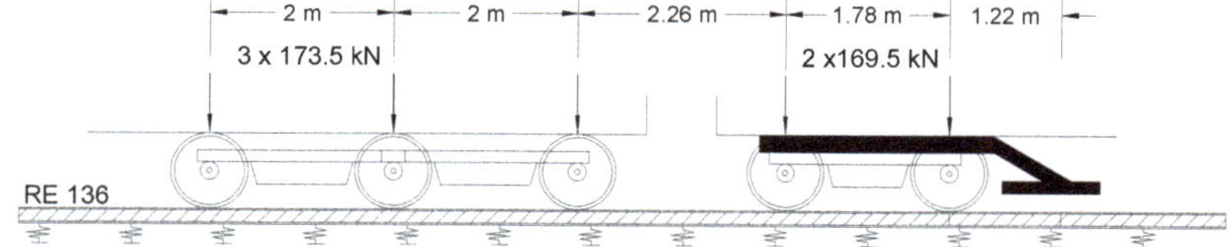

Figure 2. The loading condition in the finite element models (FEMs).

Individual Y_{rel} values are calculated from the vertical deflection profile at every 0.3048 m (≈1 ft) interval while the moving loads pass the track model. The dynamic effects of track–train interactions are not considered during the simulations due to the software's limitation. This is acceptable within the scope of this study which mostly focuses on the Canadian freight lines where speeds are most likely lower than 65 km/h.

Figure 3 shows an example of the inputted track modulus to the model and corresponding Y_{rel} output. Fallah Nafari et al., used basic statistical analysis and curve fitting approaches to study the relationship between the statistical properties of track modulus (U) and Y_{rel} [21]. The results showed that the average and standard deviation of the track modulus over a track section length can be estimated from the average and standard deviation of Y_{rel} over the same track section length. However, the estimation accuracy becomes lower by decreasing the track section length [21]. To overcome this shortcoming and increase the estimation accuracy of the track modulus, ANNs are proposed for the track modulus estimations in this study.

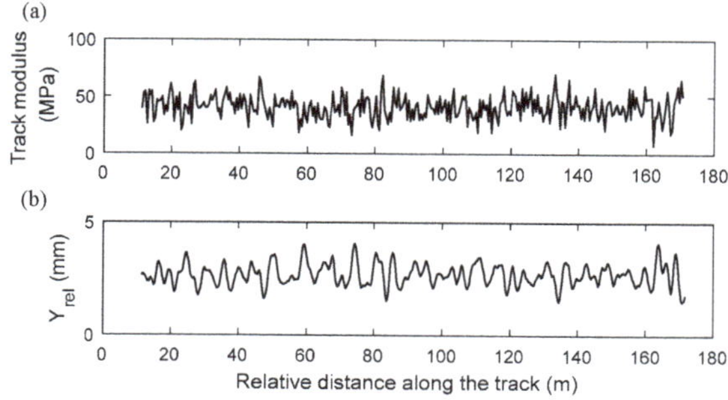

Figure 3. (**a**) Track modulus inputted to the FEM (Mean = 41.4 MPa, COV = 0.25); and (**b**) the extracted Y_{rel}.

2.4. Estimation of Track Modulus Average

2.4.1. Multilayer Perceptron Artificial Neural Networks

Multilayer perceptron neural networks (MLPNN) are typically useful for classification, and function approximation problems [34–38]. The implementation of MLPNN is operated with two stages of performance, i.e., training and testing procedures. Once the training process is successfully performed in a self-adaptive manner with all defined parameters (such as learning algorithm and network architecture including several layers, and neurons in each layer), the network can effectively approximate the input–output mapping function.

MLPNN is a network containing two or more neurons distributed in different layers, such as input layers, output layers, and hidden layers that connect the input and output layers (Figure 4a). Each neuron has a nonlinear differentiable activation function that creates real values and is highly connected to other neurons based on synaptic weights w_{ij} (n) (Figure 4b) as the level of connectivity.

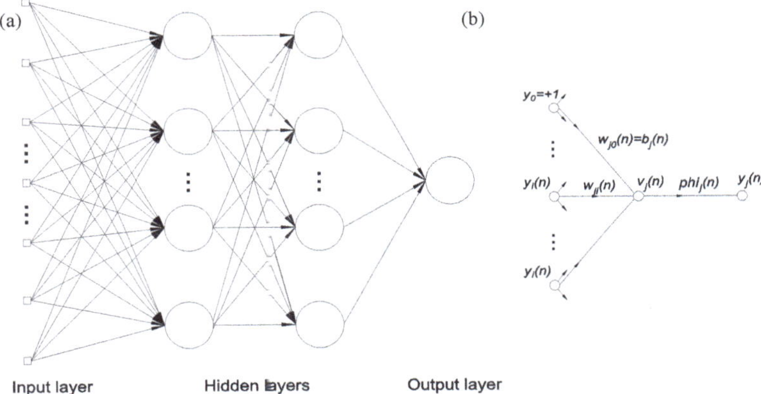

Figure 4. (**a**) Example of a two-hidden layer perceptron; (**b**) typical operation at neuron j.

One of the most complicated tasks before executing an MLPNN is that all the required parameters should be well defined to approximate the input–output relationship, which is called the learning process that contains two phases. In the forward phase, the inputs are fed into the network from left to right, and layer by layer with the fixed values of synaptic weights. In the backward phase, the error vector is first computed by subtracting the output of the network from the expected target. The error is then propagated backward from the output to the input layer. In this phase, the synaptic weights are adjusted to minimize the network error by solving the credit-assignment problems in the operation of each hidden unit. Each synaptic weight will be updated differently based upon the contribution of the corresponding hidden unit to the overall error. More information about training the network using backpropagation and gradient descent is in Haykin's book [34].

2.4.2. Estimation Procedure and Results

The inputted track modulus and the corresponding Y_{rel} data from the 180 m track models are divided into equivalent groups based on a track section length (e.g., 5 m, 10 m, etc.). Once the subgroups are defined, the average and standard deviation of Y_{rel} in each subgroup are used as the networks' inputs whereas the track modulus averages in the corresponding track segments are defined as the network's outputs.

Y_{rel} data extracted from eighty-one FEMs (out of ninety FEMs) are used to train the neural network. The accuracy of the trained network is then tested using the remaining nine (unseen) FEMs. These nine FEMs are called "unseen models" hereafter as they are not used in training the network. To test the trained network, track modulus average is estimated from Y_{rel} average and the standard deviation for the nine unseen models. The estimated track modulus average is then compared with the track modulus inputted initially into the FEMs to generate Y_{rel} data. The effectiveness of the proposed network is measured based on three parameters: the coefficient of determination (R^2), the root mean square error (RMSE), and the mean absolute percentage error (MAPE) [39]. These measures are described as follows:

$$R^2 = \left(\frac{\frac{1}{N}\sum_{i=1}^{N}\left[(o_i - \overline{o}_i)\cdot(y_i - \overline{y}_i)\right]}{\sigma_o \cdot \sigma_y} \right)^2 \quad (2)$$

$$\text{RMSE} = \sqrt{\sum_{i=1}^{N} \frac{(y_i - o_i)^2}{N}} \quad (3)$$

$$\text{MAPE} = \frac{1}{N}\sum_{i=1}^{N} 100\frac{|y_i - o_i|}{y_i} \quad (4)$$

where $\overline{o}_i$, $\overline{y}_i$, σ_o, and σ_y are the average and standard deviation of the estimated, and targeted values; N is the number of testing samples.

When a network is trained, five-fold cross-validation is employed to minimize any potential over-fitting problem and increase the network's generalization. Regarding the network architecture, a network with two hidden layers (each contains 15 hidden nodes) is used in this study. This network ensures an acceptable error range, avoids over-fitting, and optimizes the computational efficiency. From different tests, it is noted that increasing the number of hidden nodes and hidden layers, does not necessarily mean the network's performance is improved. In fact, the input configuration is the most important factor that controls the network performance.

Five networks for five different track section lengths have been fully trained to perform this study. The track modulus average over five section lengths is then estimated for the nine new models using the trained networks. Table 2 presents the accuracy level of these estimations. From the table, the network performs better when the track section length increases although the error is acceptably small even with the case of a 10 m section length. R^2 is 0.95 for the case of the 10 m section length, which means that the estimated and inputted track modulus averages are well correlated. Moreover, the RMSE and MAPE are quite small, i.e., 2.81 MPa, and 6.99% respectively, considering that range of inputted track modulus average is 12.8 to 41.4 MPa. In addition to confirming the applicability of the Y_{rel} data in indicating the track modulus information, the current methodology provides more accurate results than the other method in the literature [21]. As shown in Table 2, the R^2 value computed in the related study decreases as the length of the track segment reduces, whereas the R^2 in the current study is almost constant for cases with a 10 m track section and more.

Table 2. Estimation accuracy of the track modulus average (no noise added).

Section Length (m)	MAPE * (%)	RMSE ** (MPa)	R^2	R^2 in [21]
5	12.42	4.58	0.86	0.79
10	6.99	2.81	0.95	0.93
15	5.90	2.56	0.95	N/A
20	4.32	1.60	0.98	0.96
25	3.87	1.63	0.98	N/A

* Mean Absolute Percentage Error; ** Root Mean Square Error.

The accuracy of the estimation method for the case of the 10 and 20 m section lengths are demonstrated in Figure 5 for four models as an example. These four models had a different track modulus average and variations. From the figure, the values estimated from the networks are very close to the actual track modulus average inputted to the FEMs. Better results can be observed in the case of a 20 m section length (Figure 5b) although the performance of the estimation of track modulus over the shorter section length (Figure 5a) is still satisfactory. Most importantly, the local fluctuation of the track modulus is well captured.

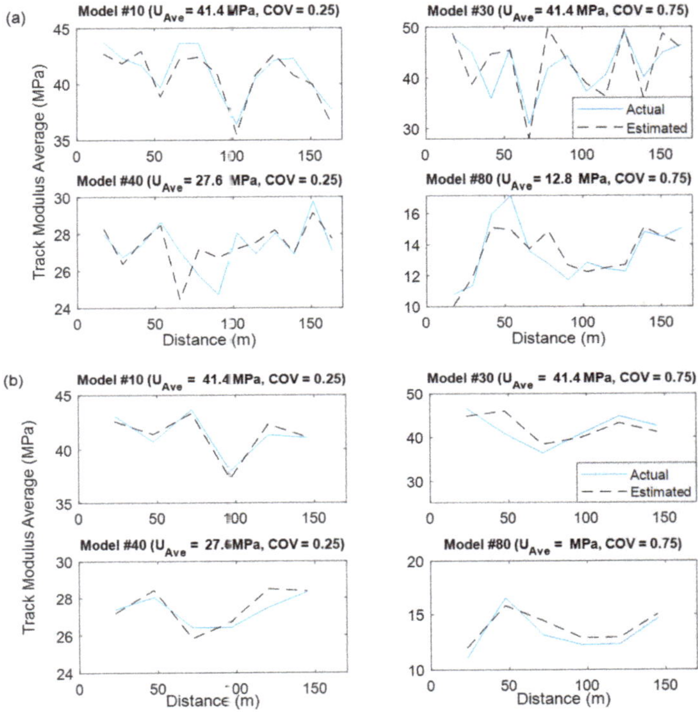

Figure 5. Moving average of the actual track modulus inputted to the FEMs vs. estimated values over: (**a**) the 10 m section length; and (**b**) the 20 m section length.

The effectiveness of the framework is further investigated by adding artificial noise to the Y_{rel} data extracted from the FEMs. This simulates the real-life condition in which the Y_{rel} measurements are affected by parameters such as the resolution of the MRail measurement system, track irregularities, etc. The artificial noise was added based on Equation (5) [40]. An example of pure vs. noise-added Y_{rel} is shown in Figure 6:

$$Y_{rel-noisy} = Y_{rel} + \alpha \cdot 0.12 + \beta \cdot 0.1 \cdot Y_{rel} \tag{5}$$

where α, and β are random numbers ranging from −1 to 1.

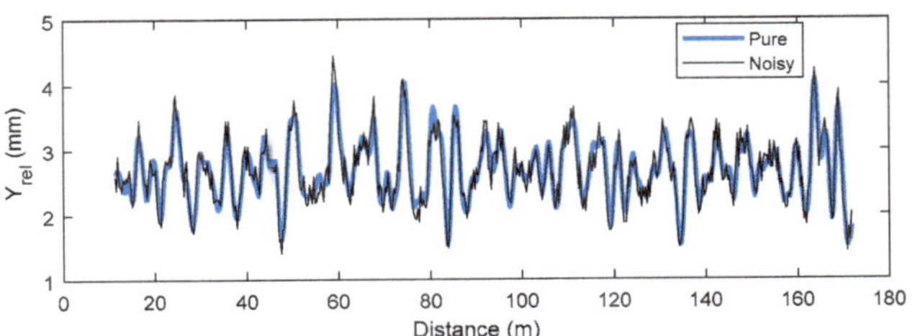

Figure 6 Demonstration of pure and noisy Y_{rel}.

The noisy Y_{rel} is used to train new networks, and then the trained networks are used to estimate the track modulus average. The estimated track modulus is then compared with the inputted track modulus for each model and the error is reported in Table 3. From the table, the estimation of the track modulus average (U_{ave}) from the noisy Y_{rel} is still successful even for the short track section length of 10 m as R^2 is 0.95, and RMSE is 2.77 MPa. This demonstrates that the framework performs effectively even when the Y_{rel} data contain noise, and thus is expected to work with real-life data.

Table 3. Estimation accuracy of the track modulus average (with added noise).

Section Length (m)	MAPE (%)	R^2	RMSE (MPa)	R^2 in [21]
5	14.09	0.83	5.07	0.79
10	7.01	0.95	2.77	0.93
15	5.93	0.96	2.36	- *
20	6.07	0.97	1.98	0.96
25	3.98	0.98	1.53	- *

* Not available for comparisons since those section lengths are not available in the previous study.

2.5. Estimation of Track Modulus Standard Deviation (U_{SD})

The estimation of the track modulus standard deviation from the Y_{rel} data using statistical methods and curve-fitting approaches has not been successful for track section lengths shorter than 80 m [21]. Therefore, frequency characteristics of the deflection data are investigated in this study to increase the estimation accuracy of track modulus standard deviation. The coefficients associated with the Y_{rel} frequency components are employed as one of the inputs to the ANNs, whose outputs are the track modulus standard deviation over different track section lengths. As demonstrated in Figure 7, Y_{rel} and the track modulus data are divided into different subgroups based on various track section lengths (similar to the procedure used for estimating the track modulus average). Then, statistical analysis, fast Fourier transform, and a liftering technique are applied on the Y_{rel} data in each subgroup to extract the average and standard deviation of the Y_{rel} and average and the standard deviation of liftering the fast Fourier transform (FFT) coefficients. These parameters are used as the inputs of ANNs.

Figure 8a shows an example of the FFT coefficients of the Y_{rel} data over a track section of 30 m for 81 models. As can be seen, the coefficients at higher orders are relatively small. This is undesirable for training the ANN due to possible bias. Therefore, the coefficients are processed using a liftering technique (Equation (6) to roughly normalize their variances) [41]:

$$X'(k) = \left(1 + \frac{L}{2}\sin\left(\frac{\pi(k+1)}{L}\right)\right) \cdot X(k),\ k = 0,\ \ldots,\ N-1 \quad (6)$$

where L is the sin lifter parameter, which is 50 in the current study, and $X(k)$ is the FFT coefficients.

Once the liftering technique is applied (Figure 8b), the average and standard deviations of the lifted FFT are calculated using Equations (7) and (8) are used as two additional inputs for ANNs.

$$P_1 = \frac{2}{N-1} \sum_{k=0}^{(N-1)/2} |X'(k)| \quad (7)$$

$$P_2 = \sqrt{\frac{2}{N-1} \sum_{k=0}^{N/2} \left(|X'(k)| - P(1)\right)^2} \quad (8)$$

The architecture used for developing the network in this section has two hidden layers and 15 hidden nodes in each layer, similar to the network's architecture for estimating the track modulus average. The trained networks are used for estimating the track modulus standard deviation over different track section lengths and three accuracy measurements are reported in Table 4. In order to show that the current input–output pair is optimized, and two network architectures are trained (ANN-1 with four inputs, i.e., the average and standard deviation of Y_{rel}, and the average and standard deviation of the lifted FFT; ANN-2 with two inputs, i.e., mean and standard deviation of Y_{rel}). In each case, the two networks are trained and tested multiple times and the mean and standard deviation of the performance parameters are computed and reported in Table 4. For the case of 5 m section length, for instance, the networks' input, and output are first extracted based on the chosen section (5 m), then ANN-1 and ANN-2 networks are trained using the training data and tested against the data extracted from nine unseen FEMs.

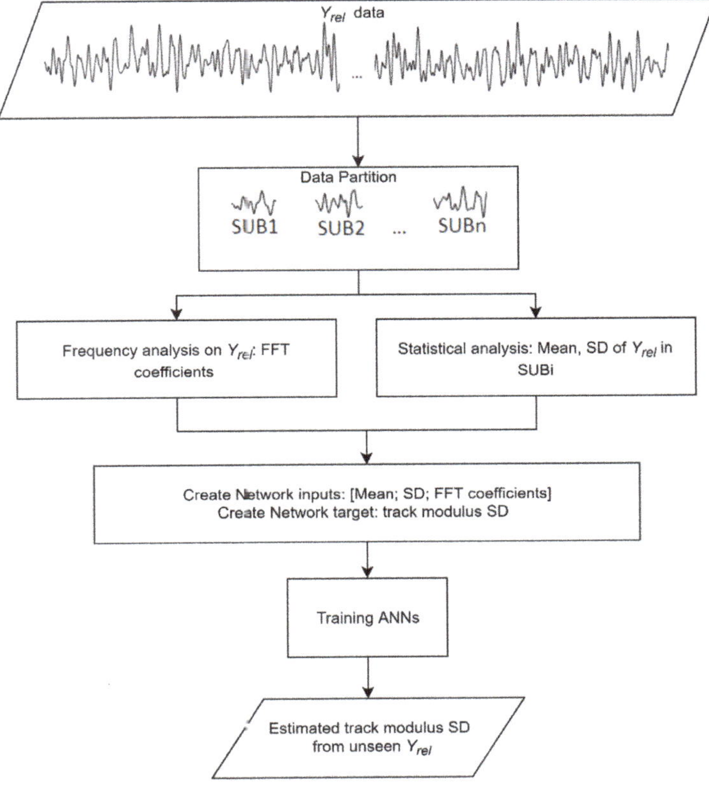

FFT: fast Fourier transform; SD: standard deviation; ANNs: artificial neural networks

Figure 7. Procedure for estimating the track modulus standard deviation (U_{SD}).

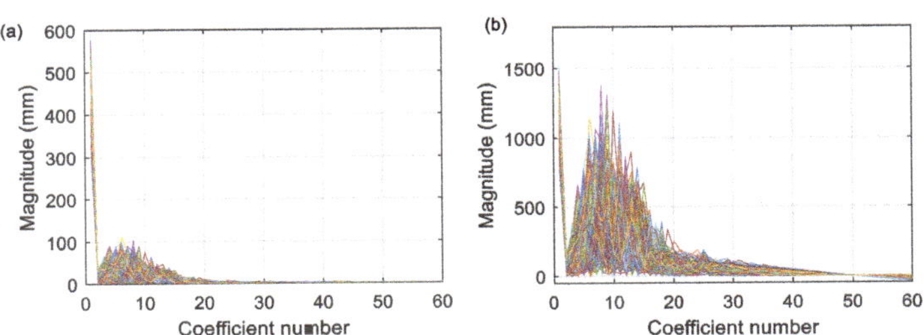

Figure 8. FFT of Y_{rel}: (**a**) before liftering; and (**b**) after liftering.

Table 4. Estimation accuracy of the U_{SD} (no noise added, the standard deviation in the parenthesis).

Section Length (m)	Network Configuration	RMSE (MPa)	MAPE (%)	R^2	R^2 in [21]
10	ANN-1	3.00 (0.16 *)	18.41	0.83	0.53
	ANN-2	3.05 (0.22)	19.12	0.82	-
15	ANN-1	2.36 (0.08)	15.01	0.89	-
	ANN-2	2.61 (0.39)	15.79	0.87	-
20	ANN-1	2.23 (0.11)	14.49	0.91	0.66
	ANN-2	2.59 (0.89)	14.47	0.88	-
25	ANN-1	1.83 (0.13)	11.96	0.94	-
	ANN-2	1.99 (0.30)	11.72	0.92	-
30	ANN-1	2.08 (0.17)	11.61	0.92	-
	ANN-2	2.14 (0.44)	11.79	0.91	-

* Standard deviation of the estimation error.

From Table 4, the error values show that the standard deviation of track modulus (U_{SD}) can be estimated satisfactorily by both network configurations (ANN-1 and ANN-2). Even for the 10-m section length case, for instance, the coefficient of correlations between the actual U_{SD} and the one estimated by the two networks are very high, e.g., 0.83 and 0.82 respectively. However, the networks with four inputs (ANN-1) slightly outperform the one with two inputs (ANN-2) regardless of the section lengths. Specifically, the RMSE and MAPE are always smaller than those arising from the trained networks whose inputs are the statistical properties of Y_{rel} only (ANN-2). Values estimated using the networks with four inputs have relatively high R^2 in all cases showing that the methodology is successful. In particular, the R^2 is as high as 0.94 for the case of the 25 m section length and the RMSE is 1.83 MPa, which is a relatively small error considering that the maximum standard deviation of the inputted track modulus in the FEMs is 31.05 MPa. Moreover, the first network (ANN-1) provides more reliable results as the standard deviation of RMSE remains stable (varying from 0.11 to 0.17 MPa) and lower than those of ANN-2. Therefore, combining FFT and statistical analysis to configure the input for the networks noticeably improves the estimation accuracy, and increases the stability of the ANNs, the mapping function between the Y_{rel} characteristics and the track modulus standard deviation (U_{SD}). Most importantly, there is a big step forward in this paper compared to the previous study, where the R^2 coefficient is 0.748 even though the 40 m section length is used [21]. The performance of this estimation can be considered ineffective as the R^2 coefficient reduced significantly in shorter track segment cases (Table 4). Hence, considering the current results, it can be claimed that neural networks are more powerful for mapping the relationship between Y_{rel} and track modulus, especially over the short track section lengths.

For more descriptive results, the strong correlation between the actual and estimated track modulus's standard deviation for the 25 m section length is demonstrated in Figure 9. As can be seen, the estimated standard deviations follow the same patterns as those of the actual values, which vary greatly from 3.2 to 31.05 MPa.

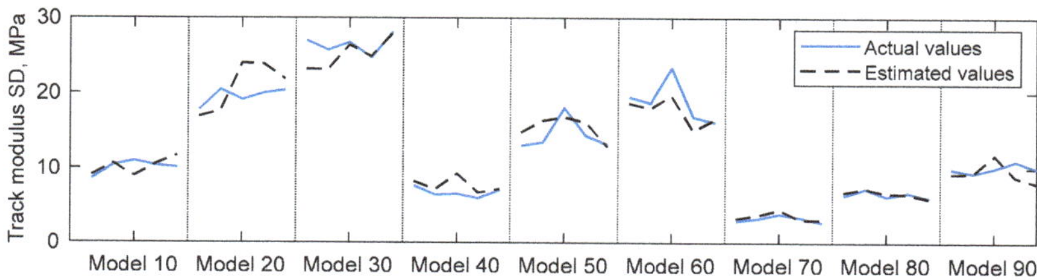

Figure 9. The actual track modulus standard deviation over the 25 m section length vs. the estimated values.

The effectiveness of the methodology is further validated by adding noise into the deflection data (Y_{rel}). Similar to the procedure mentioned in the previous section, noise is added to the Y_{rel} data from 90 models using Equation ((5). The dataset from 81 models is then used to train the networks using two approaches: networks with two inputs (average and standard deviation of Y_{rel}) and networks with four inputs (average and standard deviation of Y_{rel} and average and standard deviation of the lifted FFT). The developed networks are used to estimate the track modulus standard deviations over the different section lengths from the unseen Y_{rel} data. The estimated values are compared with the standard deviation of track modulus inputted to FEMs and results are reported in Table 5. The results show that the proposed approaches work very well even when Y_{rel} datasets are affected by noises. The R^2 is again higher than 0.90 when the 25 m or higher section lengths are utilized.

Table 5. Estimation accuracy of the U_{SD} (with noise added).

Section Length (m)	Network Configuration	R^2	RMSE (MPa)	MAPE (%)
10	ANN-1	0.82	3.06	20.12
	ANN-2	0.81	3.14	19.64
15	ANN-1	0.87	2.59	16.30
	ANN-2	0.87	2.64	16.23
20	ANN-1	0.89	2.42	16.13
	ANN-2	0.89	2.45	14.71
25	ANN-1	0.94	1.86	11.96
	ANN-2	0.93	1.88	11.73
30	ANN-1	0.94	1.84	10.43
	ANN-2	0.93	1.89	10.95

3. Conclusions

In this paper, two frameworks are proposed for estimating the track modulus average, and the standard deviation over the different track section lengths. The frameworks employed Y_{rel} data (a relative rail vertical deflection measured using the MRail system) for the track modulus estimations. The relationship between the statistical properties of the track modulus and the Y_{rel} data were investigated using artificial neural networks (ANNs). Datasets from FEMs are used to train the ANNs in which their outputs are either the track modulus average or standard deviations. Both statistical and frequency analyses were conducted to identify the optimized inputs for the ANNs from the Y_{rel} data. From the results, the track modulus average over a track section length of 10 m or longer is accurately estimated from the average and standard deviation of the Y_{rel} data within the corresponding section length. Additionally, the standard deviation of the track modulus over a section length of 25 m or longer is estimated with an acceptable level of accuracy. It is also shown that the trained ANNs work very well for the track modulus estimations even when the Y_{rel} values as the ANN inputs are affected by noise. The proposed ANNs are only applicable to a specific rail type and loading condition. Hence, a similar procedure should be followed to train the ANNs for different ranges of rail sections and loading types.

Author Contributions: Conceptualization, N.T.D., M.G. and S.F.N.; methodology, N.T.D.; software, N.T.D., S.F.N.; validation, N.T.D., M.G. and S.F.N.; formal analysis, N.T.D., M.G.; writing—original draft preparation, N.T.D.; writing—review and editing, N.T.D., M.G. and S.F.N.; supervision, M.G. All authors have read and agreed to the published version of the manuscript.

Funding: The study is funded by IC-IMPACTS (the India-Canada Centre for Innovative Multidisciplinary Partnerships to Accelerate Community Transformation and Sustainability), established through the Networks of Centres of Excellence of Canada.

Conflicts of Interest: The authors declare no conflict of interest.

References

1. Selig, E.T.; Li, D. Track modulus: Its meaning and factors influencing it. *Transp. Res. Rec.* **1994**, *1470*, 47–54.

2. Ebersohn, W.; Selig, E.T. Track modulus measurements on a heavy haul line. *Transp. Res. Rec.* **1994**, *1470*, 73.
3. Lopez Pita, A.; Teixeira, P.F.; Robuste, F. High Speed and Track Deterioration: The Role of Vertical Stiffness of the Track. *Proc. Inst. Mech. Eng. Pt. F J. Rail Rapid Transit.* **2004**, *218*, 31–40. [CrossRef]
4. Dahlberg, T. Railway Track Stiffness Variations-Consequences and Countermeasures. *Int. J. Civ. Eng.* **2010**, *8*, 1–12.
5. Sussmann, T.R.; Ebersohn, W.; Selig, E.T. Fundamental nonlinear track load-deflection behavior for condition evaluation. *Transp. Res. Rec.* **2001**, 61–67. [CrossRef]
6. Burrow, M.P.N.; Chan, A.H.C.; Shein, A. Deflectometer-based analysis of ballasted railway tracks. *Proc. Inst. Civ. Eng. Geotech. Eng.* **2007**, *160*, 169–177. [CrossRef]
7. Norman, C.; Farritor, S.; Arnold, R.; Elias, S.; Fateh, M.; El-Sibaie, M. *Design of a System to Measure Track Modulus from a Moving Railcar*; US Department of Transportation, Federal Railroad Administration, Office of Research and Development: Washington, DC, USA, 2006; paper ID 20590.
8. Eric, G.B.; Nissen, A.; Björn, S.P. Track deflection and stiffness measurements from a track recording car. *Proc. Inst. Mech. Eng. Pt. F J. Rail Rapid Transit.* **2014**, *228*, 570–580. [CrossRef]
9. Wang, P.; Wang, L.; Chen, R.; Xu, J.; Xu, J.; Gao, M. Overview and outlook on railway track stiffness measurement. *J. Mod. Transp.* **2016**, *24*, 89–102. [CrossRef]
10. Lu, S. Real-Time Vertical Track Deflection Measurement System. Ph.D. Thesis, University of Nebraska Lincoln (UNL), Lincoln, Nebraska, 2008.
11. Norman, C.D. Measurement of Track Modulus from a Moving Railcar. Master's Thesis, University of Nebraska-Lincoln, Lincoln city, NE, USA, 2004.
12. Greisen, C.J. Measurement, Simulation, and Analysis of the Mechanical Response of Railroad Track. Master's Thesis, University of Nebraska-Lincoln, Lincoln city, NE, USA, 2010.
13. Thompson, R.; Li, D. Automated vertical track strength testing using TTCI's track loading vehicle. *Technol. Digest.* **2002**, *1489*, 17–25.
14. Li, D.; Thompson, R.; Marquez, P.; Kalay, S. Development and Implementation of a Continuous Vertical Track-Support Testing Technique. *Transp. Res. Rec.* **2004**, *1863*, 68–73. [CrossRef]
15. Rasmussen, S.; Krarup, J.A.; Hildebrand, G. Non-Contact Deflection Measurement at High Speed. In Proceedings of the 6th International Conference on the Bearing Capacity of Roads, Railways and Airfields, Lisbon, Portugal, 24–26 June 2002.
16. Berggren, E.G.; Kaynia, A.M.; Dehlbom, B. Identification of substructure properties of railway tracks by dynamic stiffness measurements and simulations. *J. Sound Vib.* **2010**, *329*, 3999–4016. [CrossRef]
17. With, C.; Metrikine, A.V.; Bodare, A. Identification of effective properties of the railway substructure in the low-frequency range using a heavy oscillating unit on the track. *Arch. Appl. Mech.* **2010**, *80*, 959–968. [CrossRef]
18. Roghani, A.; Hendry, M.; Ruel, M.; Edwards, T.; Sharpe, P.; Hyslip, J. A case study of the assessment of an existing rail line for increased traffic and axle loads. In Proceedings of the IHHA 2015 Conference, Peth, Australia, 21–24 June 2015.
19. Roghani, A.; Hendry, M.T. Assessing the potential of a technology to map the subgrade stiffness under the rail tracks. In Proceedings of the Transportation research board 94th annual meeting; Transportation Research Board, Washington, DC, USA, 11–15 January 2015.
20. Roghani, A.; Macciotta, R.; Hendry, M. Combining track quality and performance measures to assess track maintenance requirements. In Proceedings of the ASME/ASCE/IEEE 2015 Joint Rail Conference, American Society of Mechanical Engineers, San Jose, CA, USA, 23–26 March 2015.
21. Fallah Nafari, S.; Gül, M.; Roghani, A.; Hendry, M.T.; Cheng, J.R. Evaluating the potential of a rolling deflection measurement system to estimate track modulus. *Proc. Inst. Mech. Eng. Pt. F J Rail Rapid Transit.* **2018**, *232*, 14–24. [CrossRef]
22. Mehrali, M.; Esmaeili, M.; Mohammadzadeh, S. Application of data mining techniques for the investigation of track geometry and stiffness variation. *Proc. Inst. Mech. Eng. Pt. F J Rail Rapid Transit.* **2019**, *234*, 439–453. [CrossRef]
23. Lu, S.; Hogan, C.; Minert, B.; Arnold, R.; Farritor, S.; GeMeiner, W.; Clark, D. Exception criteria in vertical track deflection and modulus. In *2007 ASME/IEEE Joint Rail Conference and the ASME Internal Combustion Engine Division, Spring Technical Conference (JRCICE 2007)*; American Society of Mechanical Engineers: Pueblo, CO, USA, 2007; pp. 191–198.

24. El-Sibaie, M.; GeMeiner, W.; Clark, D.; Al-Nazer, L.; Arnold, R.; Farritor, S.; Fateh, M.; Lu, S.; Carr, G. Measurement of Vertical Track Modulus: Field Testing, Repeatability, and Effects of Track Geometry. In Proceedings of the ASME/IEEE Joint Rail Conference, Wilmington, DE, USA, 22–24 April 2009; pp. 151–158.
25. Alireza, R.; Hendry, M.T. Continuous Vertical Track Deflection Measurements to Map Subgrade Condition along a Railway Line: Methodology and Case Studies. *J. Transp. Eng.* **2016**, *142*, 04016059. [CrossRef]
26. Fallah Nafari, S.; Gül, M.; Hendry, M.T.; Cheng, J.R. Estimation of vertical bending stress in rails using train-mounted vertical track deflection measurement systems. *Proc. Inst. Mech. Eng. Pt. F J Rail Rapid Transit.* **2018**, *232*, 1528–1538. [CrossRef]
27. Sadeghi, J.; Barati, P. Evaluation of conventional methods in Analysis and Design of Railway Track System. *IJCE* **2010**, *8*, 44–56.
28. Esveld, C. *Modern Railway Track*; MRT Productions: Duisburg, Germany, 1989; p. 446.
29. Feng, H. *3D-Models of Railway Track for Dynamic Analysis*; Royal Institute of Technology: Stockholm, Sweden, 2011.
30. Fallah Nafari, S.; Gül, M.; Cheng, J.J.R. Quantifying live bending moments in rail using train-mounted vertical track deflection measurements and track modulus estimations. *J. Civ. Struct. Health Monit.* **2017**, *7*, 637–643. [CrossRef]
31. Zakeri, J.A.; Abbasi, R. Field investigation on variation of rail support modulus in ballasted railway tracks. *Lat. Am. J. Solids Struct.* **2012**, *9*, 643–656. [CrossRef]
32. Sussmann, T.R.; Thompson, H.B.; Stark, T.D.; Wilk, S.T.; Ho, C.L. Use of seismic surface wave testing to assess track substructure condition. *Constr. Build. Mater.* **2017**, *155*, 1250–1255. [CrossRef]
33. Bridge, C. *Computers and Structures*; CSI: Berkeley, CA, USA, 2015.
34. Haykin, S.S. *Neural Networks and Learning Machines*; Prentice Hall: New York, NY, USA, 2009; p. 906.
35. Cybenko, G. Approximation by superpositions of a sigmoidal function. *Math. Control Signals Syst.* **1989**, *2*, 303–314. [CrossRef]
36. Guler, H. Prediction of railway track geometry deterioration using artificial neural networks: A case study for Turkish state railways. *Struct. Infrastruct. Eng.* **2014**, *10*, 614–626. [CrossRef]
37. Fink, O.; Weidmann, U. Scope and potential of applying artificial neural networks in reliability prediction with a focus on railway rolling stock. In *European Safety and Reliability Conference: Advances in Safety, Reliability and Risk Management, ESREL 2011*; Taylor and Francis Inc.: Troyes, France, 2012; pp. 508–514.
38. Sadeghi, J.; Askarinejad, H. Application of neural networks in evaluation of railway track quality condition. *J. Mech. Sci. Technol.* **2012**, *26*, 113–122. [CrossRef]
39. Hyndman, R.J.; Koehler, A.B. Another look at measures of forecast accuracy. *Int. J. Forecast* **2006**, *22*, 679–688. [CrossRef]
40. Fallah Nafari, S. *Quantifying the Distribution of Rail Bending Stresses along the Track Using Train-Mounted Deflection Measurements*; University of Alberta: Edmonton, AB, Canada, 2017.
41. Young, S.; Evermann, G.; Gales, M.; Hain, T.; Kershaw, D.; Liu, X.; Moore, G.; Odell, J.; Ollason, D.; Povey, D. *The HTK Book*; Cambridge University Engineering Department: Cambridge, UK, 2006.

Article

Experimental and Numerical Investigations into Dynamic Modal Parameters of Fiber-Reinforced Foamed Urethane Composite Beams in Railway Switches and Crossings

Pasakorn Sengsri, Chayut Ngamkhanong, Andre Luis Oliveira de Melo and Sakdirat Kaewunruen *

Laboratory for Track Engineering and Operations for Future Uncertainties (TOFU Lab), School of Engineering, The University of Birmingham, Edgbaston, Birmingham B15 2TT, UK; pxs905@student.bham.ac.uk (P.S.); cxn649@student.bham.ac.uk (C.N.); alo888@student.bham.ac.uk (A.L.O.d.M.)

* Correspondence: s.kaewunruen@bham.ac.uk

Received: 20 June 2020; Accepted: 17 July 2020; Published: 20 July 2020

Abstract: Dynamic behaviors of composite railway sleepers and bearers in railway switches and crossings are not well-known and have never been thoroughly investigated. In fact, the dynamic properties of the full-scale composite sleepers and bearers are not available in practice. Importantly, the deteriorated condition or even the failure of composite materials and components in the railway system can affect the functional limitations or serviceability of the switches and crossings. Especially, it is important to identify the dynamic modal parameters of Fiber-reinforced Foamed Urethane (FFU) composite railway sleepers and bearers so that track engineers can adequately design and optimize the structural components with their superior properties, for benchmarking with the conventional sleepers and bearers. This paper is the world's first to investigate the vibration characteristics of full-scaled FFU composite beams in healthy and damaged conditions, using the impact hammer excitation technique. This study also determines the dynamic elastic modulus of FFU composite beams from experimental dynamic measurements. It is found that the first bending mode in a vertical plane obviously is the first dominant mode of resonance under a free-free condition. The dynamic modal parameters reduce when damages occur. In this study, finite-element modeling has been used to establish a realistic dynamic model of the railway track incorporating FFU composite sleepers and bearers. Then, numerical simulations and experimental campaigns have been performed to enable new insights into the dynamic behaviors of composite sleepers and bearers. These insights are fundamental to the performance benchmarking as well as the development of vibration-based condition monitoring and inspection for predictive track maintenance.

Keywords: Fiber-reinforced Foamed Urethane (FFU); free vibration; impact hammer excitation technique

---

1. Introduction

Railway sleepers and bearers are typically made of timber, concrete, steel, and other composite materials. In traditional railway tracks, timber is normally used as railway sleepers and bearers. Due to the diverse environmental concern of noble wood leading to the high deterioration rate of timber sleepers, the need of using other materials has grown. Currently, the development and improvement of railway structure, which is economically competitive for meeting the requirements of the industry, is a key challenge for track engineers. One of the major concerns in the railway industry is the replacement of deteriorated and damaged timber sleepers in existing railway tracks [1]. Recently, polymer and composite sleepers with mostly fiber materials have been developed [2] and designed to mimic timber behavior [3,4]. This is conducted presently on a basis of a like-for-like replacement in terms of equivalent static performances (i.e., similarities of static strength, modulus of elasticity, stiffness, etc.). For example, a fiber composite system is composed of a lightweight polymer matrix with strong fibers added into the matrix [2]. These fibers can well resist forces because of their extreme strength and can be used only in the longitudinal and/or transverse direction. The static strength and elastic modulus of composites are found to be equivalent to hard timber. Recently, practitioners have strong concerns whether dynamic properties should rather be considered due to the fact that railway tracks are generally exposed to dynamic loading conditions. It is also well-known that concrete and steel are likely to have nearly no damping coefficient when compared to timber, which has an outstanding damping coefficient [5–8]. In recent reviews [9,10], it has been found that steel bearers behaved well in the short-term, but tended to have higher turnout settlements and severe ballast breakage in the long-term. In contrast, concrete tends to be an extremely good counterpart to enhance track and turnout stability in a longitudinal, vertical, and lateral direction [11,12]. However, concrete is relatively much heavier than timber and it is impractical to use concrete bearers as timber barer replacement. A major benefit of using polymer and composite sleepers and bearers is their flexibility, which results in an extreme ability to withstand vibrations induced by dynamic forces in a railway track system [13,14]. Moreover, polymer and composite sleepers and bearers are durable, simple to make, are presently cost-attractive, and need low to nearly no maintenance. Therefore, their improved lifecycle is useful for areas that are very difficult to maintain, for instance, turnouts (or referred to as 'switches and crossings'), bridges, and tunnels. Another benefit is that the utilization of the polymer and composite sleepers and bearers can handle the constant rise of concern throughout the existing environment in the present industry, because of its durability and its nearly 100% recyclability [15].

Composite railway sleepers and bearers are one of the most attractive structural elements in a railway infrastructure, acting as crosstie beams, which are placed under the rails to support track loading [16]. Their key functions are not only to transfer and distribute dynamic train loads to track substructure, but also to ensure safe rail gauge that permits the train to travel securely [17–19]. The vibration of Fiber-reinforced Foamed Urethane (FFU) composite sleepers and bearers in a railway turnout system is a key factor causing failure of FFU composite sleepers and bearers and excessive railway track maintenance costs. As such, the performance of Fiber-reinforced Foamed Urethane (FFU) composite sleepers and bearers over the entire service life and their failure modes under vibration cannot be fully identified to establish a design standard for these composite sleepers and bearers. It is important to comprehend the dynamic modal parameters of the composite sleepers and bearers to develop and design a realistic dynamic model of a railway track for predicting its responses under vibration. The essential information of the dynamic parameters can be used for dynamic performance benchmarking when a new material is manufactured for railway applications. Furthermore, the information is critical in the development of a predictive vibration-based condition assessment of the components. On this ground, it is necessary to monitor and inspect the vibration behavior of FFU composite sleepers and bearers during operation in order to prioritize and plan effective maintenance management. Note that the inspection of railway sleepers and bearers is currently carried out by visual observation. Monitoring of dynamic properties can provide an alternative technique in structural integrity assessment for track engineers.

It is noted that the use of common damage detection techniques (visual observation) is inefficient to identify any component damage in real-time and they cannot perform to completely reduce track possessions (i.e., track maintenance time). In many engineering applications, one of the reliable inspection techniques widely used in modal analysis is based on an instrumented hammer impact excitation. Modal analysis is a useful tool for comprehending the vibration characteristics of mechanical structures. This tool converts the vibration waves of excitation and response identified on a complicated structure into a range of predictive modal parameters [20]. One with the most perspectives of structural dynamics is the modal domain, which provides modal parameters (such as natural frequencies, dynamic stiffness, and dynamic damping). A structure deforms or vibrates in particular shapes, so-called 'mode shapes', when the structure is excited at its natural frequencies. It will move back and forth in a complex combination, which includes all the mode shapes under common operation conditions [20].

In a modal testing process, a Frequency Response Function (FRF) is a transfer function used for impact hammer analysis in order to determine the resonant frequencies and mode shapes, as well as damping of a vibrating structure [21]. During the design phase, the dynamic modal parameters obtained from the FRF are an important factor to consider before manufacturing a real structure to find and eliminate potential problems early [21]. In 2006, Kaewunruen and Remennikov carried out an investigation into the modal analysis of pre-stressed concrete sleepers for evaluating dynamic behaviors of the sleepers, using the impact hammer excitation technique at a particular frequency series of 0–1600 Hz [21,22]. According to their study, the PROSIG modal analysis suite was used to measure the frequency response functions (FRFs). They also used the STAR Modal analysis package to determine the natural frequencies and corresponding modal shapes of each sleeper from the FRFs. Obviously, the impact hammer excitation technique is one of the most attractive non-destructive force excitation methods to identify dynamic modal parameters of a structure under vibration. These modal parameters are helpful for the development of a realistic dynamic model of railway composite sleepers and bearers capable of predicting its dynamic responses.

In terms of mechanical properties of a common material, two independent constants called elastic modulus, E, and shear modulus, G, define the elastic properties for linearly elastic isotropic solids. The design of engineered structures has been significantly concerned about these two elastic properties. For the above reason, many experimental methods to identify E and G have been developed. These methods consist of two sets, which are static and dynamic techniques. According to studies in [23–25], the researchers were the first to determine the elastic properties of isotropic materials, based on non-destructive vibration testing. They established the formulae to calculate dynamic E and G from the natural frequencies in bending and twisting modes of cylinder and prisms, based on the Timoshenko beam theory. In fact, the base of the ASTM criterion [26] test method to characterize dynamic elastic properties was set by those researchers, using the impulse excitation technique [27]. To date, several researchers have investigated the estimation of the elastic properties of laminated composites [28–33], timber materials [34,35], or concrete materials [36], which are non-isotropic and/or inhomogeneous, using vibration-based approaches. It is important to note that the standard tests for dynamic properties gain significant supports from scientific and engineering communities in present days.

For railway applications, it is well-known that a common turnout induces high impact loads on the structural members because of its blunt geometry and mechanical connections between closure rails and switch rails. Therefore, the turnout system requires improved structural members, which use an alternative material like the FFU material, having an identical timber-like behavior. For that reason, the FFU material in switches and crossings offers its high-impact attenuation, high damping property, high UV resistance, lightweight, and long-lifecycle. The static properties of FFU bearers are presented in Table 1. However, neither of the dynamic properties of FFU bearers have been investigated before, nor are available in open literature.

In this paper, the experimental and numerical dynamic parameters of FFU composite beams in free-free conditions have been identified. The free-free condition is scientifically ideal for performance benchmarking or comparison of test results. This condition excludes uncertainties that can affect the test results, such as type of ballast, type of support, and type of fastening system. This condition is very critical when the like-for-life performance of an individual component is being assessed. In this study, two FFU composite beams have been tested using an impact hammer excitation technique over the frequency range of interest: 0 to 2100 Hz. Frequency response functions (FRFs) have been measured using the Modal Analysis Suite package to identify natural frequencies and the corresponding mode shapes, as well as damping values for the full-scale beams. The experimental results provide the correlations between modal parameters and structural damage. Then, the experimental natural frequencies are used to determine the dynamic elastic moduli of the beam in different bending modes. Therefore, the vibration parameters of FFU composite beams are inevitably required for the development of a realistic dynamic model of a railway track capable of predicting its responses to impact loads stemmed from irregularities of the rail, wheel burns, and so on.

Table 1. Static properties of Fiber-reinforced Foamed Urethane (FFU) material.

Properties [37]	Units	FFU Sleepers and Bearers [6]			
		New	After 10 Years Used	After 15 Years Used	After 30 Years Used
Elastic Modulus	GPa	8.10	8.04	8.79	8.41
Bending Strength	GPa	0.142	0.125	0.131	0.116
Shear Strength	MPa	10	9.5	9.6	7
Vertical Compressive Strength	MPa	58	66	63	55
Density	kg/m^3	740	740	740	740
Service Life	Years	50	40	35	20
Hardness	MPa	28	25	17	-

2. Materials and Methods

In this study, a non-destructive testing method is considered in order to obtain comprehensive insights into the dynamic structural behaviors of the FFU composite sleepers and bearers and the relationship between the damage and the dynamic responses obtained from the method. The method used in this investigation is the 'modal testing and analysis' to extract the dynamic Young's modulus of FFU composite beams under free vibrations from the experimental dynamic measurements.

2.1. Modal Equipment

Modal testing has been performed in conjunction with the mechanical load tests of all specimens to determine modal parameters of the specimen at various states, given in Figure 1. DATS software has been designed by PROSIG for modal analysis. The FFU composite beams have been tested using an impact hammer excitation technique over the frequency range of interest: 0 to 2100 Hz. The data is acquired using the PROSIG P8004 acquisition device for impact hammer modal testing. The modal signals have been measured and recorded using a 2-channel data acquisition for accelerometer and modal hammer connection. The FRFs, which are varied in different ways for healthy and damaged conditions, are then processed using Modal Analysis Suite package to identify natural frequencies and the corresponding mode shapes of the beam specimens.

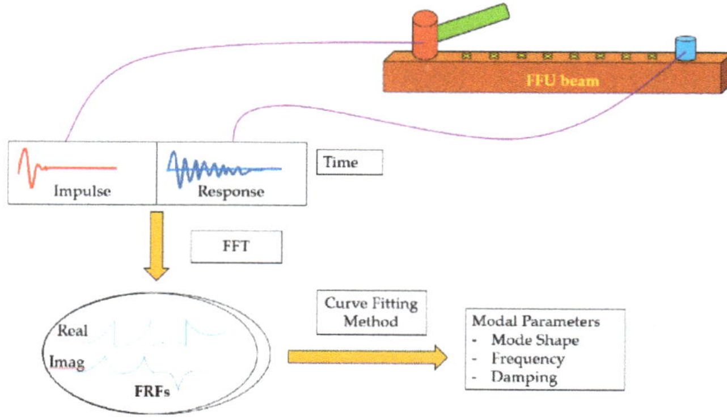

Figure 1. Modal testing and analysis.

2.2. Experimental Overview

The two full-scale composite beam specimens have been prepared for load tests, as shown in Figure 2a. The dimension of each beam is 160 mm deep × 260 mm wide × 3300 mm long, kindly provided by an industry partner. The experimental investigations are conducted in accordance with EN 13230 (Test material, specifications, support conditions, loading procedures, and other requirements needed for bending tests on railway track concrete sleepers). Note that EN 13230 has some limitations in order to detect the failure mode of flexible composites. Especially, some experimental arrangements are adapted to examine the structural damage and the failure mode of the full-scale FFU composite beams [38–41]. In this study, modal tests have been conducted using an impact excitation technique in a free-support condition (or *'free-free condition'*). The damage and failure are observed using three-point bending tests following EN 13230, in order to investigate the damage and failure of the beams. The modal parameters of FFU composite beams under different conditions are then investigated.

Figure 2. (**a**) FFU 17-06 specimens and (**b**) rubber cushions.

2.2.1. Modal Testing

The dynamic modal parameters have been identified for both healthy and damaged conditions. It should be noted that bending tests are conducted to trigger different levels of damage. Firstly, both specimens are tested under healthy condition. This test requires laying two soft rubber cushions shown in Figure 2b. These very soft cushions have been placed underneath each sample, so that the free-free boundary conditions can be incited for the modal parameters of the sole specimens. This free-free condition is imperative if the dynamic parameters of an individual component are required in any like-for-like performance comparison.

Secondly, the experimental modal analysis has been performed to identify dynamic parameters of the specimens under different severity states of damaged conditions. The equipment used for this test is a Prosig-P800 impact hammer as given in Figure 3a. The 34 uniform locations have been marked on the surface of each sample as the excitation locations of the impact hammer. The accelerometer is fixed at one corner to record the acceleration, as shown in Figure 3b. According to the EN 13230 criterion [41], the dynamic responses up to 2100 Hz are recorded. In addition, these attributes are clearly defined using a curve fitting method. Data modal analysis is a package that can create optimisation algorithms and provide relevant frequency-dependent shapes to explain the data sets. This data can be transformed into curve images; and mode shapes can be determined by the *'animation drawing suite'*.

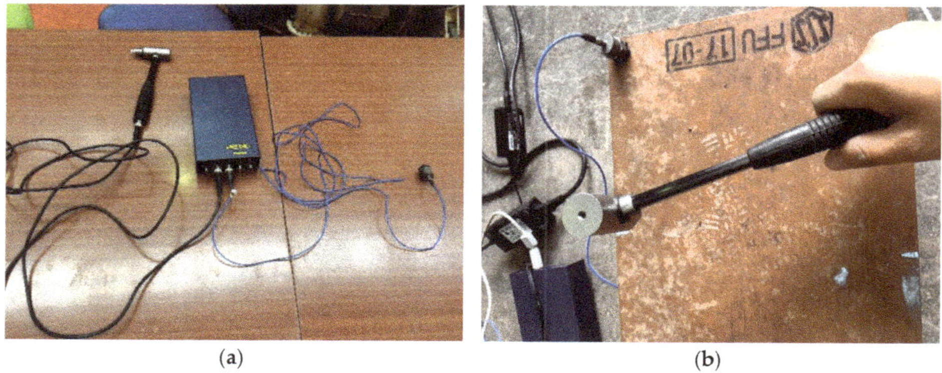

Figure 3. (**a**) PROSIG-P800 impact hammer and (**b**) excitation locations and an accelerometer position.

2.2.2. Three-Point Bending Tests

According to EN 13230-2 [41], the standard requires positive and negative three-point bending tests for sleepers at the rail seat support. Only positive bending tests have been carried out due to the symmetrical shape of the samples. This means that the samples have the same positive and negative capacity. Also, the criterion requires articulated support and must be 100 mm wide, made of steel with Brinell: HBW > 240. A static load is applied at the mid-span to cause positive 3-point bending cracks and failure. Figure 4 shows the layout of the bending load process, also illustrates the excitation locations of the impact hammer, which have been strategically installed to perform two modal tests under different bending loads. Figure 5 demonstrates two pattern tests of the samples under different bending load conditions. The investigations are sufficiently performed in order to comply with BS EN 13230-1 standard [41].

Figure 4. *Cont.*

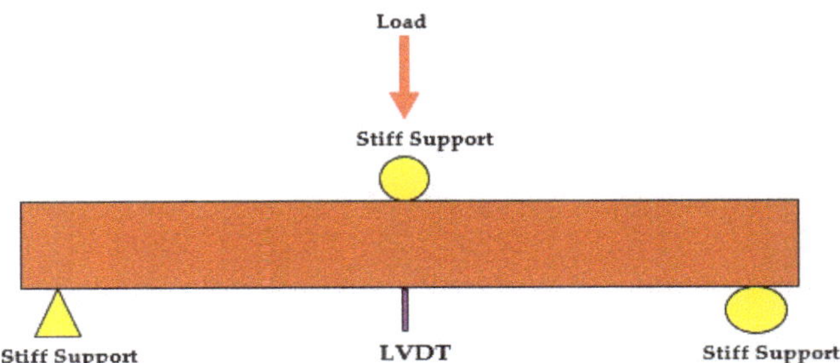

Figure 4. Testing arrangement of a full-scale FFU composite beam under bending loads.

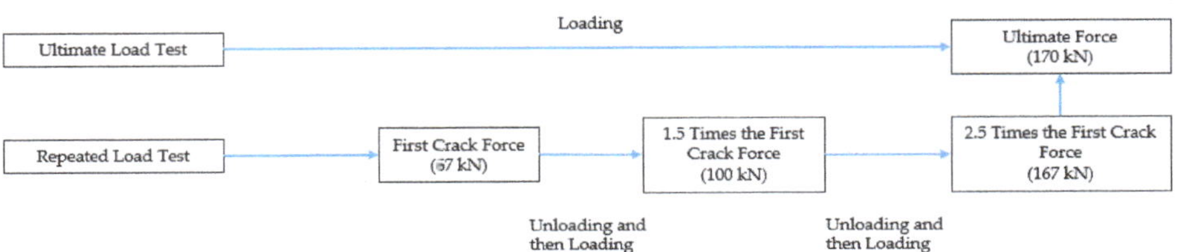

Figure 5. Testing procedure of ultimate load test and repeated load test.

2.3. Determination of Dynamic Elastic Modulus

Dynamic elastic properties of a material can thus be calculated if the mass, geometry, and mechanical resonant frequencies of the test sample can be measured. This means that the dynamic Young's modulus can be identified utilizing the resonant frequency in either the bending or longitudinal mode of vibration, as given in Equation (1) [25,42], whilst the dynamic shear modulus, also known as modulus of rigidity, can be found by employing twisting resonant vibrations [23,43]. In this section, we only focus on the determination of dynamic elastic modulus in free-free boundary conditions (for future benchmarking purpose). This is because, based on the experimental results in the following part, it can be found that the first bending mode in a vertical plane obviously dominates the first resonant mode of vibration in free-free boundary conditions. By employing bending vibration modes, slender beams based on the Euler–Bernoulli theory of bending vibrations can be applicable to the test sample. The influences of rotational inertia and shear can be negligible generally. The equations derived on these assumptions are sufficient for relatively slender beams of lower modes. Nevertheless, this theory is likely to slightly overestimate the natural frequencies. According to Euler–Bernoulli's basic equations of flexure, the dynamic elastic modulus in bending of a beam can be assessed under forced free or bending free vibrations. The dynamic elastic modulus in bending of a beam can be expressed as Equation (1):

$$\left(\frac{E_{dy}}{\rho}\right)_n = \frac{\left(2\pi L F_{f,n}\right)^2}{K_n^4 \beta}, \quad (1)$$

where E_{dy} is dynamic elastic modulus (Pa), n is mode number, L is free length (m), ρ is stabilized density (kg·m^{-3}), $F_{f,n}$ is frequency of nth mode (Hz), and K_n is a coefficient related to the beam's support condition and mode number (e.g., K_1 is equal to 4.73 for a free-free end condition and 1.785 for a fixed-free condition [44], as given in Table 2). Finally, β is the square value of gyration radius divided by free length as provided in Equation (2):

$$\beta = \left(\frac{1}{L}\sqrt{\frac{I}{A}}\right)^2 = \frac{I}{L^2 A}. \quad (2)$$

Herein, β denotes the square value of gyration radius divided by free length, L. I is the moment of inertia about the axis and A is the cross-section area. If no axis is specified, the centroidal axis is assumed.

Table 2. Dimensionless coefficients for computing the frequencies of a FFU composite beam in free-free conditions.

Mode No.	**1**	**2**	**3**	**4**	**5**	**6**
K_n	0 (Translation)	4.730	7.853	10.996	14.137	$\approx \frac{(2n-1)\pi}{2}$

It is important to note that Equation (1) is a conceptual equation of vibration, which ignores the influence of rotational deformation and shear load in a simulation. Nevertheless, for an application of using this equation, it could be dominated by L/h ratios (i.e., more than 58 in a fixed-free end condition or more than 20 in a free-free end condition) [45]. In this paper, the modeling of FFU composite beam does not take into account shear deformation and rotational bending effects (as defined by the Timoshenko theory), due to the ratio of L/h $\geq$ 20 (thin beam). Additionally, both previous equations are limited to isotropic materials. It is noted that the FFU composite beam model was considered as an isotropic material. In fact, this material would be considered to be anisotropic, but we measure its dynamic responses only in the vertical direction. Thus, the material can be considered conceptually to be isotropic. The following section presents the numerical investigations of a FFU composite beam modeling using the dynamic parameters obtained from the experiments in order to determine the dynamic elastic modulus of the beam.

2.4. A Finite-Element (FE) Model

A three-dimensional FFU composite beam model under free-free boundary conditions has been developed to study its dynamic response and compare with the experimental results. The Strand7 software [46] is used to model this 3D simulation, which employs 60 Euler–Bernoulli beam elements with 61 nodes, due to the model acting as a shallow beam. Figure 6 demonstrates the three-dimensional finite element model of a FFU composite beam. The modification for the geometric and material characteristics of these components has been based on the experimental data. The engineering properties are presented in Table 3 [47,48].

Figure 6. Finite element modeling of a FFU composite beam in free-free conditions.

Table 3. Geometric parameters employed in the dynamic simulation.

Parameter lists	**Values**	**Units**
Density	740	kg/m^3
Length	3.3	m
Cross-section area	$0.16 \times 0.26 = 0.042$	m^2

3. Results

3.1. Experimental Results

The results of the vibration tests for FFU composite beams under an ultimate load test and repeated load test for healthy and damaged conditions are presented in Tables 4 and 5 and Figures 7 and 8. The first five-mode shapes of vibration under ultimate and repeated load tests are shown in Tables 4 and 5, respectively. For all beams, the first natural bending mode in a vertical plane obviously controlled the first resonant mode of vibration both under an ultimate load test and repeated load test.

In addition, the lowest frequency corresponds to the natural bending mode, the second frequency to the lowest torsional mode, the third frequency to the second bending mode, the fourth frequency to the second torsional mode, and the fifth mode to the third bending mode. Clearly, the internal dynamic properties of FFU composite beams can be changed when damages occur.

Table 4 exhibits the results of natural frequencies and damping values of the FFU composite beams under the ultimate load test. The differences between the natural frequencies of all mode shapes in healthy and failed conditions are 16.5%, 11.4%, 15.1%, 22.46%, and 25.27%, respectively. As shown in Figure 7, the frequencies of all five modes under failed conditions are lower than those under healthy conditions. For damping values, the value of the first mode damping values under failed conditions increased by 49%, compared with those under healthy conditions. There are several transverse damages on the beam surface for the first time under failure conditions, and there are cracks (30 mm in width). Nevertheless, the beam specimens remain the same and could completely recover without any load. After measurement and unloading, the residual bending deformation level of the material is only 2 mm.

Table 4. Frequencies, damping values, and mode shapes under ultimate load test for healthy and failed conditions.

Healthy Condition		**Failed Condition**		**Difference**	
Frequency (Hz)	**Damping (%)**	**Frequency (Hz)**	**Damping (%)**	**Frequency (Hz)**	**Damping (%)**
Mode 1 (1st bending)					
68.23	3.96	56.92	5.9	11.31	1.94
Mode 2 (1st twisting)					
85.78	2.98	75.94	3.83	9.84	0.85
Mode 3 (2nd bending)					
143.61	3.37	121.87	2.8	21.74	0.57
Mode 4 (2nd twisting)					
180.14	3.85	139.68	3.99	40.46	0.14
Mode 5 (3rd bending)					
247.96	4.96	185.28	2.53	62.68	2.43

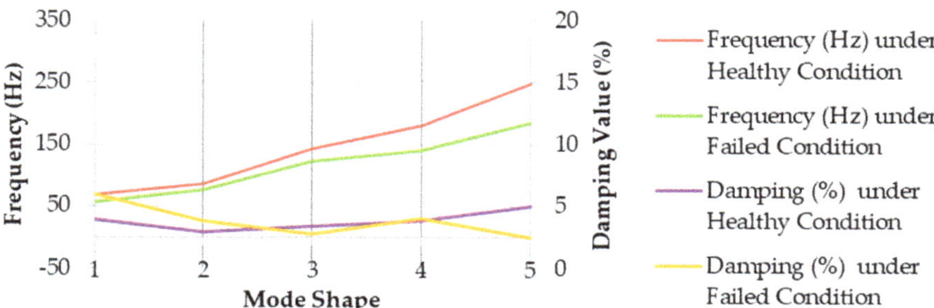

Figure 7. Frequencies against damping values over mode shapes under ultimate load test for healthy and failed conditions.

The dynamic behavior of the FFU composite beams under the repeated load test is demonstrated in Tables 5 and 6 and Figure 8. In Table 5, it is clear that all modes have no obvious deviation before the load reached 100 kN. Beyond this load to the ultimate load, the frequencies of all five modes tend to reduce with percent variations of different mode shapes. The maximum difference of frequency is found in the first mode, approximately 27%, compared with the frequency under healthy conditions. Surprisingly, the frequencies of the fourth mode are unchanged under the different loading conditions. A comparison of natural frequencies and damping values between healthy and damaged conditions in all the five modes is presented in Table 6, which shows that there were maximum differences in natural frequencies and damping values between the healthy condition and the ultimate loading condition (170 kN). We note that the minimum difference in frequencies and damping values between the healthy condition and the damaged condition could be found under a load of 67 kN.

However, the different frequencies of the other modes reduce dramatically, as shown in Figure 8. In regards to damping values in Table 5, all the modes except the first and fifth mode are scant. Obviously, the difference of damping values between under 100 kN and 167 kN loading in the first mode is significantly high and increased two-fold from 3.94 to 8.24. In addition, the difference of damping values between healthy conditions and 67 kN loading in the fifth mode is considerable, which decreased by 36% from 2.39 to 1.53. It is clear that the dynamic modal parameters of FFU beams decrease when damages appear. These beams could reduce with the damage severity.

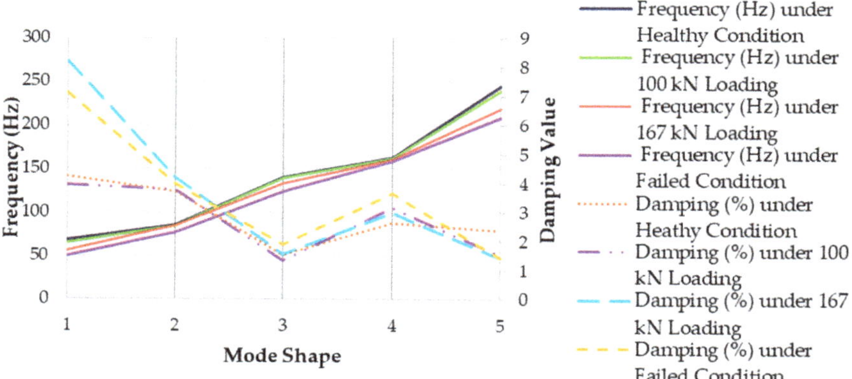

Figure 8. Frequencies against damping values over mode shapes under repeated load test for healthy and damaged conditions.

Table 5. Frequencies, damping values, and mode shapes under repeated load test for healthy and damaged conditions.

Healthy	67 kN	100 kN	167 kN	170 kN (Failed)
Frequency (Hz)/Damping (%)				
Mode 1 (1st bending)				

69.2/4.23	68.03/5.04	65.84/3.94	55.99/8.24	50.44/7.12
Mode 2 (1st twisting)				

85.63/1.55	84.8/3.84	85.24/3.78	85.54/4.22	77.31/3.97
Mode 3 (2nd bending)				

140.09/1.55	139.88/1.6	139.04/1.35	132.98/1.58	124.08/1.9
Mode 4 (2nd twisting)				

163.3/2.66	162.43/3.16	162.13/3.16	160.82/3.00	159.07/3.66
Mode 5 (3rd bending)				

244.77/2.39	241.69/1.53	239.30/1.54	219.40/1.44	209.17/1.44

Table 6. Relative values of frequencies and damping to healthy condition.

	Difference in Frequencies (Hz)/Damping (%)			
No. Mode	67 kN	100 kN	167 kN	170 kN (Failed)
1	1.17/0.81	3.36/0.29	13.21/4.01	18.76/2.89
2	0.83/2.29	0.39/2.23	0.09/2.67	8.32/2.42
3	0.21/0.05	1.05/0.20	7.11/0.03	16.01/0.35
4	0.87/0.50	1.17/0.50	2.48/0.34	4.23/1.00
5	3.08/0.86	5.47/0.85	25.37/0.95	35.60/0.95

3.2. Numerical Results

Based on the frequencies experimentally obtained by the impact hammer excitation technique, dynamic Young's modulus, E, can be computed using Equation (1). As shown in Table 7, the big difference of the Young's modulus values is significant in the first bending mode and relatively small when compared with the Young's modulus for FFU composite sleepers, E = 8.1 GPa, according to the reviews in [48,49].

In order to verify the model, the natural frequencies of a full-scale FFU beam in free-free conditions are calibrated against the existing experiments. The values of a dynamic elastic modulus in different bending modes obtained from Table 7 have been used in the finite element analysis. A comparison between numerical and experimental investigations for frequencies and mode shapes are given in Table 8, especially the experimental data based on the ultimate load test. The results are found to be in a very good agreement in all the first five modes. The maximum difference of frequencies between the numerical and experimental data is less than 4% in the second twisting mode, because of the effects of experimental disturbances in our laboratory. Additionally, there is a satisfied correlation between both results for the shifts in natural frequencies under free vibration. It is important to note that the numerical modal analysis of a FFU composite beam can only be achieved under free-free boundary conditions. This free-free boundary condition is commonly used for performance benchmarking of an individual component (i.e., like-for-like comparison), especially for railway sleepers and bearers, which are safety-critical components [50]. In the near future, the situ investigation into modal parameters of FFU composite beams can be further carried out in order to determine the effect of different boundary conditions (e.g., type of ballast aggregate, or resultant effects of particle size distribution, tamping) on vibration properties of the beams.

Table 7. Determination of dynamic Young's modulus from dynamic measurements in free-free end conditions.

Mode No.	Experimental Frequency (Hz)	Dynamic Young's Modulus, E (GPa)
1 (bending)	68.23	15.34
2 (twisting)	85.78	-
3 (bending)	143.61	8.81
4 (twisting)	180.14	-
5 (bending)	247.96	6.83

Table 8. Natural frequencies of a conceptual FFU composite beam (Hz) in the free-free conditions.

Mode No.	Mode Shape	Numerical (Hz)	Experimental (Hz)	Difference (%)
1 (First Bending)		68.70	68.23	0.69
2 (First Twisting)		86.82	85.78	1.21
3 (Second Bending)		143.44	143.61	0.12
4 (Second Twisting)		173.58	180.14	3.64
5 (Third Bending)		247.44	247.96	0.21

4. Conclusions

Dynamic modal parameters of FFU composites are extremely significant for the development of a realistic dynamic model of a railway track capable of predicting its dynamic responses for predictive and preventative maintenance to ensure railway safety. The results of the experimental and numerical modal analysis for Fiber-reinforced Foamed Urethane composite beams in free-free boundary conditions are indicated in this study. For the purpose of like-for-like performance benchmarking for a particular component, the free-free boundary condition is considered to be more suitable since the test results will not be affected by uncertainties stemmed from supports (e.g., dimension and particle size distribution of ballast, tamping technique, ballast geological properties). Full-scaled experiments have been performed to artificially create damage and failure in accordance with European standards. Dynamic parameter tests have been conducted by using an impact hammer excitation technique throughout the frequency range of interest: from 0 to 2100 Hz. According to experimental results, it provides the correlations between modal parameters and structural damage. Furthermore, the dynamic parameters obtained are later used to extract the dynamic elastic moduli. The results of frequency parameters under free-free conditions are in a very good agreement between experimental and numerical data with less than 4% discrepancy. Further research could be conducted to investigate the vibration characteristics of Fiber-reinforced Foamed Urethane composite beams in situ conditions in order to consider the influence of various ballast conditions on the natural frequencies, modal damping values, and vibration mode shapes of FFU composite beams under the in-situ boundary conditions. Some interesting novel findings from this research can be concluded as follows:

- The first bending mode in a vertical plane obviously dominates the first resonant mode of vibration under a free-free condition;
- The dynamic modal parameters of full-scale FFU composite beams reduce when damages occur. Thus, they decrease with damage severity;
- The highest dynamic Young's modulus of FFU composite beams is found in the resonant frequency of the first bending mode and also reduces when the second and third bending modes appear.

Author Contributions: Conceptualization, P.S. and S.K.; data curation, P.S. and A.L.O.d.M.; formal analysis, P.S., C.N., A.L.O.d.M., and S.K.; funding acquisition, S.K.; investigation, P.S., C.N., A.L.O.d.M., and S.K.; methodology, P.S., C.N., and S.K.; project administration, S.K.; resources, S.K.; software, S.K.; supervision, S.K.; validation, P.S., C.N., and A.L.O.d.M.; visualization, P.S., C.N., and A.L.O.d.M.; writing—original draft, P.S., C.N., and A.L.O.d.M.; writing—review & editing, S.K. All authors have read and agreed to the published version of the manuscript.

Funding: The authors wish to gratefully acknowledge the European Commission for the financial sponsorship of the H2020-MSCA-RISE Project No. 691135 "RISEN: Rail Infrastructure Systems Engineering Network," which enables a global research network that tackles the grand challenge in railway infrastructure resilience and advanced sensing in extreme environments (www.risen2rail.eu). In addition, this project is partially supported by the European Commission's Shift2Rail, H2020-S2R Project No. 730849 "S-Code: Switch and Crossing Optimal Design and Evaluation". The APC is kindly sponsored by MDPI's Invited Paper Program.

Conflicts of Interest: The authors declare no conflict of interest.

References

1. Van Erp, G.; McKay, M. Recent Australian Developments in Fibre Composite Railway Sleepers. *Electron. J. Struct. Eng.* **2013**, *13*, 62–66.
2. Manalo, A. Behaviour of Fibre Composite Sandwich Structures: A Case Study on Railway Sleeper Application. Ph.D. Thesis, Centre of Excellence in Engineered Fibre Composites Faculty of Engineering and Surveying University of Southern Queensland Toowoomba, Toowoomba, Australia, 2011.
3. Ngamkhanong, C.; Kaewunruen, S.; Costa, B.J.A. State-of-the-Art Review of Railway Track Resilience Monitoring. *Infrastructures* **2018**, *3*, 3. [CrossRef]
4. Gamage, E.K.; Kaewunruen, S.; Remennikov, A.M.; Ishida, T. Toughness of Railroad Concrete Crossties with Holes and Web Openings. *Infrastructures* **2017**, *2*, 3. [CrossRef]
5. Kaewunruen, S. Monitoring in-service performance of fibre-reinforced foamed urethane material as timber-replacement sleepers/bearers in railway urban turnout systems. *Struct. Monit. Maint.* **2014**, *1*, 131–157.
6. Kaewunruen, S. In situ performance of a complex urban turnout grillage system using fibre-reinforced foamed urethane (FFU) bearers. In Proceedings of the 10th World Congress on Rail Research, Sydney, Australia, 25–28 November 2013.

7. Kaewunruen, S. Monitoring structural deterioration of railway turnout systems via dynamic wheel/rail interaction. *Case Stud. Nondestr. Test. Eval.* **2014**, *1*, 19–24. [CrossRef]
8. Indraratna, B.; Salim, W.; Rujikiatkamjorn, C. *Advanced Rail Geotechnology—Ballasted Track*; CRC Press/Balkema: Leiden, The Netherlands, 2011.
9. RailCorp. *Timber Sleepers & Bearers; Engineering Specification SPC 231*; RailCorp: Sydney, Australia, 2012.
10. Kaewunruen, S.; Remennikov, A.M. Dynamic flexural influence on a railway concrete sleeper in track system due to a single wheel impact. *Eng. Fail. Anal.* **2009**, *16*, 705–712. [CrossRef]
11. Remennikov, A.M.; Kaewunruen, S. A review of loading conditions for railway track structures due to train and track vertical interaction. *Struct. Control. Health Monit.* **2008**, *15*, 207–234. [CrossRef]
12. Standards Australia. *Australian Standards: AS3818.2 Timber*; Standards Australia: Sydney, Australia, 2001.
13. Dindar, S.; Kaewunruen, S.; An, M. Identification of Appropriate Risk Analysis Techniques for Railway Turnout Systems. *J. Risk Res.* **2016**, *21*, 974–995. [CrossRef]
14. Pen, L.L. Track Behaviour: The Importance of the Sleeper to the Ballast Interface. Ph.D. Thesis, University of Southampton, Southampton, UK, 2008.
15. Griffin, D.W.P.; Mirza, O.; Kwok, K.; Kaewunruen, S. Finite element modelling of modular precast composites for railway track support structure: A battle to save Sydney Harbour Bridge. *Aust. J. Struct. Eng.* **2015**, *16*, 150–168. [CrossRef]
16. Kaewunruen, S.; Sengsri, P.; de Melo, A.L.O. Experimental and Numerical Investigations of Flexural Behaviour of Composite Bearers in Railway Switches and Crossings. In *Sustainable Issues in Transportation Engineering*; Mohammad, L., Abd El-Hakim, R., Eds.; Sustainable Civil Infrastructures; GeoMEast; Springer: Cham, Switzerland, 2019.
17. Silva, E.A.; Pokropski, D.; You, R.; Kaewunruen, S. Comparison of structural design methods for railway composites and plastic sleepers and bearers. *Aust. J. Struct. Eng.* **2017**, *18*, 160–177. [CrossRef]
18. Kaewunruen, S. Discussion of "Evaluation of an Innovative Composite Railway Sleeper for a Narrow-Gauge Track under Static Load" by Wahid Ferdous, Allan Manalo, Gerard Van Erp, Thiru Aravinthan, and Kazem Ghabraie. *J. Compos. Constr.* **2019**, *23*, 07018001. [CrossRef]
19. Kaewunruen, S.; Remennikov, A.M.; Murray, M.H. Introducing a new limit states design concept to railway concrete sleepers: An Australian experience. *Front. Mater.* **2014**, *1*, 8. [CrossRef]
20. Crystal Instruments. Basics of Structural Vibration Testing and Analysis. Available online: https://www.crystalinstruments.com/basics-of-modal-testing-and-analysis (accessed on 12 January 2020).
21. Kaewunruen, S. Experimental and Numerical Studies for Evaluating Dynamic Behaviour of Prestressed Concrete Sleepers Subject to Severe Impact Loading. The University of Wollongong. Available online: http://ro.uow.edu.au/theses/277 (accessed on 8 January 2020).
22. Remennikov, A.; Kaewunruen, S. Experimental Investigation on Dynamic Railway Sleeper/Ballast Interaction. *Exp. Mech.* **2006**, *46*, 57–66. [CrossRef]
23. Pickett, G. Equations for computing elastic constants from flexural and torsional resonant frequencies of vibrating prisms and cylinders. *ASTM Proc.* **1945**, *45*, 846–865.
24. Spinner, S.; Reichard, T.W.; Tefft, W.E. A Comparison of Experimental and Theoretical Relations Between Young's Modulus and the Flexural and Longitudinal Resonance Frequencies of Uniform Bars. *J. Res. Natl. Bur. Stand. Sect. A Phys. Chem.* **1960**, *64*, 2–147. [CrossRef] [PubMed]
25. Spinner, S.; Tefft, W.E. A Method for Determining Mechanical Resonance Frequencies and for Calculating Elastic Moduli from These Frequencies. *ASTM Proc.* **1961**, *61*, 1221–1238.
26. ASTM Standard E 1876–07. *Standard Test Method for Dynamic Young's Modulus, Shear Modulus, and Poisson's Ratio by Impulse Excitation of Vibration*; ASTM International: West Conshohocken, PA, USA, 1876.
27. Wesolowski, M.; Barkanov, E.; Rucevskis, S.; Chate, A.; La Delfa, G. Characterisation of elastic properties of laminated composites by non-destructive techniques. In Proceedings of the 17th International Conference on Composite Materials, Edinburgh, UK, 27–31 July 2009.
28. Yu, L.; Wang, J.; Xia, D. Vibration Method for Elastic Modulus of Glued Laminated Beams. In Proceedings of the 3rd International Conference on Material, Mechanical and Manufacturing Engineering, IC3ME 2015, Guangzhou, China, 27–28 June 2015; Atlantis Press: Amsterdam, The Netherlands, 2015; pp. 1603–1606.
29. Giaccu, G.F.; Meloni, D.; Valdès, M.; Fragiacomo, M. Dynamic determination of the modulus of elasticity of maritime pine cross-laminated panels using vibration methods. *WIT Trans. Ecol. Environ.* **2017**, *226*, 571–579.

30. Ivanova, Y.; Partalin, T.; Georgiev, I. Characterisation of elastic properties of laminated composites by ultrasound and vibration. *Sci. Proc. Sci. Tech. Union Mech. Eng.* **2016**, *1*, 418–425.
31. Gillich, G.R.; Samoilescu, G.; Berinde, F.; Chioncel, C.P. Experimental determination of the rubber dynamic rigidity and elasticity module by time-frequency measurements. *Mater. Plast.* **2007**, *44*, 18–21.
32. Iancu, V.; Vasile, O.; Gillich, G.R. Modelling and Characterization of Hybrid Rubber-Based Earthquake Isolation Systems. *Mater. Plast.* **2012**, *49*, 237–241.
33. Nedelcu, D.; Gillich, G.R.; Cziple, F.; Ciuca, I.; Padurean, I. Considerations about using polymers in adaptive guardrails construction. *Mater. Plast.* **2008**, *45*, 47–52.
34. Roohnia, M. An Estimation of Dynamic Modulus of Elasticity in Cantilever Flexural Timber Beams. *Drv. Ind.* **2014**, *65*, 1–10. [CrossRef]
35. Kubojima, Y.; Tonosaki, M.; Yoshihara, H. Young's modulus obtained by flexural vibration test of a wooden beam with inhomogeneity of density. *J. Wood Sci.* **2006**, *52*, 20–24. [CrossRef]
36. Lee, K.M.; Kim, D.S.; Kim, J.S. Determination of Dynamic Young's Modulus of Concrete at Early Ages by Impact Resonance Test. *KSCE J. Civ. Eng.* **1997**, *1*, 11–18. [CrossRef]
37. Kaewunruen, S.; Lewandrowski, T.; Chamniprasart, K. Dynamic responses of interspersed railway tracks to moving train loads. *Int. J. Struct. Stab. Dyn.* **2017**, *18*, 1850011. [CrossRef]
38. Kaewunruen, S.; Ngamkhanong, C.; Papaelias, M.; Roberts, C. Wet/dry influence on behaviors of closed-cell polymeric cross-linked foams under static, dynamic and impact loads. *Constr. Build. Mater.* **2018**, *187*, 1092–1102. [CrossRef]
39. Krezo, S.; Mirza, O.; Kaewunruen, S.; Sussman, J.M. Evaluation of CO_2 emissions from railway resurfacing maintenance activities. *Transp. Res. Part. D Transp. Environ.* **2018**, *65*, 458–465. [CrossRef]
40. Kaewunruen, S.; Lopes, L.M.C.; Papaelias, M.P. Georisks in railway systems under climate uncertainties by different types of sleeper/crosstie materials. *Lowl. Technol. Int.* **2018**, *20*, 77–86.
41. BSI. *BSI Standards Publication Railway Applications—Track—Concrete Sleepers and Bearers Part 2: Prestressed Monoblock Sleepers*; BSI: London, UK 2016.
42. Kaewunruen, S. Systemic values of enhanced dynamic damping in concrete sleepers – Comments on the paper: Ahn S, Kwon S, Hwang Y-T, Koh H-I, Kim H-S, Park J. Complex structured polymer concrete sleeper for rolling noise reduction of high-speed train system. *Comp, Struct.* **2020**, *234*, 111711. [CrossRef]
43. Resonance, S. *Standard Test Method for Dynamic Young's Modulus, Shear Modulus, and Poisson's Ratio by Impulse Excitation of Vibration 1*; ASTM International: West Conshohocken, PA, USA, 2005.
44. Stokey, W.F. Chapter 7 Vibration of Systems Having Distributed Mass and Elasticity. In *Shock and Vibration Handbook*. 22 October 2001. Available online: https://www.globalspec.com/reference/64452/203279/chapter-7-vibration-of-systems-having-distributed-mass-and-elasticity (accessed on 1 February 2020).
45. Kaewunruen, S.; Janeliukstis, R.; Ngamkhanong, C. Dynamic properties of fibre reinforced foamed urethane composites in wet and dry conditions. *Mat. Today. Procs.* **2020**, *5*, 690. [CrossRef]
46. G+D Computing. *Using Strand7 Introduction to the Strand7 Finite Element Analysis System*; G+D Computing Pty Ltd.: Sydney, Australia, 2002.
47. Kaewunruen, S.; Goto, K.; Xie, L. Failure modes of fibre reinforced foamed urethane composite beams: Full-scale experimental determination. *Mat. Today. Procs.* **2020**, *5*, 691. [CrossRef]
48. Kaewunruen, S.; You, R.; Ishida, M. Composites for Timber-Replacement Bearers in Railway Switches and Crossings. *Infrastructures* **2017**, *2*, 13. [CrossRef]
49. Sengsri, P.; Ngamkhanong, C.; Melo, A.L.O.; Papaelias, M.; Kaewunruen, S. Damage Detection in Fiber-Reinforced Foamed Urethane Composite Railway Bearers Using Acoustic Emissions. *Infrastructures* **2020**, *5*, 50. [CrossRef]
50. Kaewunruen, S.; Sussman, J.M.; Matsumoto, A. Grand Challenges in Transportation and Transit Systems. *Front. Built Environ.* **2016**, *2*, 4. [CrossRef]

Article

Effect of Impeller Diameter on Dynamic Response of a Centrifugal Pump Rotor

Alireza Shooshtari [1,*], Mahdi Karimi [1], Mehrdad Shemshadi [1] and Sareh Seraj [2]

1 Department of Mechanical Engineering, Bu-Ali Sina University, Hamedan 65175-4161, Iran; karimi_mh@yahoo.com (M.K.); m.shemshadi@eng.basu.ac.ir (M.S.)
2 Automotive Engineering Department, Tehran 16846-13114, Iran; Sare_Seraj@yahoo.com
* Correspondence: shooshta@basu.ac.ir

Abstract: This paper investigates the effect of impeller diameter on the dynamic response of a centrifugal pump using an inverse dynamic method. For this purpose, the equations of motion of the shaft and the impeller are derived based on Timoshenko beam theory considering the impeller as a concentrated mass disk. For practical modeling, the model of Jones and Harris is added to the equation to include the effect of bearings. As a case study, the model is applied to a process pump used in an oil refinery. Computing the eigenvalues of the model and comparing them with the natural frequencies of the structure, the model updating of the problem is performed through an indirect method. Three impellers with different diameters are applied to the updated model. The results show that increasing the diameter of the pump impeller can increase the amplitude of vibration up to 52% at critical speeds of the rotor. It is found that in addition to the hydraulic condition and efficiency, the impeller diameter should be considered as an important factor in the selection of centrifugal pumps.

Keywords: rotor dynamic; bearing; centrifugal pump; impeller diameter; Lagrangian equations

Citation: Shooshtari, A.; Karimi, M.; Shemshadi, M.; Seraj, S. Effect of Impeller Diameter on Dynamic Response of a Centrifugal Pump Rotor. *Vibration* **2021**, *4*, 117–129. https://doi.org/10.3390/vibration4010010

Academic Editors: Hamed Kalhori and Aleksandar Pavic

Received: 26 November 2020
Accepted: 26 January 2021
Published: 9 February 2021

1. Introduction

Rotating machineries such as pumps, compressors, turbines, etc., play an important role in many different industries. Accurate predictions of rotor system dynamic characteristics are very important in the design of any type of machines. There have been many studies relating to the field of rotor dynamic systems during the recent years. Engineers have developed several new techniques about the dynamics of rotating machines.

The first recorded theory of rotor dynamics was in a classic paper of Jeffcott [1]. The Jeffcott rotor model has been used to explain the whirling effect. It consists of a simply supported flexible massless shaft with a rigid disc mounted at the mid-span. In the Jeffcott model, the moments of inertia I_p and I_t are not considered. This is because there are no gyroscopic moments exerted on the shaft. The disc is assumed to move in a plane that is perpendicular to the shaft spin axis. By developments in the technology of rotating machines, the rotational speed of rotors became higher, and so, the non-conservative forces generated through the bearings of the rotor become considerable. To determine the critical speeds in which resonance has occurred, it is necessary to know the natural frequencies, mode shapes, and forced responses, which are caused by unbalancing in rotor systems. Prohl studied the critical speed evaluation of a turbine shaft, and he suggested the transfer matrix method [2]. The first application of the finite element method for a rotor system was made by Ruhl and Booker [3]. In their study, the influences of the rotary inertia, gyroscopic moment, bending, shear deformation, axial load, and internal damping were neglected to simplify the model. The theory has been developed by considering the rotary inertia, gyroscopic moment, and axial forces. Nelson and McVaugh extended this to include gyroscopic effects. They derived the equations of motion for the shaft and the effects of translational and rotary inertia and gyroscopic moments on it [4]. Erturka et al.

presented an analytical method based on Timoshenko beam theory for calculating the frequency response function (FRF) of a spindle–holder–tool combination. They proposed a mathematical model and obtained the point FRF for a tool [5]. Subbiah et al. showed that a rotor has certain speed ranges in which a large amplitude of vibration could occur. These speed ranges are known as critical speed, which results in excessive rotor deflection [6]. Phadatare and Barun described a step forward in calculating the nonlinear frequencies and resultant dynamic behavior of a high-speed rotor bearing system with unbalanced mass. In this study, Fast Fourier Transform (FFT) analysis was established for finding the fundamental frequencies of the rotor according to the variation of shaft diameter and location of unbalanced mass [7]. Metsebo et al. have focused on the influence of the rotating shaft on the dynamic of a rotor ball bearing system. They carried out a mathematical modeling for the system considering the shaft as a Timoshenko beam [8].

Ball bearings are the essential elements of rotating machineries. So, the influence of bearings on the performance of rotor-bearing systems is very important. El-Sayed derived a set of equations for the stiffness of bearings using the Hertz theory and determined the total deflections of inner and outer races caused by an applied load [9]. Tamura and Tsuda performed a theoretical study about fluctuations of radial spring characteristics of a ball bearing due to ball revolutions [10]. Many researchers also estimated bearing stiffness by carrying out some experiments. Stone and Walford developed a rotor-bearing test rig to estimate the bearing's radial stiffness and damping by measuring the response of the rotor [11]. Jairo et al. presented an experimental validation for a mathematical modeling of ball bearing. In this model, the bearing was considered as a mass-spring-damper system based on Hertz equations for contact deformation [12]. Xia sheng et al. proposed a mathematical model for the stiffness of bearings, which is varying by speed. They explained that the speed of a rolling bearing varies the stiffness of the bearing [13]. Zhang et al. investigated the effect of ring misalignment on the service characteristics of ball bearing and rotor systems [14]. They improved a quasi-static model of ball bearing considering the misalignment in its ring. Neisi et al. worked on the effect of off-sized balls on contact stresses in a touchdown bearing [15]. Yi Qin et al. developed a dynamics model for deep groove ball bearings with local faults based on coupled and segmented displacement excitation [16]. Chandrasekaran et al. used computational fluid dynamic (CFD) methods and mathematical modeling to investigate the impeller design parameters on the effect of fluid follows in the centrifugal pumps [17].

The objective of this paper is investigation on the effect of impeller diameter on the amplitude of vibration at critical speeds in an overhung centrifugal pump. Modeling of the shaft and impeller is based on Timoshenko theory, and modeling of the bearing is based on the work of Jones and Harris, which is added to the model of shaft simultaneously. Then, numerical analysis for a real centrifugal pump in the oil refinery has been done based on the proposed model. It has shown that in addition of the effect of the geometrical parameter of the shaft, the effect of the diameter of the impeller on the dynamical behavior of the pump is very important.

The novelty of this research is the development of the mathematical model of the shaft, impeller, and bearing simultaneously. Using this developed model, the effect of impeller size diameter on the dynamic behavior of a centrifugal pump has been investigated, and it has been shown that it is very considerable.

This paper includes four sections. In the first section, the literature has been reviewed and the necessity of research has been recognized. In the second section, the model of the system has been introduced and the equation of motion based on the energy method has been derived by calculating the potential and kinetic energy of the system. In the third section, the model of the bearing has been presented, and the governing equations of motion for the ball bearing have been introduced. Finally, in the last section, the derived equations have been used and solved.

For an actual centrifugal pump, the effect of impeller diameter on the amplitude of vibration has been investigated.

2. System Modeling

The rotor of a centrifugal pump is approximated by a simple model as shown in Figure 1a. The model is composed of a shaft of length L and supported by two bearings located at L_1 and L_2 along the shaft. u, v, and w reflect the displacement in the e_x, e_y, and e_z directions, respectively.

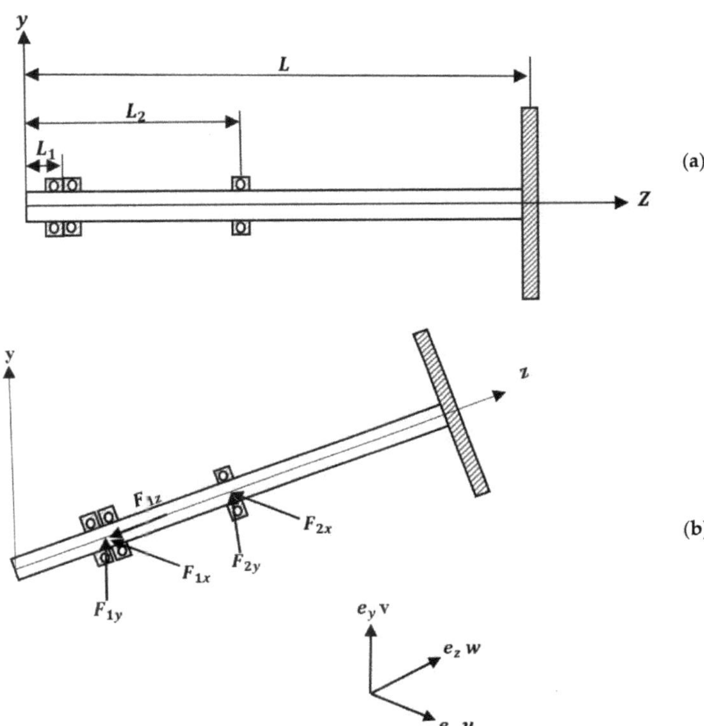

Figure 1. Model of centrifugal pump (**a**) definitions; (**b**) force of the bearing on the rotor system.

The shaft is modeled as a Timoshenko beam. In this model, first-order shear deformation theory with rotary inertia and gyroscopic effect has been considered. The shaft rotates at a constant speed around its longitudinal axis. In addition, it has a uniform, circular cross-section.

The equations of motion of the system are obtained using the Lagrangian equations as:

$$\frac{d}{dt}\frac{\partial T}{\partial \dot{q}_n} - \frac{\partial T}{\partial q_n} + \frac{\partial U}{\partial q_n} = (Q_{nc})^T \tag{1}$$

where T is the total kinetic energy of the system, U is the total potential energy of the system, q_n is the generalized coordinate and Q_{nc} represents non-conservative forces that are not directly related to the potential energy of the system. So, to derive the governing equations of motion using Equation (1), one must calculate the kinetic and potential energy of the system.

The total kinetic energy of a rotor system is estimated by the dynamic motion of the shaft and disk [14].

$$T = T_{disk} + T_{shaft} \tag{2}$$

The kinetic energy due to the rotation of the disk is difficult to calculate. Therefore, we assume that the disk is symmetric and rigid and has been fixed at the end of the shaft. The motion of the disk can be defined as two superimposed rotations θ_x, θ_y and two translational deflections u, v in directions e_x, e_y. So, the kinetic energy of the disk can be expressed as:

$$T_{disk} = \frac{1}{2}M\left(\dot{u}^2 + \dot{v}^2\right) + \frac{1}{2}I_t\left(\dot{\theta}_x^2 + \dot{\theta}_y^2\right) - I_p\Omega\dot{\theta}_x\theta_y + \frac{1}{2}\,I_pL\Omega^2 \quad (3)$$

where M is the mass of the rigid disk, I_t is the diametral mass moment of inertia, and I_p is the polar mass moment of inertia. The term $I_p\Omega\dot{\theta}_x\theta_y$ represents the gyroscopic effect, and finally, the last term defines the kinetic energy due to the rotation of the disk.

The kinetic energy of the shaft involves the kinetic energy due to bending of the shaft, effect of rotatory inertia, and gyroscopic effect. The kinetic energy of the shaft can be derived as:

$$T_{shaft} = \int_0^l \frac{1}{2}\rho_{shaft}A(\dot{u}^2 + \dot{v}^2)dz + \int_0^l \frac{1}{2}J_t(\dot{\theta}_x^2 + \dot{\theta}_y^2)dz + \frac{1}{2}J_pL\Omega^2 - \int_0^l I_p\Omega\dot{\theta}_x\theta_y\,dz \quad (4)$$

where A is the cross-sectional area, ρ_{shaft} is the density of the shaft, and J_t and J_p are the diametric and polar inertia of the shaft, respectively. So, substituting Equations (3) and (4) in Equation (2), the total kinetic energy of system has been obtained.

The potential energy of the system includes the strain energy due to the deformation of the shaft (U_1) and the strain potential energy due to the deflection of the bearing installed on the shaft (U_2) [12]:

$$U = U_1 + U_2. \quad (5)$$

The strain energy of the shaft can be expressed as:

$$U_1 = \frac{1}{2}\int_0^l EI((\frac{\partial^2 u}{\partial^2 z})^2 + (\frac{\partial^2 v}{\partial^2 z})^2)dz + \frac{1}{2}\int_0^l kGA\left(\Phi_x^2 + \Phi_y^2\right)dz \quad (6)$$

where E represents the modulus of elasticity, G is the shear modulus, and k is the shear coefficient. In addition, A and I are the cross-section area and moment of inertia of the shaft receptivity. In the above equation, the first term is related to the shaft bending, and the second term is due to shear deformation. In addition, the potential energy caused by bearing forces can be expressed as:

$$U_2 = (-F_{1x}u - F_{1y}v)_{l_1} + (-F_{2x}u - F_{2y}v)_{l_2}. \quad (7)$$

Different forces work on the impeller as a consequence of the fluid. These forces are non-conservative forces and are unknown, and they have been ignored for simplicity. However, mass unbalance generates an additional centrifugal force that makes it possible to calculate this non-conservative force. If the impeller is out of balance, the resulting centrifugal force will induce the rotor to vibrate. When the shaft rotates at a speed equal to the natural frequency, this vibration becomes large. Unbalance is defined by a small mass m_u situated at a distance d_u from the geometric center of the impeller, as shown in Figure 2 [13].

The out of balance force at the end of the rotor is:

$$F_u = m_u\,d_u\,\Omega^2. \quad (8)$$

This rotating force can be resolved into two components in the x and y direction as:

$$\begin{aligned} F_{u-x} &= m_u\,d_u\,\Omega^2 cos\Omega t \\ F_{u-x} &= m_u\,d_u\,\Omega^2 sin\Omega t. \end{aligned} \quad (9)$$

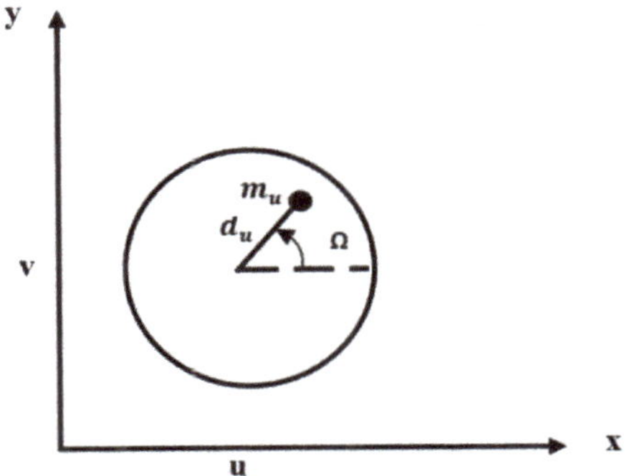

Figure 2. Mass unbalance.

It is assumed that the unbalance mass is in the x direction in the initial state. The deflections in the x and y directions are expressed as:

$$\begin{aligned} u(z.t) &= f(z)q_1(t) = f(z)q_1 \\ v(z.t) &= f(z)q_2(t) = f(z)q_2 \end{aligned} \tag{10}$$

where q_1 and q_2 are generalized independent coordinates and $f(z)$ is the displacement function that satisfies the boundary conditions of the system. As the rotor of the centrifugal pump has simply supported at both ends, $f(z)$ has been selected as [12]:

$$f(z) = sin\left(\frac{n.\pi}{L_2 - L_1}z - L_1\right). \tag{11}$$

The angular displacements θ_x and θ_y are assumed to be small. So, they are approximated by the derivative of u and v with respect to the z-direction. Therefore, using Equations (10) and (11), θ_x and θ_y have been expressed as:

$$\begin{aligned} \theta_x(z.t) &= -\frac{\partial u}{\partial z} = -\frac{df(z)}{dz}q_1 = -g(z)q_1 \\ \theta_y(z.t) &= \frac{\partial v}{\partial z} = \frac{df(z)}{dz}q_2 = g(z)q_2. \end{aligned} \tag{12}$$

To derive the equations of motion, the kinetic energy and the potential energy are specified by generalized coordinates. So, using Equation (3), the kinetic energy of the disk in generalized coordinates is expressed as:

$$T_{disk} = \frac{1}{2}Mf^2(L)\left(\dot{q}_1^2 + \dot{q}_2^2\right) + \frac{1}{2}I_t g^2(L)\left(\dot{q}_1^2 + \dot{q}_2^2\right) + \frac{1}{2}I_p\Omega^2 + I_p\Omega g^2(L)(\dot{q}_1 q_2) \tag{13}$$

where

$$g(z) = \frac{\partial f(z)}{\partial z} = \frac{n\,\pi}{(L_2 - L_1)}cos\left(\frac{n\,\pi}{(L_2 - L_1)}z - L_1\right). \tag{14}$$

In addition, Equation (4) in the generalized equation becomes:

$$T_{shaft} = \frac{1}{2}\rho A\int_0^L f^2(z)dz\left(\dot{q}_1^2 + \dot{q}_2^2\right) + \frac{1}{2}J_t\int_0^L g^2(z)dz\left(\dot{q}_1^2 + \dot{q}_2^2\right) + \frac{1}{2}J_p\Omega^2 + J_p\Omega\int_0^L g^2(z)dz\,(\dot{q}_1 q_2). \tag{15}$$

Likewise, using the generalized coordinates, the potential energy in the generalized coordinate will be:

$$\begin{aligned} U = \tfrac{1}{2}EI\textstyle\int_0^L h^2(z)dz\left(\dot{q}_1^2 + \dot{q}_2^2\right) + \tfrac{1}{2}KGA\textstyle\int_0^L\left(\beta_x^2 + \beta_y^2\right)dz + (-F_{1x}\,q_1 - F_{1y}\,q_2)\,\delta(z - L_1) \\ + (-F_{2x}\,q_1 - F_{2y}\,q_2)\delta(z - L_2) \end{aligned} \tag{16}$$

where δ represnts the dirac delta function and

$$h(z) = \frac{\partial f(z)}{\partial z} = -\left(\frac{n\,\pi}{(L_2 - L_1)}\right)^2 sin\left(\frac{n\,\pi}{(L_2 - L_1)}z - L_1\right). \quad (17)$$

In addition, substituting the displacement function into the kinetic energy of the mass unbalance expression of Equation (11) gives:

$$T_u \cong m_u d\Omega\, f(L)(\dot{q_1}\ cos\Omega t - \dot{q_2}\ sin\Omega t). \quad (18)$$

Ball Bearing Model

In this section, a mathematical model for calculating bearing stiffness is proposed by analyzing the equations in the bearing dynamic model, which is based on Jones and Harris's efforts [14]. This mathematical model aims to give a comprehensive consideration of the nonlinear stiffness of the ball bearing, and it can be seen that the bearing stiffness is critically dependent on the preloading, g, of the rolling elements. In case of rolling element bearings, the elastic deformation takes place at both the inner raceway and the outer raceway with the rolling element. Based on the Hertzian contact theory, the relation between load F and deflection is [15]:

$$F = k_p \delta_p^{3/2} \quad (19)$$

where δ_p is the contact deformation and k_p is a load–deformation constant for a single point contact (either at the inner or outer raceway). If the ball and raceway are made of steel, then [16,17]:

$$k_p = 2.15 \times 10^5 \left(\sum \rho\right)^{-\frac{1}{2}} (\delta^*)^{-3/2} \qquad \frac{N}{mm^{1.5}} \quad (20)$$

where δ^* is the dimensionless contact deformation and $\sum \rho$ is the curvature sum [14]. So, one can write the stiffness of the inner and outer ring contact as:

$$\begin{aligned} k_{pi} &= 2.15 \times 10^5 (\sum \rho_i)^{-\frac{1}{2}} ({\delta^*}_i)^{-3/2} \qquad \frac{N}{mm^{1.5}} \\ k_{po} &= 2.15 \times 10^5 (\sum \rho_o)^{-\frac{1}{2}} ({\delta^*}_o)^{-3/2} \qquad \frac{N}{mm^{1.5}} \end{aligned} \quad (21)$$

where

$$\begin{aligned} \sum \rho_i &= \tfrac{1}{D_b}\left(4 - \tfrac{1}{f_i} + \tfrac{2\gamma}{1-\gamma}\right) \\ \sum \rho_o &= \tfrac{1}{D_b}\left(4 - \tfrac{1}{f_o} + \tfrac{2\gamma}{1-\gamma}\right) \end{aligned}$$

and

$$f_i = \frac{r_i}{D_b} \quad f_o = \frac{r_o}{D_b} \quad \gamma = \frac{D}{D_m} \quad D_m = \frac{1}{2}(d_i + d_o) \cong \frac{1}{2}(D + d).$$

In the above equations, subscripts i and o represent inner and outer raceways and D_b is the ball diameter. In addition, D_m is the pitch diameter, d_i and d_o are the inner and outer ring raceway contact diameter, and r_i and r_o are the inner and outer raceway groove radius, respectively.

The total deformation, δ, at a single rolling element location is given by:

$$\delta = \delta_{pi} + \delta_{po} = \left(\frac{1}{k_{pi}} + \frac{1}{k_{po}}\right) F^{\frac{2}{3}}. \quad (22)$$

Equation (19) can be rewritten as:

$$F = k_{pio} \delta^{\frac{3}{2}} \quad (23)$$

where

$$k_{pio} = \frac{k_{pi}k_{po}}{(k_{pi}^{\frac{2}{3}} + k_{po}^{\frac{2}{3}})^{3/2}}.$$

k_{pio} is the load–deformation constant for two point contacts of a ball with raceways.

To calculate the deformations, the conventions shown in Figure 3 are used. Considering the presence of radial clearance, the force produced by every spring in the horizontal and vertical direction can be obtained [18]:

$$\begin{aligned} F_x &= k_{pio}(g + ucos\psi_i + vsin\psi_i)^{\frac{3}{2}} cos\psi_i \\ F_y &= k_{pio}(g + vcos\psi_i + vsin\psi_i)^{\frac{3}{2}} cos\psi_i \end{aligned} \tag{24}$$

where g is the radial preload between the ball and races, u and v are the displacements of the moving ring in the x and y directions, respectively, and ψ_i is the angular position of the i th element.

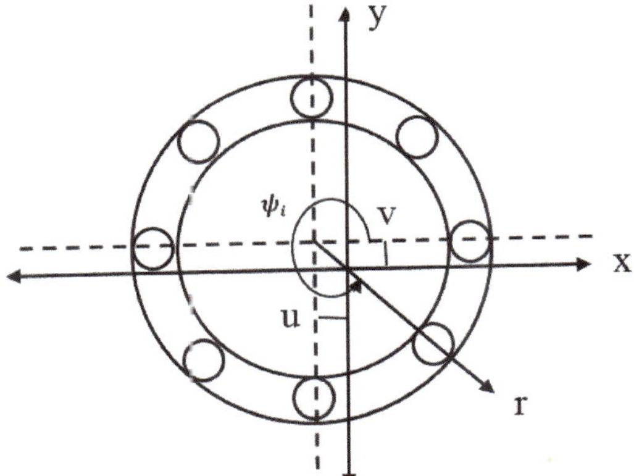

Displacement in r direction is:

$u\ cos\psi_i + v\ sin\psi_i$

Figure 3. Bearing model.

Finally, the bearing stiffness can be simplified to [15]:

$$\begin{aligned} k(u) &= k_{pio} \sum_{i=1}^{z} \{g + ucos\psi_i\}^{\frac{1}{2}} \{cos\psi_i - \tfrac{B}{1.5A} sin\psi_i\} cos\psi_i \\ A &= \sum_{i=1}^{z} (g + ucos\psi_i)^{\frac{1}{2}} sin^2\psi_i \\ B &= \sum_{i=1}^{z} (g + ucos\psi_i)^{1/2} sin\psi_i\ cos\psi_i \\ \psi_i &= \tfrac{\pi}{Z}(2i - 1) \qquad i = 1, 2, 3, \ldots .z \end{aligned} \tag{25}$$

where Z is the number of rolling elements.

It can be seen that the bearing stiffness is critically dependent on value of the preloading g of the rolling elements.

3. Numerical Solution

The model is applied for an installed pump (P-502B) in a Kermanshah oil refinery in Iran. This centrifugal pump and its impeller are shown in Figure 4.

Figure 4. Rotor pump P-502B in a Kermanshah oil refinery.

The schematic of the shaft and impeller of the above pump is shown in Figure 5. The flowing fluid is light petroleum gases including propane and butane.

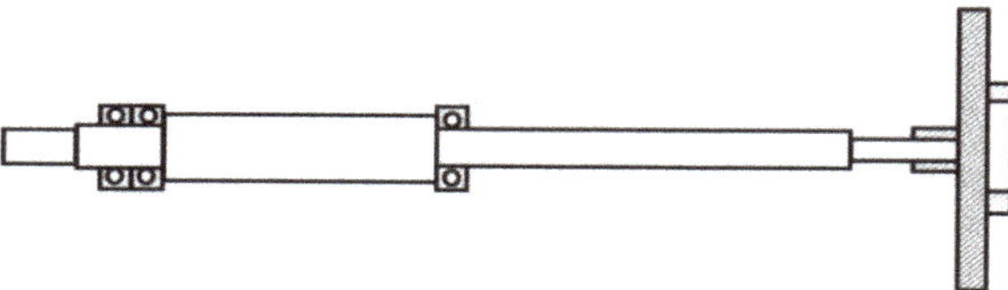

Figure 5. The schematic diagram of the pump rotor (P-502B).

The shaft of the mentioned pump is supported by two bearings SKF7308 and SKF6307, which have been installed in distance $L_1 = 0.125$ m and $L_2 = 0.521$ m respectively, see Figure 1a. In addition, three impellers with different diameters have been installed at the end of the shaft.

The characteristics of shaft and impeller are shown in Tables 1 and 2, respectively.

Table 1. Characteristics of shaft P-502B.

Parameter	Value
Material	AISI4140
Length (m)	0.7
Equivalent diameter (m)	0.036
Young's modulus (N/m^2)	2.10×10^{11}
Density (kg/m^2)	7850
Mass (kg)	5.6
Polar inertia of mass (kg-m^2)	0.000939
Diametric inertia of mass (kg-m^2)	0.2373
Moment of inertia (m^4)	8.24×10^8

Table 2. Characteristics of impeller P-502B.

Parameter	Value
Material	GG25
Type	closed
Equivalent thickness (m)	0.12
Diameters (m)	0.254–0.281–0.31
Density (kg/m^2)	7150
Mass (kg)	4.3–5.2–6
Polar inertia of mass (kg-m^2)	0.03467–0.05231–0.06153
Diametric inertia of mass (kg-m^2)	0.017338–0.02422–0.03283

Bearing Stiffness of P-502B

To obtain the stiffness of the bearing, knowledge of the properties is necessary. The general properties of these ball bearings are shown in Table 3.

Table 3. Dimensional specifications and extracted parameters in bearings.

	SKF6307	SKF7308
Parameter	Value	value
D_m (mm)	57.5	65
D_b (mm)	13.5	15
Z	8	8
g (mm)	0.051	0.053
r_i r_o (mm)	41.2	46.35
γ	0.23	0.23
f_i	3.05	3.08
f_o	3.05	3.08
$\sum\rho_i$ (mm^{-1})	0.31	0.28
$\sum\rho_o$ (mm^{-1})	0.24	0.22
$F(\rho)_i$	0.216	0.215
$F(\rho)_o$	0.212	0.211
δ_i^*	0.9865	0.9866
δ_o^*	0.9869	0.9871
k_{pi} (N/mm$^{1.5}$)	3.940×10^5	4.146×10^5
k_{po} (N/mm$^{1.5}$)	4.472×10^5	4.673×10^5
k_{pio} (N/mm$^{1.5}$)	1.301×10^5	1.342×10^5
Limited Speed (RPM)	12,000	10,000

It can be seen that the bearing stiffness is critically dependent on the value of preloading force g for the rolling elements.

Substituting the above values in Equation (25), the values of stiffness in the x and y directions have been obtained as:

Bearing stiffness for SKF6307:

$$k(x) = k(y) = \left(1.101 - 94.6x^2\right) \times 10^5 \ \frac{N}{m}. \tag{26}$$

Bearing stiffness for SKF7308:

$$k(x) = k(y) = \left(1.13 - 97.8x^2\right) \times 10^5 \ \frac{N}{m}. \tag{27}$$

Finally, using the Equations (3), (4), (6), (7) and Lagrangian Equation (1), the mathematical equation for the case study can be expressed as:

$$\begin{bmatrix} 10.86 & 0 \\ 0 & 10.86 \end{bmatrix}\begin{bmatrix} \ddot{q_1} \\ \ddot{q_2} \end{bmatrix} + \begin{bmatrix} 0 & 0.902\Omega \\ -0.902\Omega & 0 \end{bmatrix}\begin{bmatrix} \dot{q_1} \\ \dot{q_2} \end{bmatrix} +$$
$$\begin{bmatrix} 2.44 \times 10^7 + 125 \times 10^5 {q_1}^2 & 0 \\ 0 & 2.44 \times 10^7 + 125 \times 10^5 {q_2}^2 \end{bmatrix}\begin{bmatrix} q_1 \\ q_2 \end{bmatrix} = \tag{28}$$
$$\begin{bmatrix} 28.5 \times 10^{-5} sin\Omega t & 0 \\ 0 & 28.5 \times 10^{-5} cos\Omega t \end{bmatrix}.$$

The above equation is a lumped parameter model that can be described as:

$$\widetilde{M}\ddot{q} - \Omega\widetilde{D}\dot{q} + \widetilde{K}q = \widetilde{Q}^{nc}. \tag{29}$$

Looking at Equation (28), one can see that the damping matrix is dependent on the angular velocity of the shaft. The stiffness matrix is implicitly dependent on the amount of amplitude of vibration and so, the equations of motion are nonlinear equations. This will lead to a complicated model. So, for simplicity, normal clearance has been considered to make the value of the stiffness coefficient independent from the deflection, and so, the equations of motion will become linear ordinary differential equations. According to the SKF General catalog, hence $g = 0.051$ and $g = 0.053$ used for SKF6307 and SKF7308 respectively, the equivalent stiffness for bearings in a normal clearance case have been derived as follows:

$$\boldsymbol{k(r) = 1.040 \times 10^5 \ \tfrac{N}{m}} \text{ for SKF6307}$$

$$\boldsymbol{k(r) = 1.066 \times 10^5 \ \tfrac{N}{m}} \text{ for SKF7308}$$

To obtain the critical speed of the rotor, the eigenvalues and thus the natural frequencies of the system must be determined. To do this, a slightly different but equivalent approach can be used to combine the pair in Equation (29). By letting $\alpha = q_2 - jq_1$ and subtracting $j\times$ in the second Equation (29) from the first Equation (29), we have [19]:

$$\widetilde{M}\ddot{\alpha} - j\Omega\widetilde{D}\dot{\alpha} + \widetilde{K}\alpha = 0. \tag{30}$$

Now, we can solve this equation by letting $\alpha = \alpha_0 e^{j\Omega t}$ and taking the positive root to obtain backward critical speed. If we let $\alpha = \alpha_0 e^{-j\Omega t}$ in which Ω is the forward critical speed, one must add the out of balance moments to the system to determine the response. Thus:

$$\widetilde{M}\ddot{\alpha} - j\Omega\widetilde{D}\dot{\alpha} + \widetilde{K}\alpha = F_u. \tag{31}$$

Likewise, by $\alpha = q_2 - jq_1$ and subtracting $j\times$ in the second equation from the first, the rotor amplitude is calculated [19]. In addition, to determine the effect of impeller diameter, the various values of this parameter have been considered. To do this procedure, a code in MATLAB software has been written, and the following outputs shown in Figures 6–8 have been derived.

The amplitudes of the rotor response plots for all pump impellers that are recommended by pump manufacture are shown in Figures 6–8. The summary of the results is shown in Table 4.

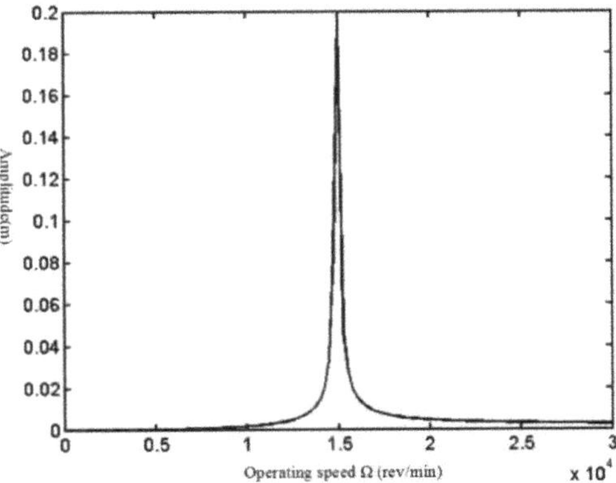

Figure 6. Calculatec response at impeller diameter = 0.254 m for API unbalance, normal bearing clearance.

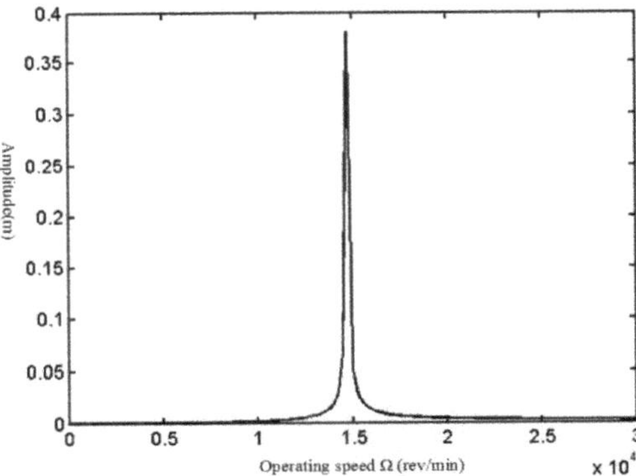

Figure 7. Calculated response at impeller diameter = 0.281 m for API unbalance, normal bearing clearance.

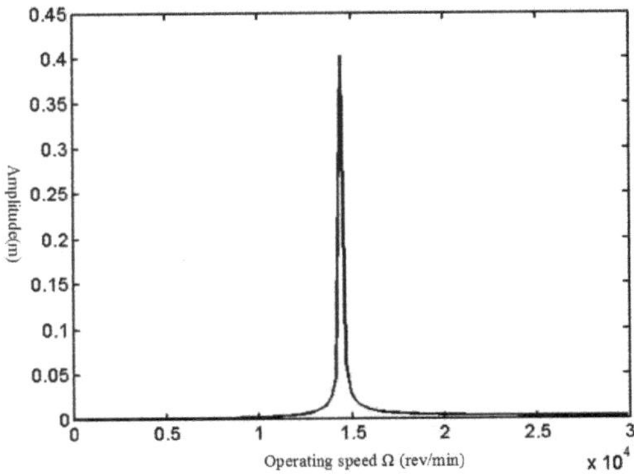

Figure 8. Calculated response at impeller diameter = 0.31 m for API unbalance, normal bearing clearance.

Table 4. Effect of impeller diameter on the amplitude of vibration and critical speed of the rotor.

Diameter of Impeller (mm)	Amplitude of Vibration (m)	Critical Speed (Backward) (Rpm)	Critical Speed (Forward) (Rpm)
254	0.198	13,480	14,646
281	0.380	12,909	14,437
310	0.401	12,493	14,184

So, one can see that by increasing the impeller diameter, the amplitude of vibration at critical speed will be increased up to 52%, while the value of critical speed will decrease. The value of amplitude of vibration shown in this table has been obtained without considering the structural damping of the shaft, bearings, and impeller and also the effects of fluid flow in the pump. So, in reality, these values are much smaller than the tabulated values in Table 4.

4. Conclusions

In this paper, the equations of motion for the shaft of a centrifugal pump have been derived using Lagrangian equations and based on Timoshenko beam theory and Jones and Harris [20,21] modeling for bearings based on Hertzian theory for rolling element [22,23]. By these equations, it is possible to define a lumped parameter model for the system and determine the amplitude of rotor in critical speeds [12,24]. The solutions of these equations for an applied case study show that as the impeller diameter increases, the amplitude of vibration at critical speeds increases, too. In this case study, by replacing the impeller diameter in the equations with a larger size (impeller diameter 0.256 m replaced with 0.32 m), it is shown that the amplitude is approximately increased by 52% in the critical speed. So, as the impeller trimming or impeller exchange is a general method for the maintenance in plants, it must be considered that the bigger diameter will yield more amplitude of vibration at critical speed.

It must be considered that modeling of the shaft, bearings, and impellers with details and assembling them and then conducting finite element analysis is very complex, and as the case study is a process pump in an oil refinery that works 24 h a day, stopping the production line and then measuring, modeling, and conducting FEM analysis is very expensive. On the other hand, the purpose of this manuscript is to introduce an analytical model for all the essential rotary parts of a centrifugal pump in detail and then conduct a sensitive analysis for the effect of impeller diameter on the critical speed of it inversely. For this purpose, the dynamic inverse solution is very suitable when there are practical limitations for FEM analysis, such as it being very complex and expensive.

Author Contributions: Investigation, A.S., M.K. and M.S.; Methodology, A.S. and S.S. All authors have read and agreed to the published version of the manuscript.

Funding: This research received no external funding.

Institutional Review Board Statement: Not applicable.

Informed Consent Statement: Not applicable.

Data Availability Statement: Not applicable.

Conflicts of Interest: The authors declare no conflict of interest.

References

1. Jeffcott, H.H. The lateral vibration of loaded shafts in neighborhood of a whirling speed: The effect of want of balance. *Philos. Mag.* **1919**, *6*, 304–314. [CrossRef]
2. Prohl, M.A. A general method for calculating the critical speeds of flexible rotors. *J. Appl. Mech.* **1945**, *12*, 142–148.
3. Ruhl, R.L.; Booker, J.F. A finite element model for distributed parameter turbo generator system. *J. Eng. Ind.* **1972**, *94*, 126–132. [CrossRef]

4. Nelson, H.D.; McVaugh, J.M. The dynamics of rotor-bearing systems using finite elements. *J. Eng. Ind. May* **1976**, *98*, 593–600. [CrossRef]
5. Erturk, A.; Ozguven, H.N.; Budak, E. Analytical modeling of spindle–tool dynamics on machine tools using Timoshenko beam model and receptance coupling for the prediction of tool point FRF. *Int. J. Mach. Tools Manuf.* **2006**, *46*, 1901–1912. [CrossRef]
6. Subbiah, R.; Rieger, N. On the transient analysis of rotor bearing system. *J. Vib. Acoust. Stress Reliab. Des.* **1988**, *110*, 515–520. [CrossRef]
7. Phadatare, H.P.; Pratiher, B. Nonlinear Frequencies and Unbalanced Response Analysis of High Speed Rotor-Bearing Systems. In Proceedings of the International Conference on Vibration Problems, Warangal, India, 18–20 February 2015.
8. Metsebo, J.; Upadhgay, N.; Kankar, P.K.; Nbendjo, B.R.N. Modeling of a rotor-ball bearing system using timoshenko beam and effect of rotating shaft on their dynamics. *J. Mech. Sci. Technol.* **2016**, *30*, 5339–5350. [CrossRef]
9. El-Sayed, H.R. Stiffness of deep-groove ball bearings. *Wear* **1980**, *63*, 89–94. [CrossRef]
10. Tamura, H.; Tsuda, Y. On the spring characteristics of ball bearing. *Bull. JSME* **1980**, *23*, 961–969. [CrossRef]
11. Stone, B.J.; Walford, T. The measurement of the radial stiffness of rolling element bearings under oscillating conditions. *J. Mech. Eng. Sci.* **1980**, *22*, 175–181.
12. Grajales, J.A.; Lopez, J.F.; Quintero, H.F. Ball bearing vibration model: Development and experimental validation. *J. Ing. Compet.* **2014**, *16*, 279–288.
13. Xia, S.; Li, B.; Wu, Z.; Li, H. Calculation of ball bearing speed-varying stifness. *Mech. Mach. Theory* **2014**, *81*, 166–180.
14. Zhang, Y.; Fang, B.; Kong, L.; Li, Y. Effect of the ring misalignment on the services characteristics of ball bearing and rotor system. *Mech. Mach. Theory* **2020**, *151*, 103889. [CrossRef]
15. Neisi, N.; Sikanen, E.; Heikkinen, J.E.; Sopanen, J. Effect of off-sized balls on contact stresses in a touchdown bearing. *Tribol. Int.* **2018**, *120*, 340–349. [CrossRef]
16. Qin, Y.; Cao, F.; Wang, Y.; Chen, W.; Chen, H. Dynamics modelling for deep groove ball bearings with local faults based on coupled and segmented displacement excitation. *J. Sound Vib.* **2019**, *447*, 1–19. [CrossRef]
17. Chandrasekaran, M.; Santhanam, V.; Venkashwaran, N. Impeller design and CFD analysis of fluidflow in rotordynamic pumps. *Mater. Today* **2020**. [CrossRef]
18. Van Esch, B.P.M. Rotor Dynamcis of a Centrifugal Pump. Master's Thesis, Technical University Eindhoven, Eindhoven, The Netherlands, February 2006.
19. Atepor, L. Vibration Analysis and Intelligent Control of Flexible Rotor Systems Using Smart Materials. Ph.D. Thesis, University of Glasgow, Glasgow, UK, October 2008.
20. Harris, T.A. *Rolling Bearing Analysis*, 4th ed.; Willey: New York, NY, USA, 2001.
21. Hertz, H.; Jones., D.E.; Schott., G.A. *On the Contact of Rigid Elastic Solids and on Hardness*; Miscellaneous Papers; Macmillan: London, UK, 1896; pp. 163–183.
22. Tiwari, R. *Analysis and Identification in Rotor-Bearing System*; Indian Institute of Technology: Guwahati, India, 2005.
23. Harris, T.; Kotzalas, M. *Rolling Bearing Analysis—Essential Concepts of Bearing Technology*, 5th ed.; Taylor and Francis: Abingdon, UK, 2007.
24. Friswell, M.I.; Penny, J.E.; Garvey, S.D.; Lees, A.W. Dynamic of Rotating Machines Solution Manual. Version 1. July 2011. Available online: http://www.rotordynamics.info/DRM_solutions.pdf (accessed on 8 February 2021).

 *vibration*

Article

$\mathcal{H}_2$ and $\mathcal{H}_\infty$ Optimal Control Strategies for Energy Harvesting by Regenerative Shock Absorbers in Cars †

Alessandro Casavola ‡, Francesco Tedesco ‡ and Pasquale Vaglica *

Modeling, Electronics and Systems (DIMES), Department of Informatics, University of Calabria, Via P. Bucci, 42/C, 87036 Rende (CS), Italy; a.casavola@dimes.unical.it (A.C.); ftedesco@dimes.unical.it (F.T.)

* Correspondence: pasquale.vgl@gmail.com

† This paper is an extended version of our paper published in 2020 IEEE American Control Conference, Denver, CO, USA, 1–3 July, 2020.

‡ These authors contributed equally to this work.

Received: 17 April 2020; Accepted: 20 May 2020; Published: 22 May 2020

Abstract: Regenerative suspension systems, unlike traditional passive, semi-active or active setups, are able to convert the traditionally wasted kinetic energy into electricity. This paper discusses flexible multi-objective control design strategies based on LMI formulations to suitably trade-off between the usual road handling and ride comfort performance and the amount of energy to be harvested. An electromechanical regenerative vehicle suspension system is considered where the shock absorber of each wheel is replaced by a linear electrical motor which is actively governed. It is shown by simulations that multivariable centralized control laws designed on the basis of a full-car model of the suspension system are able to achieve larger amount of harvested energy under identical ride comfort prescriptions with respect to scalar decentralized control strategies, designed on the basis of a single quarter-car model and implemented independently on each wheel in a decentralized way. Improvements up to 40% and 20% of harvested energy are respectively achievable by the centralized multivariable H_2 and H_∞ optimal controllers under the same test conditions.

Keywords: regenerative shock absorbers; energy harvesting; active control of automobile suspension systems

---

1. Introduction

Research on regenerative suspension systems has gained increasing interest in recent years for the potential energy savings achievable in implementing active control strategies that ensure enhanced dynamic performance and the ability to convert wasted kinetic energy in electrical power for both energy harvesting and self-powered implementation. The state-of-the-art in the field has been recently reviewed in [1,2].

In a typical regenerative setup the viscous shock absorber is usually replaced by an electrical actuator that can be regulated to mimic a standard shock absorber (virtual shock absorber) or to provide a more general dynamical behavior in order to better trade-off between the conflicting requirements of harvesting large amounts of energy and ensuring good road handling and ride comfort performance.

In [3] a multiobjective H_∞ control design methodology has been recently proposed for actively regulating regenerative suspension systems and it was shown to be much more flexible in trading-off among conflicting requirements and capable to dramatically improve the amount of harvested energy with respect to PI controllers implementing virtual shock absorbers where the only design knob is the dumping parameter.

The specificity of the control strategy proposed in [3], there referred to as Maximum Induced Power Control (MIPC) strategy, is that it may directly consider the amount of energy to be harvested amongst its objectives. This is done by imposing a model-reference prescription in closed-loop to the kinetic energy coming from the road that makes it arbitrarily large and sign-defined. The rationale is that, in the average, the amount of harvested electrical energy cannot be greater than the energy drained into the system from the road unevenness. Thus, maximizing the energy induced by the road and making it sign-defined may be a more effective control design objective in order to increase the amount of harvestable energy. Simulations and comparisons undertaken in [3] have confirmed the flexibility and the potentiality of the MIPC approach. A gain-scheduling version based on a LPV system formulation can be found in [4]. Similar conclusions on the superiority of the MIPC approach in terms of energy requirements for implementing the control strategy in comparison with standard passive control strategies were also drawn out in [5], where experimental comparisons of several control laws were undertaken on a lab regenerative shock absorber prototype.

One of the limitations of the MIPC approach described in [3] is that it is based on a quarter-car model and is limited to regulate the behavior of the regenerative suspension system of a single wheel. This approach obviously produces sub-optimal results because the four MIPC control laws are individually applied to the four wheels of the vehicles in a decentralized way, and the pitch and roll motions have not directly been taken into account in the design phase. As a consequence, each regulated suspension has the same dynamic behavior and provides the same amount of energy. Then, the total amount is simply given by four times the energy harvested by a single wheel. This approach, although suboptimal, is anyway of interest here because it will be used as a baseline solution for comparison purposes.

The main contribution of this paper is to present complete multivariable centralized MIPC approach is presented for the four wheels of a vehicle based on a full-car model of the regenerative suspension system. Both multi-objective H_∞ and H_2 state-feedback control strategies are presented based on standard LMI control design formulations. Dynamical output feedback control syntheses are also possible but the details are not presented here for space limitation.

A final example is provided where comparisons among the decentralized state-feedback H_∞ solution of [3], and the centralized state-feedback H_∞ and H_2 solutions presented here are reported. From these simulations, as expected, it clearly results that the multivariable centralized H_∞ and H_2 MIPC approaches proposed here overcome the decentralized implementation because of the better system description, the coordinated implementation of the four controllers and the extra degrees of freedom available for the optimization. Moreover, it is also found that the centralized H_2 approach allows one to harvest larger amount of energy for the same ride comfort requirement than the H_∞ MIPC approaches.

2. The Model and the Overall Control Architecture

2.1. The Full-Car Regenerative Suspension Model

The full-car regenerative suspension schematic under consideration is depicted in Figure 1. In such a system, the passive viscous dampers, usually present in any passive suspension systems for all four wheels, are here replaced by electromechanical actuators that generate forces $f_i(t)$, $i = 1, \ldots, 4$ so as to suitably dampen the vertical, pitch and roll motions of the car body.

The 7DOF mathematical model proposed in [6] is used here to describe the pitch θ_s and roll φ_s rotational motions (with I_p and I_s the corresponding moments of inertia) and the vertical motion of the sprung mass m_s along its center of mass z_s. They consist of the following coupled linear differential equations (the dependance on the time is omitted for simplicity)

$$\begin{aligned} I_r\ddot{\varphi}_s = & -k_f t_f(z_{s1}-z_{u1})+k_f t_f(z_{s2}-z_{u2})-k_r t_r(z_{s3}-z_{u3})+k_r t_r(z_{s4}-z_{u4}) \\ & +t_f f_1 - t_f f_2 + t_r f_3 - t_r f_4 \end{aligned} \tag{1}$$

$$\begin{aligned} I_p\ddot{\theta}_s = & -k_f a(z_{s1}-z_{u1})-k_f a(z_{s2}-z_{u2})+k_r b(z_{s3}-z_{u3})+k_r b(z_{s4}-z_{u4}) \\ & +af_1+af_2-bf_3-bf_4 \end{aligned} \tag{2}$$

$$m_s\ddot{z}_s = -k_f(z_{s1}-z_{u1})-k_f(z_{s2}-z_{u2})-k_r(z_{s3}-z_{u3})-k_r(z_{s4}-z_{u4})+f_1+f_2+f_3+f_4 \tag{3}$$

where z_{ui}, $i = 1,\ldots,4$ denote the vertical motions of the four unsprung masses described by

$$m_{u1}\ddot{z}_{u1} = k_f(z_{s1}-z_{u1})-k_{tf}(z_{u1}-z_{r1})-f_1 \tag{4}$$

$$m_{u2}\ddot{z}_{u2} = k_f(z_{s2}-z_{u2})-k_{tf}(z_{u2}-z_{r2})-f_2 \tag{5}$$

$$m_{u3}\ddot{z}_{u3} = k_r(z_{s3}-z_{u3})-k_{tr}(z_{u3}-z_{r3})-f_3 \tag{6}$$

$$m_{u4}\ddot{z}_{u4} = k_r(z_{s4}-z_{u4})-k_{tr}(z_{u4}-z_{r4})-f_4 \tag{7}$$

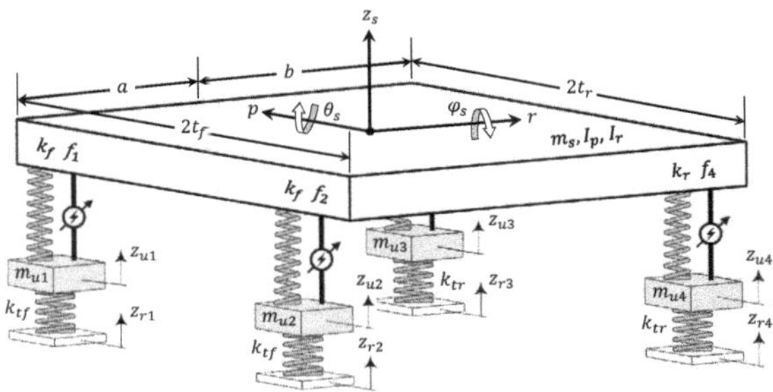

Figure 1. Full-car schematic for a regenerative setup.

In the above equations m_{u1} and m_{u2} represent the front wheels whereas m_{u3} and m_{u4} the rear ones, whose coupling with the road is simply modeled by the elastic stiffnesses k_{tf} and k_{tr} (for the front and rear couples of tires respectively). Moreover, k_f and k_r are the front and rear suspension stiffnesses respectively. Notice in particular that z_{si}, $i = 1,\ldots,4$ represent the vertical motions of the car body corners that, under a usual small pitch and roll angles assumption, are related to φ_s, θ_s and z_s via the following linear expressions

$$z_{s1} = t_f\varphi_s + a\theta_s + z_s \quad \dot{z}_{s1} = t_f\dot{\varphi}_s + a\dot{\theta}_s + \dot{z}_s \tag{8}$$

$$z_{s2} = -t_f\varphi_s + a\theta_s + z_s \quad \dot{z}_{s2} = -t_f\dot{\varphi}_s + a\dot{\theta}_s + \dot{z}_s \tag{9}$$

$$z_{s3} = \quad t_r\varphi_s - b\theta_s + z_s \quad \dot{z}_{s3} = \quad t_r\dot{\varphi}_s - b\dot{\theta}_s + \dot{z}_s \tag{10}$$

$$z_{s4} = \quad -t_r\varphi_s - b\theta_s + z_s \quad \dot{z}_{s4} = \quad -t_r\dot{\varphi}_s - b\dot{\theta}_s + \dot{z}_s \tag{11}$$

Finally, the parameters a, b, t_f and t_r characterize geometrically the car body, z_{ri}, $i = 1, \ldots, 4$ denote the road profiles, $z_{si} - z_{ui}$, $i = 1, \ldots, 4$ the suspension strokes and $z_{ui} - z_{ir}$, $i = 1, \ldots, 4$ the tire deflections.

2.2. Power Flow Analysis

In order to arrive to consistent guidelines on how to maximize the harvested electrical power, a power flows analysis of the regenerative suspension system is accomplished in this subsection. To this end, with reference to Figure 1, one notes that the only exogenous signals that provide power to the regenerative suspension system are the four road profiles z_{ri} and no dissipative units, like viscous shock absorbers are present.

Then, by writing down the total energy $E = K + V$ of the system, as the sum of the kinetic energy K of the masses and the potential energy V of the springs, one has

$$K = \frac{1}{2}I_r\dot{\varphi}_s^2 + \frac{1}{2}I_p\dot{\theta}_s^2 + \frac{1}{2}m_s\dot{z}_s^2 + \frac{1}{2}m_{uf}(\dot{z}_{u1}^2 + \dot{z}_{u2}^2) + \frac{1}{2}m_{ur}(\dot{z}_{u3}^2 + \dot{z}_{u4}^2) \tag{12}$$

$$\begin{aligned} V = & \frac{1}{2}k_f[(z_{s1} - z_{u1})^2 + (z_{s2} - z_{u2})^2] + \frac{1}{2}k_r[(z_{s3} - z_{u3})^2 + (z_{s4} - z_{u4})^2] \\ & + \frac{1}{2}k_{tf}[(z_{u1} - z_{r1})^2 + (z_{u2} - z_{r2})^2] + \frac{1}{2}k_{tr}[(z_{u3} - z_{r3})^2 + (z_{u4} - z_{r4})^2] \end{aligned} \tag{13}$$

Because power cannot be dissipated within the suspension system of Figure 1, it can only be exchanged with the road via its irregularities z_{ri} and with the electromagnetic devices via the exchanged forces f_i. In particular, such an exchanged power can be positive or negative. In fact, the road can introduce kinetic energy via the irregularities and absorbs part of the potential energy of the suspension during their discharging. On the other hand, the electromagnetic devices can act either as motors or generators by introducing or absorbing energy from the system.

Then, it makes sense to consider the power exchanged with the road P_s and the mechanical power exchanged with the actuators P_m and express the total instantaneous power $\dot{E}$ in terms of the above two terms

$$\dot{E}(t) = P_s(t) + P_m(t) \tag{14}$$

By exploiting (1)–(7) and after direct mathematical manipulations one arrives to

$$\begin{aligned} \dot{E} = & -k_{tf}[(z_{u1} - z_{r1})\dot{z}_{r1} + (z_{u2} - z_{r2})\dot{z}_{r2}] - k_{tr}[(z_{u3} - z_{r3})\dot{z}_{r3} + (z_{u4} - z_{r4})\dot{z}_{r4}] \\ & + f_1(\dot{z}_{s1} - \dot{z}_{u1}) + f_2(\dot{z}_{s2} - \dot{z}_{u2}) + f_3(\dot{z}_{s3} - \dot{z}_{u3}) + f_4(\dot{z}_{s4} - \dot{z}_{u4}) \end{aligned} \tag{15}$$

and, on the basis of the above considerations, one can recognize that the power contribution from the road profiles and the actuators are clearly indetifiable

$$P_s(t) = -k_{tf}\sum_{j=1}^{2}(z_{uj} - z_{rj})\dot{z}_{rj} - k_{tr}\sum_{j=3}^{4}(z_{uj} - z_{rj})\dot{z}_{rj} \tag{16}$$

$$P_m(t) = \sum_{i=1}^{4} f_i(\dot{z}_{si} - \dot{z}_{ui}) \tag{17}$$

As a result, as far as the exchanged power flows are concerned, the system can be considered as a series of two-port subsystems, as depicted in Figure 2, where the power P_s exchanged with the road acts at one terminal of the first subsystem, the mechanical power P_m is then exchanged with the electromechanical actuators that provide/request the electrical power P_e. In all terminals of the two-port systems we adopt the convention that the power is entering into the system when positive.

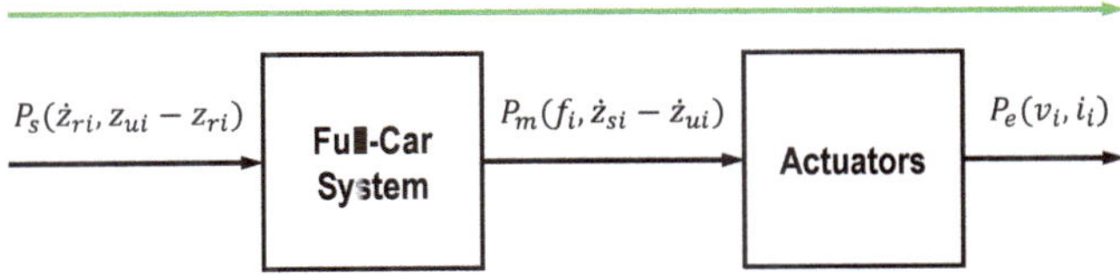

Figure 2. Power flows with the corresponding dual variables.

The above expressions for P_m and P_s generalized the ones achieved in [3] for a single regenerative suspension system derived from a quarter-car model and the same considerations can be drawn out here, briefly recalled hereafter for the reader convenience:

1. Ideally one would like to dispose of a control action capable to make both quantities large and sign defined for all time:
$$P_m(t) << 0, \quad P_s(t) >> 0 \text{ and } P_m(t) = -P_s(t), \tag{18}$$
If this were possible it would imply $\dot{E}(t) = 0$, which would ensure that the system has a perfect transfer of energy from the road to the electrical batteries.
2. Because $f_i(t)$ are directly manipulable variables, it is quite easy to make $P_m(t) < 0$ sign-defined. However,
$$\int_0^T |P_m(t)|dt \leq \int_0^T |P_s(t)|dt \tag{19}$$
expresses an obvious constraint on the energy that can be harvested in any interval $[0, T]$ of interest, which cannot be larger than the energy provided by the road. Thus, it might make nonsense to try to maximize P_m by suitably design the control actions if P_s were small without disposing of any control degree of freedom to increase its amount any further.
3. Observe first that P_s is zero when the road profiles are all completely flat ($\dot{z}_{ri} = 0$, $i = 1, \ldots, 4$) or when the tire deflections are all zero, $(z_{ui}(t) - z_{ri}(t) = 0)$, $i = 1, \ldots, 4$. Thus, high levels of energy harvesting on flat roads are incompatible with good road handling performance. From a control perspective, observe also that P_s depends on the regulated variables z_{ui}. Thus, if P_s were made sign-defined and as large as possible, viz. $P_s >> 0$ via a suitable control law this would increase the harvested energy regardless of the behavior (the sign) of P_m. Roughly speaking, making $P_s > 0$ it would provide a barrier for the internal energy of the system from flowing back towards the road.

The above considerations were at the basis of the Maximum Induced Power Control (MIPC) decentralized design approach of [3] that will be extended here to the multivariable centralized case. It is expected that the more degrees of freedom arising in driving all four electromagnetic actuators by a single multivariable centralized controllers make easier the achievement of the above control requirements with respect to the use of four decentralized and no coordinated control actions.

2.3. Electromechanical Actuator Model

In this paper we consider four identical linear permanent-magnets electromechanical devices depicted in Figure 3 as a force actuators (LPMA) for the four wheels. Simple linear models of the generic LPMA actuator are given by

$$v_{ai}(t) = R_{ai}i_i(t) + L_{ai}\frac{di_i(t)}{dt} + K_i\dot{z}_i(t) \tag{20}$$

$$\dot{z}_i(t) = \dot{z}_{si}(t) - \dot{z}_{ui}(t) \tag{21}$$

$$f_i(t) = K_i i_i(t) \tag{22}$$

for $i = 1, \ldots, 4$ where f_i are the damping forces, i_i the armature currents, v_{ai} the armature voltages and $\dot{z}_i$ the speed of the suspension stroke. The parameters R_{ai} and L_{ai} represent respectively the armature resistances and inductances while K_i are the force constants. The terms $K_i\dot{z}_i(t)$ represent the back EMF voltages whereas the forces are proportional to the corresponding armature currents. The electrical power $P_{ei}(t) = i_i(t)v_{ai}(t)$ results to be given by

$$P_{ei}(t) = R_a i_i^2(t) + L_a\frac{di_i(t)}{dt}i_i(t) + K_i\dot{z}_i(t)i_i(t), \tag{23}$$

where: $P_{di} = R_{ai}i_i^2(t)$ represents the power dissipated by the Joule effect, $P_{ci}(t) = L_{ai}\frac{di_i(t)}{dt}i_i(t)$ the electrical power stored in the inductance L_{ai} and $P_{mi}(t) = K_i\dot{z}_i(t)i_i(t)$ the mechanical power acting on the i-th actuators. For a more accurate analysis of the dynamics of electromechanical devices well suited for energy harvesting purposes and their models see e.g., [7], where a discussion of the validity of assuming constant the coupling coefficients between the electrical and mechanical parts of the device is reported.

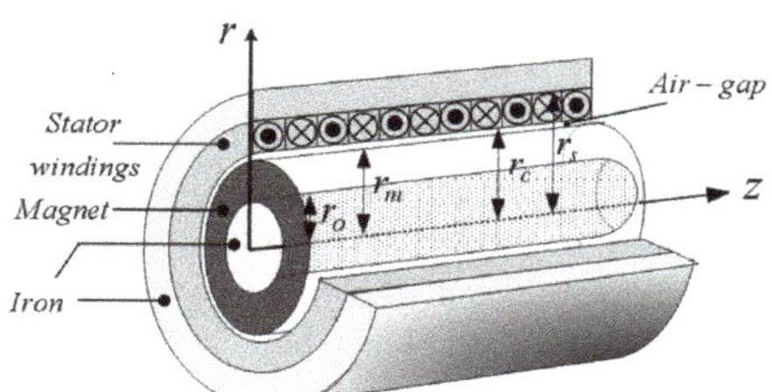

Figure 3. Permanent magnets linear electrical actuator.

Thus, Equation (23) suggests to work with electrical machines characterized by high voltages and low currents, in order to have small Joule losses and, in turn, a good efficiency. This requires actuators with large values of force constants K_i and small values for R_{ai}. Other losses are relevant only for high currents and are here neglected for simplicity.

2.4. The Overall Control Architecture

Figure 4 depicts the overall control architecture where two nested control loops are present. The inner one is governed by the *Current Controller* (CC) while the outer controller is termed *Regenerative Vibration Controller* (RVC) and it is in charge to provide the currents set-points to the inner CC controller.

The CC controller aims at regulating the armature current loops in the bank of four bidirectional DC-DC converters which establish the bidirectional electrical power flows between the battery and the four actuators. Its design is not considered here while we focus on the design of the RVC controller which is expected to provide the usual control performance of standard active suspension systems plus the ability to harvest as much energy as possibile. Among many, for brevity, only optimal H_∞ and H_2 centralized and decentralized static state-feedback control actions will be considered for the design of the RVC controller. Dynamic output feedback approaches will be considered in a future work.

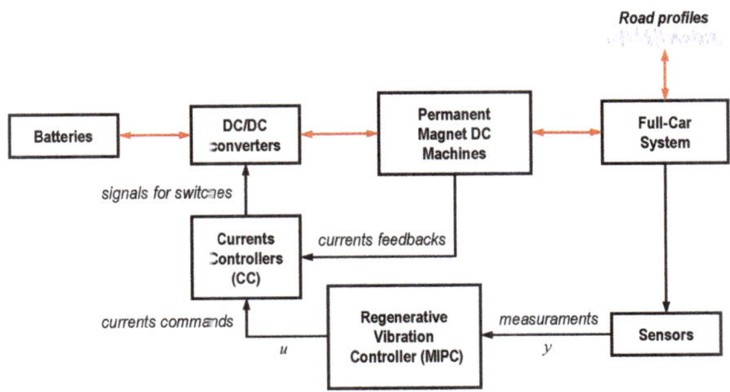

Figure 4. Overall control architecture.

3. RVC Control Design Specifications and LMI Based Designs

3.1. Ride Comfort and Road Handling Control Specifications

Traditional control specifications in classical passive and active suspension systems mainly consists of finding a suitable trade-off between the *ride comfort* and *road handling* requirements, see e.g., [8,9].

Ride comfort is related to the passengers perception of vibrations at various frequency (0.5–80 Hz) that greatly depends on the capability of the suspension systems to attenuate the perceived acceleration levels as much as possibile, especially at those frequencies more dangerous for the human body (0.5–5 Hz). Studies have shown that ride comfort may be directly related to sprung mass vertical accelerations $\ddot{z}_s$ and/or $\ddot{z}_{si}$.

In order to avoid subjectivity, the ISO-2631 standard (see [10]) defines an index, referred to as the *Ride Index* (RI), that quantifies the human exposure to vibration by weighting the perceived acceleration by a human sensitivity curve defined by a band-pass filter:

$$Ride\ \ Index = \sqrt{a_{wx}^2 + a_{wy}^2 + a_{wz}^2} \tag{24}$$

$$a_{i,RMS} = \sqrt{\frac{1}{T}\int_0^T a_{wi}^2(t)dt} \quad i = x, y, z \tag{25}$$

where a_{wi} is the acceleration along the i-th axis weighed by the following filter:

$$W_k(s) = \frac{81.89s^3 + 796.6s^22 + 1937s + 0.14}{s^4 + 80s^3 + 2264s^2 + 7172s + 21196} \tag{26}$$

Lower values of the ride index imply better vibrations attenuation.

Road handling is another important objective, related to vehicle and passengers' safety. It becomes extremely important in sportive/racing cars. It can usually be represented by the tire deflection $z_u - z_r$ that has to be minimized in some norm.

3.2. Energy Harvesting Control Specification

A further requirement of a regenerative suspension system is that to harvest energy from the road unevenness. Based on the above considerations, this requirements is here optimized by making P_s sign-defined and as larger as possible.

This condition could be achieved by trying to enforce the following conditions on the tire deflections in closed-loop

$$z_{ui}(t) - z_{ri}(t) = -\alpha \dot{z}_{ri}(t), \; i = 1,2,3,4 \tag{27}$$

where $\alpha > 0$ is a free constant design parameter. If all such conditions were satisfied for all time instants, the power P_s would result

$$\begin{aligned} P_s(t) &= -\left[k_{tf}\sum_{j=1}^{2}(z_{uj} - z_{rj})\dot{z}_{rj} + k_{tr}\sum_{j=3}^{4}(z_{uj} - z_{rj})\dot{z}_{rj}\right] \\ &= \alpha k_{tf}(\dot{z}_{r1}^2(t) + \dot{z}_{r2}^2(t)) + \alpha k_{tr}(\dot{z}_{r3}^2(t) + \dot{z}_{r4}^2(t)) \end{aligned} \tag{28}$$

In this way, in principle, $P_s(t)$ would be always positive (the power would always flow from the road to the suspension system) and large as desired by choosing a suitable large α. However, it is worth pointing out that making the tire deflections large is in contrast with good road handling performance. Moreover, high levels of harvested energy require high armature currents that increase the electrical losses in the actuators. Then, depending on the application at hands, a suitable trade-off among the above conflicting control specifications has to be addressed in the design the RVC controller that can naturally formulated as a multi-objective optimal control design problem.

In fact, conditions (27) can be reformulated as the following model reference errors $z_{pi}(t$ to be minimized

$$z_{pi}(t) = z_{ui}(t) - z_{ri}(t) + \alpha \dot{z}_{ri}(t), \; i = 1,2,3,4 \tag{29}$$

3.3. State-Space Realization for the RVC Control Synthesis

Because the need of having a system description which uses the derivatives $\dot{z}_{ri}$ instead of z_{ri} as exogenous signals for causally realizing z_{pi} in (29) we extend to the full-car case the alternative state-space representations used in [11,12] for the quarter-car model. The following state-space representation results

$$\dot{x}(t) = Ax(t) + Bu(t) + E\dot{z}_r(t) \tag{30}$$

corresponding to the following system vectors

$$x = \begin{bmatrix} z_{s1} - z_{u1} \\ z_{s2} - z_{u2} \\ z_{s3} - z_{u3} \\ z_{s4} - z_{u4} \\ \dot{\varphi}_s \\ \dot{\theta}_s \\ \dot{z}_s \\ z_{u1} - z_{r1} \\ z_{u2} - z_{r2} \\ z_{u3} - z_{r3} \\ z_{u4} - z_{r4} \\ \dot{z}_{u1} \\ \dot{z}_{u2} \\ \dot{z}_{u3} \\ \dot{z}_{u4} \end{bmatrix} \qquad u = \begin{bmatrix} i_1 \\ i_2 \\ i_3 \\ i_4 \end{bmatrix} \qquad \dot{z}_r = \begin{bmatrix} \dot{z}_{r1} \\ \dot{z}_{r2} \\ \dot{z}_{r3} \\ \dot{z}_{r4} \end{bmatrix} \tag{31}$$

for certain matrices A, B and C detailed in [13].

It is worth pointing out that the above state-space realization is not minimal in that the state has dimension 15 while the 7DOF full-car model (1)–(7) could be realized by a state of dimension 14. This choice is dictated by the greater easiness in specifying the various control objectives for the design of the controller that this realization offers.

The extra dimension has as a consequence that $\mathrm{rank}\{A\} = 14$. In fact, it can be observed that A has one dominant real eigenvalue in zero and seven couple of pure imaginary coniugate eigenvalues. The eigenvalue in zero expresses the fact a linear combinations of state components remains constant during the free evolutions of the systems. In particular, it is found that the following linear dependence arises

$$t_r(z_{s1}(t) - z_{s2}(t)) - t_f(z_{s3}(t) - z_{s4}(t)) = 0, \ \forall t \tag{32}$$

along all the free evolutions of the system ($z_{ri}(t) \equiv 0, i = 1, \ldots, 4$). This condition trivially results by considering (8)–(11) and expresses the fact that in a rigid body the positions of three points are sufficient to characterize the positions of all other points of the body.

The state-space realization (30) is fully controllable. Thus, if the state is fully measurable ($y(t) = x(t)$) linear state-feedback control laws can be freely designed. On the contrary, if only an output is available

$$y(t) = C_y x(t) + D_{yu} u(t) + D_{yw} z_r(t) \tag{33}$$

it is also full observable for many choices of sensors. For example, in [13] it is shown that the pair (A, C_y) corresponding to the following output

$$y = \begin{bmatrix} z_{s1} - z_{u1} & z_{s2} - z_{u2} & z_{s3} - z_{u3} & z_{s4} - z_{u4} & \ddot{z}_{s1} & \ddot{z}_{s2} & \ddot{z}_{s3} & \ddot{z}_{s4} \end{bmatrix}^T \tag{34}$$

is full observable. In this second case one has full freedom in designing any form of dynamic output feedback control laws.

Finally, for control design purposes it is convenient to introduce the following performance vector

$$z(t) = Cx(t) + Du(t) + F\dot{z}_r(t) \tag{35}$$

$$z = \begin{bmatrix} \ddot{z}_{s1} \ \ddot{z}_{s2} \ \ddot{z}_{s3} \ \ddot{z}_{s4} \ i_1 \ i_2 \ i_3 \ i_4 \ z_{p1} \ z_{p2} \ z_{p3} \ z_{p4} \end{bmatrix}^T \tag{36}$$

where all conflicting objective variables of interest are considered. In particular, $z(t)$ collects the four vertical accelerations $\ddot{z}_{si}$ accounting for the drive comfort requirements, the four control actions i_i used to moderate the control energy so as to mitigate the electrical losses and the four variables z_{pi} defined in (29) related to the energy harvesting specifications. Again, details on the C, D and F matrices can be found in [13].

3.4. Frequency Shaping of the State-Space Realization

A common practice in the design of optimal control laws is that of shaping the control objectives in frequency in order to penalize/enhance the performance in certain frequency bands of interest. This can be done by filtering the signals of the control systems.

By denoting with $W_{ai}(s)$ the scalar, stable and proper filter shaping the vertical acceleration of the body corner $\ddot{z}_{si}(t)$ and with $W_{ui}(s)$ the one related to the current $i_i(t)$, the objective vector $z(t)$ results filtered by the diagonal multivariable filter $W_z(s)$ defined as:

$$W_z(s) = diag\left\{W_a(s), W_u(s), I_4\right\} \tag{37}$$

In particular, the aggregate filters $W_a(s)$ e $W_u(s)$ have the following structure

$$\begin{aligned} W_a(s) &= diag\left\{W_{a1}(s), W_{a2}(s), W_{a3}(s), W_{a4}(s)\right\} \\ W_u(s) &= diag\left\{W_{u1}(s), W_{u2}(s), W_{u3}(s), W_{u4}(s)\right\} \end{aligned} \tag{38}$$

and for simplicity are assumed identical

$$\begin{aligned} W_{a1}(s) &= W_{a2}(s) = W_{a3}(s) = W_{a4}(s) \\ W_{u1}(s) &= W_{u2}(s) = W_{u3}(s) = W_{u4}(s) \end{aligned} \tag{39}$$

It is convenient to select the $W_{ai}(s)$ filters according to the ISO-2641 recommendations and using (26) to shape the accelerations in order to have the objectives directly related to the Ride Index. On the contrary, $W_{ui}(s)$ are usually selected as high-pass filters in order to penalize the control actions at high frequency.

The energy of the road profiles are optionally shaped by a suitable bank of filters as well

$$W_d(s) = diag\left\{W_{d1}(s), W_{d2}(s), W_{d3}(s), W_{d4}(s)\right\} \tag{40}$$

$$W_{d1}(s) = W_{d2}(s) = W_{d3}(s) = W_{d4}(s) \tag{41}$$

Thus, by denoting with (A_z, B_z, C_z, D_z) e (A_d, B_d, C_d, D_d) the state-space representations of the filters $W_z(s)$ and $W_d(s)$, with x_z and x_d the corresponding states, on the basis of the following extended state

$$x_g = \begin{bmatrix} x^T & x_z^T & x_d^T \end{bmatrix}^T \tag{42}$$

one gets the extended state-space representation

$$\begin{cases} \dot{x}_g(t) = A_g x_g(t) + B_{gu} u(t) + B_{gw} \dot{z}_r(t) \\ \tilde{z}(t) = C_{gz} x_g(t) + D_{gzu} u(t) + D_{gzw} \dot{z}_r(t) \end{cases} \tag{43}$$

with

$$A_g = \begin{bmatrix} A & \varnothing_{15\times 20} & EC_d \\ B_z C & A_z & B_z F C_d \\ \varnothing_{4\times 15} & \varnothing_{4\times 20} & A_d \end{bmatrix} \quad B_{gu} = \begin{bmatrix} B \\ B_z D \\ \varnothing_{4\times 4} \end{bmatrix}$$

$$B_{gw} = \begin{bmatrix} ED_d \\ E_z F D_d \\ B_d \end{bmatrix} \quad C_{gz} = \begin{bmatrix} D_z C & C_z & D_z F C_d \end{bmatrix} \tag{44}$$

$$D_{gzu} = D_z D \quad D_{gzw} = D_z F D_d$$

3.5. H_2 and H_∞ Optimal State Feedback Designs

We assume hereafter that the state is measurable and that the system (43) is fully controllable. This happens to be true when no loss of controllability arises from the introduction of the shaping filters (poles/zeros cancellations). Thus, state-feedback control laws of the form

$$u(t) = K x_g(t) \tag{45}$$

can be arbitrarily designed. In particular, the following closed-loop system results

$$\begin{cases} \dot{x}_g(t) = (A_g + B_{gu} K) x_g(t) + B_{gw} \dot{z}_r(t) \\ \tilde{z}(t) = (C_{gz} + D_{gzu} K) x_g(t) + D_{gzw} \dot{z}_r(t) \end{cases} \tag{46}$$

with

$$T(s) = (C_{gz} + D_{gzu} K)(sI - (A_g + B_{gu} K))^{-1} B_{gw} + D_{gzw} \tag{47}$$

being the closed-loop transfer matrix between $\dot{z}(t)$ and $\tilde{z}(t)$.

3.5.1. Optimal H_∞ Control Synthesis

The H_∞ optimal state-feedback control law

$$u(t) = K_\infty x_g(t) \tag{48}$$

that minimizes the H_∞ norm of (47), that is

$$K_\infty := \arg\min_K \|T(s)\|_{H_\infty} \tag{49}$$

can be achieved by solving the following LMI optimization problem [14]

$$\begin{cases} \min\limits_{X,Y,\gamma} \gamma \\ \text{s.t.} \\ \begin{bmatrix} A_{cl} + A_{cl}^T & B_{gw} & (C_{gz}X + D_{gzu}Y)^T \\ * & -\gamma I & D_{gzw}^T \\ * & * & -\gamma I \end{bmatrix} < 0 \\ X = X^T > 0 \end{cases} \tag{50}$$

where $A_{cl} := A_g X + B_{gu} Y$. If the problem is feasible, the optimal H_∞ state-feedback gain is given by

$$K_\infty = YX^{-1} \tag{51}$$

3.5.2. Optimal H_2 Control Synthesis

The H_2 optimal state-feedback control law

$$u(t) = K_2 x_g(t) \tag{52}$$

that minimizes the H_2 norm of (47), that is

$$K_2 := \arg\min_K \|T(s)\|_{H_2}^2 \tag{53}$$

can be achieved by solving the following LMI optimization problem [14]

$$\begin{cases} \min\limits_{X,Y,Q,\nu} \nu \\ \text{s.t.} \\ \begin{bmatrix} A_{cl} + A_{cl}^T & B_{gw} \\ * & -I \end{bmatrix} < 0 \\ \begin{bmatrix} X & (C_{gz}X + D_{gzu}Y)^T \\ * & Q \end{bmatrix} > 0 \\ X = X^T > 0 \\ Q = Q^T > 0 \\ Tr(Q) \leq \nu, \end{cases} \tag{54}$$

where $A_{cl} := A_g X + B_{gu} Y$. If the problem is feasible, the optimal H_2 state-feedback gain is given by

$$K_2 = YX^{-1} \tag{55}$$

4. Simulation Results

Several Matlab/Simulink simulations have been undertaken for assessing the full-car H_2 and H_∞ MIPC approaches presented here and compare them with the quarter-car H_∞ state-feedback solution described in [3]. The latter control law, designed for the suspension systems of a single wheel, will be then applied to all four suspension systems in a decentralized way.

The three control approaches, namely H_∞ decentralized, H_∞ centralized and H_2 centralized, will be compared on the same car, actuators and driving scenario with the available design knobs tuned to achieve the same Ride Index for the three control strategies.

In particular, three values of the Ride Index will be considered: RI = 0.25, RI = 0.47 and RI = 0.70 that, according to the ISO 2631 RI classification, correspond respectively to *not uncomfortable*, *a little uncomfortable* and *fairly uncomfortable* likely passengers reactions

Road's profiles complying with the ISO-8608 standard [15] have been used in the simulations. In particular, all simulations have been undertaken by assuming to drive on a **C** straight road at 70 Km/h. Figure 5 depicts the corresponding profiles

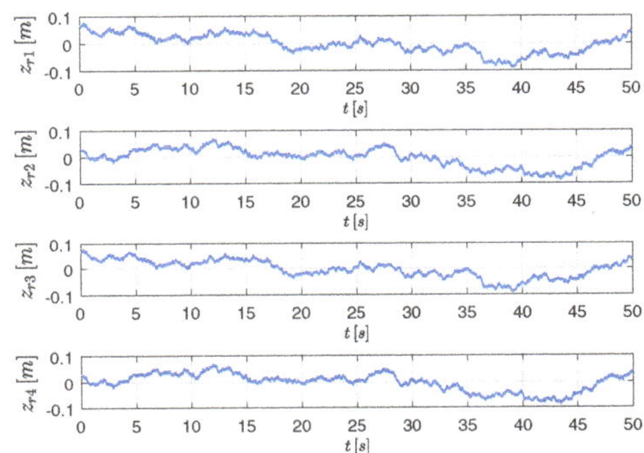

Figure 5. Road profiles $z_{r1}(t)$, $z_{r2}(t)$, $z_{r3}(t)$ and $z_{r4}(t)$ used in the simulations.

The same standard class-C vehicle and actuator considered in [3] (where all relevant parameters are listed) is used here.

The following dynamical weights have been used in the three control strategies

$$W_{ai}(s) = \rho\, W_k(s) \;\; \text{(defined in (26))} \tag{56}$$

$$W_{ui}(s) = \beta\, \frac{s+1}{s+10} \tag{57}$$

$$W_{di}(s) = \gamma\, \frac{10}{s+1000} \tag{58}$$

with ρ, β, γ and α in (29) as free design knobs. The fact that the regulation of the Ride Index can be simply achieved by tuning the few control design knobs testifies favorably on the flexibility of the proposed MIPC approach.

4.1. Simulations for Ride Index = 0.25

The values of design knobs values used in the RI = 0.25 simulations are reported in Table 1 whereas the plots of the acceleration $\ddot{z}_{s1}$, actuation current $i_1(t)$ and the harvested electrical power $P_{e1}(t)$ under the three control laws are reported in Figure 6

Table 1. Knobs tuning—RI = 0.25.

RI = 0.25 (Excellent Comfort/Smallest Harvesting)				
Control	ρ	β	γ	α
$H_{\infty,dec}$	0.2067	100	10	30
$H_{\infty,cen}$	0.42	3.45	1	27
$H_{2,cen}$	0.5	2	1	29

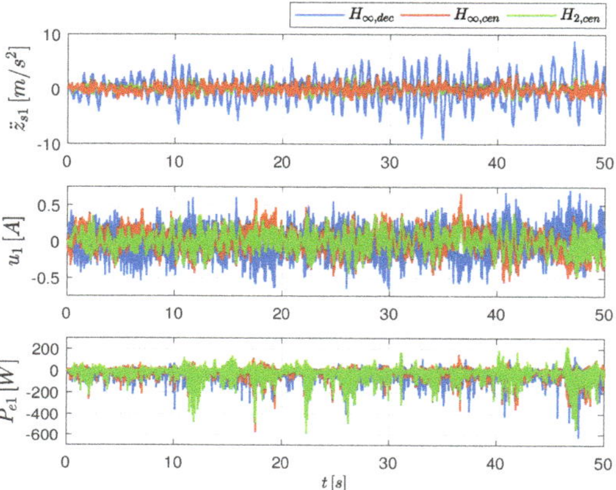

Figure 6. (**top**) Accelerations $\ddot{z}_{s1}(t)$, (**middle**) Actuation current $i_1(t)$, (**down**) Instantaneous harvested electrical power $P_{e1}(t)$.

4.2. Simulations for Ride Index = 0.47

In Table 2 the design knobs values used in the RI = 0.47 simulations are reported while the plots of the corresponding acceleration $\ddot{z}_{s1}$, actuation current $i_1(t)$and the electrical power $P_{e1}(t)$ under the three control laws are reported in Figure 7.

Table 2. Knobs tuning—RI = 0.47.

RI = 0.47 (Trade-off between Comfort/Harvesting)				
Control	ρ	β	γ	α
$H_{\infty,dec}$	0.323	100	10	42
$H_{\infty,cen}$	1.67	2.25	1	30
$H_{2,cen}$	0.82578	3.764	1	31

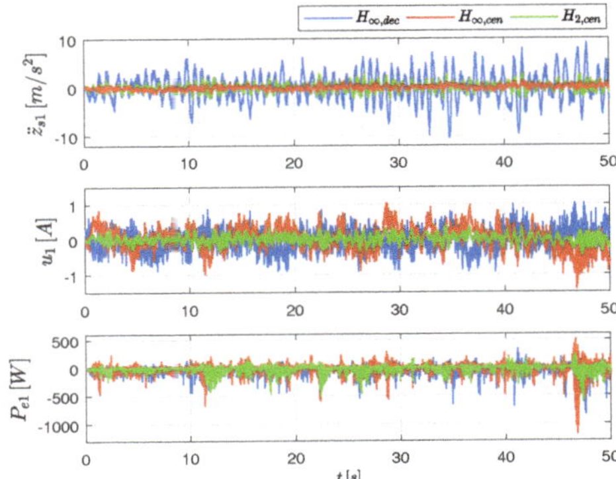

Figure 7. (**top**) Accelerations $\ddot{z}_{s1}(t)$, (**middle**) Actuation current $i_1(t)$, (**down**) Instantaneous harvested electrical power $P_{e1}(t)$.

4.3. Simulations for Ride Index = 0.70

Finally, in Table 3 the values of the design knobs used for the case RI = 0.70 are reported while the plots of the acceleration $\ddot{z}_{s1}$, actuation current $i_1(t)$ and the harvested electrical power $P_{e1}(t)$ under the three control laws are reported in Figure 8.

Table 3. Knobs tuning—RI = 0.70.

RI = 0.25 (Perceivable Discomfort/Largest Harvesting)				
Control	ρ	β	γ	α
$H_{\infty,dec}$	0.42	100	10	47
$H_{\infty,cen}$	2.82	3	1	33
$H_{2,cen}$	2.167	4–69	1	33

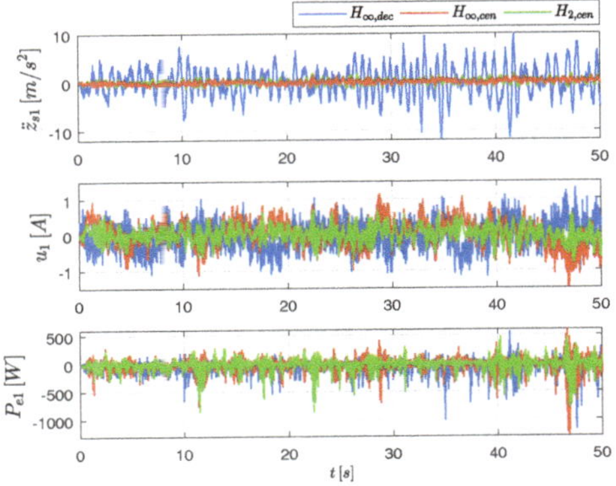

Figure 8. (**top**) Accelerations $\ddot{z}_{s1}(t)$, (**middle**) Actuation current $i_1(t)$, (**down**) Instantaneous harvested electrical power $P_{e1}(t)$.

4.4. Average Harvested Electrical Power

Finally, next Table 4 summarize the average harvest electrical power during the simulations.

From the simulation it results that the $H_{\infty,cen}$ and $H_{2,cen}$ regulators, designated on the basis of the full-car suspension model, guarantee good rider performance and higher levels of harvested energy with respect to the $H_{\infty,dec}$ controller in all the situations tested. Moreover, the ride comfort performance of the $H_{\infty,dec}$ regulator, designated on the basis of the quarter-car suspension model and implemented in a decentralized way, degrades remarkably at the increase of the energy harvesting requirements. This can be seen from the accelerations $\ddot{z}_{s1}(t)$ plots but similar conclusions can be drawn out by observing the sprung mass accelerations $\ddot{z}_s(t)$ and the pitch and roll angles evolutions (non reported here for space limitations but available in [13]).

As far as the amount of harvested energy during the simulations, Table 4 allows one to observe that the centralized $H_{\infty,cen}$ and $H_{2,cen}$ regulators always are able to recover more energy than the decentralized $H_{\infty,dec}$ controller, the more at lower values of the RI index. Moreover, it can be seen that the $H_{2,cen}$ regulator recovers the largest amount of electric power, while the $H_{\infty,cen}$ regulator is more robust in rejecting the exogenous signals acting to the system.

Table 4. Average harvest electrical power. The percentages express the improvements with respect to the $H_{\infty,dec}$ control achievements for the same RI.

Average Harvest Electrical Power			
Control	**RI = 0.25**	**RI = 0.47**	**RI = 0.70**
$H_{\infty,dec}$	4 × 100 W	4 × 124 W	4 × 153 W
$H_{\infty,cen}$	4 × 120 W (+20%)	4 × 141 W (+13%)	4 × 160 W (+4.5%)
$H_{2,cen}$	4 × 141 W (+40%)	4 × 154 W (+24%)	4 × 168 W (+10%)

5. Conclusions

The usage of regenerative suspension systems in modern electrical/hybrid cars could contribute to the vehicle's autonomy with a modest degradation to the usual ride comfort and road handling performance. The integration of such systems with other existing energy harvesting devices, such as regenerative brakes, may help the diffusion of this kind of cars, providing together a total amount of many tens/hundreds watts.

This paper has complemented the results achieved in [3] on the design of active control laws for the regulation of regenerative suspension systems by extending the scalar MIPC approach there presented for a quarter-car system to the general multivariable solution achieved on the basis of a full-car model. Moreover, both the H_∞ and H_2 state-feedback solutions have been considered based on a novel and *ad-hoc* state-space realization and have been shown to be enough flexible and powerful to easily trade-off amongst conflicting energetic and dynamic requirements.

From the simulations it results that the centralized solutions have to be preferred with respect to the decentralized one. In fact, from Table 4 it results that the improvements on the harvested energy are increasingly higher (up to 40% for the centralized H_2 and up to 20% for the centralized H_∞ optimal controllers) as the ride index RI decreases. This is especially important in the large part of cars usage, where maintaining low values of the ride index is of paramount importance for ensuring a good ride comfort. It is also found that the centralized H_2 control is able to gain the double of the energy harvested by the centralized H_∞ control for the same value of the ride index.

An important contribution of this work is the demonstration that the coordination of the control actions achievable by a multivariable design has to be preferred than a simpler decentralized implementation. The usage of modern H_2 and H_∞ multi-objective control design methodologies make the extra numerical complexity for the multivariable design negligible.

Author Contributions: Methodology, supervision and writing-review and editing A.C.; Investigation and formal analysis F.T.; Software, data curation, visualization and writing-original draft preparation P.V. All authors have read and agreed to the published version of the manuscript.

Funding: This research received no external funding.

Conflicts of Interest: The authors declare no conflict of interest.

References

1. Abdelkareema, M.A.A.; Xua, L.; Alia, M.K.A.; Elagouza, A.; Mia, J.; Guoa, S.; Liuc, Y.; Zuoc, L. Vibration energy harvesting in automotive suspension system: A detailed review. *Appl. Energy* **2018**, *229*, 672–699. [CrossRef]
2. Zhang, R.; Wang, X.; John, S. A Comprehensive Review of the Techniques on Regenerative Shock Absorber Systems. *Energies* **2018**, *11*, 1167. [CrossRef]
3. Casavola, A.; di Iorio, F.; Tedesco, F.; Multiobjective, A. H_∞ Control Strategy for Energy Harvesting in Regenerative Vehicle Suspension Systems. *Int. J. Control* **2018**, *91*, 741–754. [CrossRef]
4. Casavola, A.; di Iorio, F.; Tedesco, F. Gain-scheduling control of electromagnetic regenerative shock absorbers for energy harvesting by road unevenness. In Proceedings of the 2014 IEEE American Control Conference, Portland, OR, USA, 4–6 June 2014.
5. Sultoni, A.I.; Sutantra, I.N.; Pramono, A.S. H_∞ Multi-Objective Implementation for Energy Control of Electromagnetic Suspension. *Int. Rev. Mech. Eng.* **2015**, *9*, 542–547. [CrossRef]
6. Kim, C.; Ro, P.I. An Accurate Full Car Ride Model Using Model Reducing Techniques. *J. Mech. Des.* **2002**, *124*, 697–705. [CrossRef]
7. Kecik, K.; Mitura, A.; Lenci, S.; Warminski, J. Energy harvesting from a magnetic levitation system. *Int. J. Non-Linear Mech.* **2017**, *94*, 200–206. [CrossRef]
8. Sammier, D.; Sename, O.; Dugard, L. $\mathcal{H}_\infty$ Control of Active Vehicle Suspensions. In Proceedings of the IEEE International Conference on Control Applications, Anchorage, AK, USA, 25–27 September 2000.
9. Poussot-Vassal, C.; Sename, O.; Dugard, L.; Gáspár, P.; Szabó, Z.; Bokor, J. A new semi-active suspension control strategy through LPV technique. *Control Eng. Pract.* **2008**, *16*, 1519–1534. [CrossRef]
10. ISO. *ISO 2631-4: Human Exposure to Mechanical Vibration and Shock. Part 4: Guidelines for the Evaluation of the Effects of Vibration and Rotational Motion on Passenger and Crew Comfort in Fixed-Guideway Transport Systems*; International Organization for Standardization: Geneva, Switzerland, 2001.
11. Chen, H.; Guo, K.H. Constrained Control of Active Suspensions: An LMI Approach. *IEEE Trans. Control Syst. Technol.* **2005**, *13*, 412–421. [CrossRef]
12. Rajamani, R. *Vehicle Dynamics and Control*, 2nd ed.; Springer: Berlin, Germany, 2012.
13. Vaglica, P. Modeling and Control of Regenerative Suspension Systems for Automobiles. Master's Thesis, University of Calabria, Rende, Italy, 2019.
14. Boyd, S.; El Ghaoui, L.; Feron, E.; Balakrishnan, V. *Linear Matrix Inequalities in System and Control Theory*; Society for Industrial and Applied Mathematics: Philadelphia, PA, USA, 1994; Volume 15.
15. ISO. *ISO 8608:1995(E): Mechanical Vibration—Road Surface Profiles—Reporting of Measured Data*; International Organization for Standardization: Geneva, Switzerland, 1995.

MDPI
St. Alban-Anlage 66
4052 Basel
Switzerland
Tel. +41 61 683 77 34
Fax +41 61 302 89 18
www.mdpi.com

Vibration Editorial Office
E-mail: vibration@mdpi.com
www.mdpi.com/journal/vibration